T0205904

Analytics and Optimization for Renewable Energy Integration

Analytics and Optimization for Renewable Energy Integration

Ning Zhang
Chongqing Kang
Ershun Du
Yi Wang

CRC Press
Taylor & Francis Group
Boca Raton London New York

CRC Press is an imprint of the
Taylor & Francis Group, an **informa** business

CRC Press
Taylor & Francis Group
6000 Broken Sound Parkway NW, Suite 300
Boca Raton, FL 33487-2742

First issued in paperback 2022

© 2019 by Taylor & Francis Group, LLC
CRC Press is an imprint of Taylor & Francis Group, an Informa business

No claim to original U.S. Government works

Version Date: 20190123

ISBN 13: 978-1-03-240163-8 (pbk)
ISBN 13: 978-1-138-31682-9 (hbk)

DOI: 10.1201/9780429455094

**Visit the Taylor & Francis Web site at
http://www.taylorandfrancis.com**

**and the CRC Press Web site at
http://www.crcpress.com**

To Our Alma Mater
--Tsinghua University

Contents

III   Short-Term Operation Optimization   159

Preface

In recent years, wind power and photovoltaics (PV) have experienced a rapid growth around the world. As representatives of renewable energy, large number of wind power and PV have been integrated into power systems because of their low-carbon superiority and continuously decreasing investment cost. Countries around the world set aggressive goals for high penetration of renewables in future power systems to decarbonize the energy system. Large scale centralized renewable generations are connected to transmission grids in countries such as China and United States. In some other countries, distributed renewables are employed to partially supply the load in distribution systems. According to the estimate of the *Energy Transition Outlook: Renewables, Power and Energy Use Report*, 85% of the world electricity will be supplied by renewable energy in 2050 (http://www.ourenergypolicy.org/wp-content/uploads/2017/09/DNV-GL_-Energy-Transition-Outlook-2017_renewables_lowres-single_0109.pdf).

The fast development of renewable energy continuously changes the behavior of the power system. Uncertainty and variability are the two most significant features of renewable energy. However, power systems have always been designed for controllable generators such as thermal and hydro-power generators instead of intermittent energy for more than a century. The uncertainty and variability of renewables will fundamentally change the way that the power system balances the generation and load, in terms of both planning for the long term and operation the short term, and in both transmission and distribution networks. The aim of many emerging power system technologies is exactly to increase the flexibility of the power system to accommodate more renewable energy. To face this ever-changing power system trend, new theories and methodologies for analytics and optimization are required for the planning and operation of the future power system.

Stochastic mathematics made great progress in recent decades. It not only helps us understand the uncertainties in our everyday life but also provides wisdom and skills to handle the uncertainty. More and more stochastic mathematics are starting to be used in power system analytics and decision making. It is evident that stochastic mathematics analytic and operation techniques will benefit the power system with renewable energy integration, in terms of both security and economy.

The scope of this book covers the modeling and forecasting of renewable energy as well as operation and planning of power systems with renewable energy integration. The book can be divided into four parts as shown in Figure 1. The first part presents some stochastic mathematics theories as a preparation for reading the main part of the book. The second part presents various modeling and analytic techniques for renewable energy generation, including how to model long- and short-term renewable energy uncertainty, what the dependencies among multiple renewable power plants will be, how to simulate renewable energy generation, and how to find representative output time series. The third part provides solutions of how to handle the uncertainty of renewable energy in power system operation, including a fast and efficient algorithm to calculate the probabilistic power flow, risk-based stochastic unit commitment models to acquire the optimal generation schedule under uncertainty, an efficient operation strategy for renewable power plants participating in electricity markets, and how to make the tie-line scheduling under uncertainty. From the perspective of long-term planning, the last part introduces a series of efficient approaches on how to simulate long-term power system operation, how to calculate the capacity credit of renewable energy, how to optimize the location, the capacity, and the sequence for renewable investments, and how to optimize the generation and transmission expansion planning of power systems with renewable energy integration.

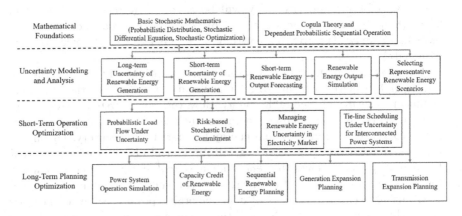

FIGURE 1 Research Framework of Analytics and Optimization for Renewable Energy Integration.

The aim of this book is threefold: **Firstly, the book provides a comprehensive understanding of the uncertain behavior of renewable energy generation.** For the power industry, the transition of deterministic thinking to stochastic is crucial for renewable energy integration. This book not only explains it from a theoretical prospective but also describes how to understand this problem from an engineering point of view. **Secondly, the book provides state-of-art mathematics and methodologies for power systems under uncertainty.** We think stochastic mathematics will play an important role in the future power system. Therefore this book proposes a whole range of models, from basic models to deal with the uncertainty, to small math skills to make the calculation tractable. It opens the door to develop practical decision-making tools for power system operation and planning with high renewable energy penetration. **Thirdly, this book helps the reader to understand the impact of renewable energy integration on power systems.** The book provides many case studies to demonstrate the cost of hedging against renewable energy uncertainty and the benefit of integrating renewable energy, how the renewable energy integration affects the decision-making process, and to what extent the power system economy, reliability, and environmental impact will be changed with renewable energy integration. Especially, some case studies use real-world data and provide some practical information and experiences from industry.

The content of this book is as follows:

Chapter 1 provides several fundamental stochastic mathematics techniques for uncertainty analysis and modeling including probability theory, stochastic differential equations, and the stochastic process and scenarios. These basic theories and techniques will be used in the analytic and optimization of renewable energy in this book.

Chapter 2 introduces the method to model and calculate multiple stochastic dependent variables using copula theory and dependent probabilistic sequence operation (DPSO). DPSO can be viewed as discrete convolution that is able to handle dependent variables. It has wide applications in power system reliability assessment, reserve quantification, probabilistic forecasting, etc.

Chapter 3 proposes the long-term uncertainty models for wind power and PV generation. We use the copula function to model the distribution of multiple wind farms or PV stations in order to capture their spatiotemporal dependencies. Their long-term fluctuation is modeled by autocorrelation functions. Compared with wind power, the output of PV shows obvious periodicity. We show how to divide the PV output into the deterministic part and stochastic part.

Chapter 4 addresses the issue of renewable energy forecasting. Considering the fact that the accuracy of short-term wind power forecast highly depends on numerical weather prediction (NWP), we propose an improved forecasting approach by mining the outliers of NWP, where outliers are detected by feature extraction and K-means clustering.

Chapter 5 proposes short-term uncertainty models for wind power and PV generation. The short-term uncertainty of renewable energy is reflected as the forecast error. We use the copula function to model the dependencies between the real output and its forecasted value among different renewable energy sites. The conditional forecast error of renewable energy can then be obtained, which provides the uncertainty of the wind power / PV forecast varying with the level of its output.

Chapter 6 introduces renewable energy operation simulation techniques. The need for such techniques is that there usually is not enough clean historical data to be used in power system long-term planning. We propose a sequenced differential equation-based method to simulate the wind speed and the shedding of solar radiation to further obtain long-term output of wind power and solar radiation. The proposed method is able to simulate massive renewable energy output data that retain the basic characteristics of wind power and PV output, including their intermittence, variety, and dependency.

Chapter 7 proposes the method of finding representative renewable energy scenarios. Several scenario reduction methods are proposed and are compared using several evaluation metrics. The obtained representative renewable energy scenarios act as the basis of the scenario-based stochastic optimization problems.

Chapter 8 conducts probabilistic load flow analysis with high penetrated renewable energy. It combines dependent probabilistic sequence operation and a novel linear power flow model to boost the computational efficiency. Since there are so many dependent uncertain injections in the networks, dependent probabilistic sequence operation suffers from the curse of dimensionality. An effective high-dimensional dependent probabilistic sequence operation calculation method is provided.

Chapter 9 proposes a risk-based stochastic unit commitment model to address the renewable energy uncertainty in short-term power system operations. The risks of the loss of load, renewable energy curtailment, and branch overflow caused by renewable energy uncertainty are analytically formulated in the unit commitment model. The model is proven to be able to be transformed into a mixed integer linear programing and shows efficiency improvements compared with the scenario-based unit commitment model.

Chapter 10 discusses the operation strategy of renewable energy producers participating in electricity markets. The renewable energy producer is allowed to participate in both energy and reserve markets to hedge the generation uncertainty. The optimal bidding strategy for renewable energy producers under probabilistic forecast to maximize their expected revenue in the market is analytically derived.

Chapter 11 addresses the issue of tie-line scheduling of inter-connected power systems with large-scale renewable energy integration. We propose an improved tie-line scheduling approach that is able to optimize the electricity exchange and share reserves among different balancing areas by considering the renewable energy integration quota in each area.

Chapter 12 introduces the power system operation simulation technique considering large-scale renewable energy integration. The operation simulation considers the operation characteristics of different kinds of generation such as renewable energy, coal / gas thermal power, hydro power, pumped hydro storage, nuclear power and combined heat and power generators. Detailed power system topology and the transmission capacity of AC/DC transmission lines are considered. Calculation acceleration techniques are proposed so that the model is able to carry out 8760-hour year-round operation simulation in a reasonable time. We provide several case studies using real-world data to demonstrate the application of this technique in thermal generation cost variation analysis due to renewable energy integration, wind power accommodation analysis, and pump storage planning.

Chapter 13 discusses the issue of capacity credit of renewable energy from both a numerical calculation prospective and analytical derivation. A dependent probabilistic sequence operation-based method is proposed that can quickly and accurately evaluate the capacity credit of renewable energy. An analytical model based on a reliability function is proposed that uncovers the factors that influence the value of the capacity credit.

Chapter 14 proposes a sequence and location planning method for renewable energy. The model optimizes the sequence and location of the wind farm to maximize the overall energy production and capacity value of wind power. The wind resource and the dependencies between different sites are considered. Since the optimization model is un-analytical, ordinal optimization theory is applied to identify suboptimal decisions.

Chapter 15 proposes a stochastic long-term generation planning model to efficiently integrate wind power. The uncertainty of wind power generation is represented by scenarios, including both typical and extreme scenarios. Since it is difficult to directly solve the proposed two-stage stochastic planning model considering multiple scenarios, an improved Benders decomposition algorithm is introduced to accelerate the model solving.

Chapter 16 proposes a stochastic long-term transmission planning model to efficiently integrate renewable energy. Both long-term and short-term uncertainties are considered in this model, which are described via scenarios and operating conditions, respectively.

We would like to conclude this preface by expressing our gratitude to a number of institutions and people. We are thankful to the Tsinghua University for providing us with an invaluable research environment. Some chapters of the book come from the PhD thesis of Qianyao Xu and the M. Sc. thesis of Xi Zhang and Weijia Zhao. The authors are indebted to PhD candidates Zhenyu Zhuo, Hai Li, Hongjie He, and Haiyang Jiang at Tsinghua University for their assistance in preparing the material, editing the manuscript, and proofreading the whole book. This book is supported by the National Key R&D Program of China (No. 2016YFB0900100), and National Natural Science Foundation of China (No. 51677096, 51620105007). The authors really appreciate their financial supports.

Tsinghua University, Beijing   *Ning Zhang*
November 2018   *Chongqing Kang*
         Ershun Du
         Yi Wang

List of Abbreviations

ANN	Artificial Neural Network
ARIMA	Auto Regressive Integrated and Moving Average
ARMA	Auto Regressive and Moving Average
ATC	Available Transmission Capacity
AWPF	Abnormal Wind Power Forecast
BIC	Bayes Information Criterion
BP	Back Propagation
BSC	Basic System Constraints
CCGT	Combined Cycle Gas Turbine
CCV	Constrained Cost Variable
CDF	Conditional Density Function
CHP	Combined Heat and Power
COPT	Capacity Outage Probability Table
CPDF	Conditional Probability Density Function
CVaR	Conditional Value of Risk
DA	Day-ahead
DEENS	Derivative of EENS
DPSO	Dependent Probability Sequence Operation
ECGC	Equivalent Convectional Generation Capacity
EENS	Expected Energy not Supplied
EFC	Equivalent Firm Capacity
ELCC	Effective Load-carrying Capability
ERCOT	Electric Reliability Council of Texas
ESS	Energy Storage System
EV	Electrical Vehicle
FOR	Forced Outage Rate
FRC	Frequency Regulation Constraints
GC	Guaranteed Capacity
GEP	Generation Expansion Planning
GOPT	Grid Optimization Planning Tool
GP	Gaussian Process
GTC	Grid Transmission Constraints
GWPC	Grid-Accommodable Wind Power Capacity
IEEE	Institute of Electrical and Electronics Engineers
ISO	Independent System Operator
KF	Kalman Filtering

LFC	Load-Following Constraints
LMP	Locational Marginal Prices
LOLE	Loss of Load Expectation
LOLP	Loss of Load Probability
MAPS	Multi Area Production Simulation Software
MLP	Multi-Layer Perception
NCG	Northwestern China Grid
NN	Neural Network
NPV	Net Present Value
NWP	Numerical Weather Prediction
NWPF	Normal Wind Power Forecast
PDF	Probability Density Function
PLF	probabilistic Load Flow
PM	Persistence Method
PRC	Peak-Valley Regulation Constraints
PV	Photovoltaic
RF	Reliability Function
RMSE	Root Mean Square Error
RPS	Renewable Portfolio Standard
RT	Real-time
RUC	Risk-Based Unit Commitment
SCUC	Security-Constrained Unit Commitment
SRC	Spinning Reserve Constraints
SUC	Stochastic Unit Commitment
SVM	Support Vector Machine
TLC	Tie-Line Constraints
TSO	Transmission System Operator
UC	Unit Commitment
UCED	Unit Commitment and Economic Dispatch
VaR	Value of Risk
VOLL	Value of Loss Load
WA	Wavelet Analysis
WPF	Wind Power Forecast
WPO	Wind Power Output
WPP	Wind Power Producer
WSF	Wind Speed Forecast

Part I

Mathematical Foundations

1

Basic Stochastic Mathematics

The integration of high penetration of renewable energy presents greater uncertainties for power systems. These uncertainties, i.e., outputs of wind power and PV, electrical and heat demands, energy price, and charging behaviors of electrical vehicles (EVs), are related and result in complex dependencies based on relevant factors such as weather and time of the day. The modeling and analysis of high-dimensional dependent uncertainties are of great significance for the reliability and security of future power systems. Stochastic mathematics is a branch of Mathematics. There are several mature textbooks about this topic. In this chapter, we only provide the most fundamental mathematics that will be used in this book for renewable energy analytics and optimization.

1.1 Random Variables, Probability Distribution, and Scenarios

1.1.1 Random Variables

Random variables represent real single-value functions of various results of random tests. The possible values of the random variables are outcomes of a random phenomenon. A random variable X can be viewed as a measurable function from a set of possible outcomes Ω to a measurable space E. Mathematically, the probability that X takes on a value in a measurable set $S \subseteq E$ is written as:

$$Pro(X \in S) = Pro(\{\omega \in \Omega | X(\omega) \in S\}) \tag{1.1}$$

where Pro is the probability measure equipped with Ω.

Random variables can be divided into discrete variables and continuous variables. For example, whether there is an outage for one generator can be modeled as a discrete random variable. The set of possible outcomes $\Omega = \{0,1\}$, where $X = 0$ denotes the generator is in an outage state; $X = 1$ denotes the generator works well. If the outage rate of the generator is 0.02, the probability

$$Pro(X = 0) = 0.02, \quad Pro(X = 1) = 0.98,$$
$$Pro(X = 0) + Pro(X = 1) = 1. \tag{1.2}$$

The output of one wind farm is a continuous random variable. The set of possible outcomes $\Omega = [0, P_{max}]$, where P_{max} denotes the maximum output of the wind farm.

1.1.2 Probability Distribution

A probability distribution is a mathematical function that provides the probabilities of occurrence of different possible outcomes in an experiment. The sum of all the probabilities for all possible outcome is equal to 1:

$$\sum_{i=-\infty}^{+\infty} p_i = 1, \quad \text{or} \quad \int_{x=-\infty}^{+\infty} p(x) = 1. \tag{1.3}$$

The typical probabilistic distributions include Gaussian distribution, Beta distribution, Weibull distribution, etc.

1.1.3 Scenario

In addition to the probabilistic distribution, the scenario is another way to describe the uncertainty of a random variable. A scenario can be viewed as a single realization of a stochastic process. From a computational viewpoint, a convenient manner to characterize stochastic processes is through scenarios. Scenario generation and reduction are research hot topics in presenting the uncertainty of renewable energy. More details about scenario reduction can be found in Chapter 7.

1.2 Multivariate Probabilistic Distributions

Since there are complex uncertainties within the power system, we need to extend single random variable analysis to multiple random variable analysis. There are three important probabilistic distributions to model the uncertainties and their dependencies of multiple random variables: joint distribution, marginal distribution, and conditional distribution.

In this section, we would like to use a very simple example to illustrate the differences among three very important concepts of probabilistic distributions. Here, we have two wind farms, the output of each wind farm is either 1, 2, or 3. We denote the outputs of the two wind farms as X and Y, and the probabilities of each output combination (X, Y) are presented in Table 1.1.

TABLE 1.1 Probabilities for each output combination of two wind farms

X	1	1	1	2	2	2	3	3	3
Y	1	2	3	1	2	3	1	2	3
Probabilities	0.18	0.1	0.12	0.1	0.05	0.12	0.05	0.1	0.18

1.2.1 Joint Distribution

Joint distribution gives the probability of two or more random variables. The joint probability mass function of two discrete random variables X, Y is:

$$P(X = x_i, Y = y_i) = p_{ij}$$

$$\text{where} \quad p_{ij} \geq 0, \quad \sum_i \sum_j p_{ij} = 1. \tag{1.4}$$

The joint probability density function are denoted as $f_{X,Y}(x,y)$ for two continuous random variables. We have:

$$\int_{-\infty}^{+\infty} \int_{-\infty}^{+\infty} f_{X,Y}(x,y) dx dy = 1. \tag{1.5}$$

For our example, Table 1.1 provides the possibilities of all the combinations of X and Y. It can be viewed as the discrete joint distribution.

1.2.2 Marginal Distribution

The marginal distribution of a subset of a collection of random variables is the probability distribution of the variables contained in the subset. It gives the probabilities of various values of the variables in the subset without reference to the values of the other variables. It can be viewed as a kind of dimension reduction.

The distribution is denoted as "marginal" because they can be found by summing values in a table along rows or columns, and writing the sum in the margins of the table. For discrete random variables, the marginal distribution of X and Y can be calculated as follows:

$$p_{i\cdot} = \sum_{j=1}^{+\infty} p_{ij}, \quad p_{\cdot j} = \sum_{i=1}^{+\infty} p_{ij}. \tag{1.6}$$

For continuous random variables, the marginal distribution of X and Y can be calculated as follows:

$$f_X(x) = \int_{-\infty}^{+\infty} f_{X,Y}(x,y) dy, \quad f_Y(y) = \int_{-\infty}^{+\infty} f_{X,Y}(x,y) dx. \tag{1.7}$$

For our example, the marginal distribution of X is:

$$p(x = 1) = 0.4; \quad p(x = 2) = 0.27; \quad p(x = 3) = 0.33. \tag{1.8}$$

1.2.3 Conditional Distribution

For two random variables, the conditional distribution describes the probability distribution of one random variable when the other variable is known to be a particular value. The concept can be extended to multiple random variables.

For discrete random variables, for a given value $Y = y_i$, the conditional distribution of X can be calculated as follows:

$$p_{i|j} = Pro(X = x_i | Y = y_j) = \frac{Pro(X = x_i, Y = y_i)}{Pro(Y = y_i)} = \frac{p_{ij}}{p \cdot j} p \cdot j. \qquad (1.9)$$

For continue random variables, for a given value $Y = y$, the conditional distribution of X can be calculated as follows:

$$f_X(x|Y = y) = \frac{f_{(X,Y)}(x,y)}{f_Y(y)}. \qquad (1.10)$$

For our example, the conditional distribution of X for a give $Y = 2$ is:

$$p(x = 1 | y = 1) = \frac{0.18}{0.18 + 0.1 + 0.05};$$

$$p(x = 2 | y = 1) = \frac{0.1}{0.18 + 0.1 + 0.05}; \qquad (1.11)$$

$$p(x = 3 | y = 1) = \frac{0.05}{0.18 + 0.1 + 0.05}.$$

Based on the conditional distribution, we can deduce the total probability formula:

$$f_Y(y) = \int_{-\infty}^{+\infty} f(y|x) f_X(x) dx, \quad f_X(x) = \int_{-\infty}^{+\infty} f(x|y) f_Y(y) dy, \qquad (1.12)$$

and the Bayes rule:

$$f(x|y) = \frac{f(x,y)}{f_Y(y)}, \quad f(y|x) = \frac{f(x,y)}{f_X(x)}. \qquad (1.13)$$

We can see that both the marginal distribution and conditional distribution can be deduced from the joint distribution. However, constructing the joint distribution of two or more random variables is not a trivial task because of their complex dependencies. We provide more deep insight into this issue in the next chapter.

1.3 Stochastic Process

A stochastic process is can be defined as a collection of random variables which are associated with or indexed by a set of numbers, usually viewed as points in time. For example, the output of a wind farm can be viewed as a stochastic process.

Mathematically, a stochastic process is defined as a collection of random variables defined on a common probability space $(\Omega, \mathcal{F}, P)$, where Ω is a sample space, $\mathcal{F}$ is a σ-algebra, and P is a probability measure. The random variables X_t are indexed by some set T and all take values in the same mathematical space S, which must be measurable. A stochastic process can be written as:

$$X_t : t \in T. \tag{1.14}$$

A stochastic process consists of four basic parts:

1. **Index set**: The set T is called the index set or parameter set of the stochastic process. Often this set is some subset of the real line, such as the natural numbers or an interval, giving the set T the interpretation of time.

2. **State space**: The mathematical space S of a stochastic process is called its state space. This mathematical space can be defined using integers, real lines, n-dimensional Euclidean spaces, complex planes, or more abstract mathematical spaces.

3. **Sample function**: A sample function is a single outcome of a stochastic process, so it is formed by taking a single possible value of each random variable of the stochastic process.

4. **Increment**: An increment of a stochastic process is the difference between two random variables of the same stochastic process. For a stochastic process with an index set that can be interpreted as time, an increment is how much the stochastic process changes over a certain time period.

Typical stochastic processes include the Bernoulli process, random walk, Wiener process, Poisson process, Markov processes, etc.

1.4 Stochastic Differential Equation

A type of stochastic differential equation is developed in this chapter that can fit both the marginal distribution and the correlation structure. Suppose the probability density f is continuous and strictly positive on (l, u), zero outside (l, u), bounded, and has finite variance. The following stochastic differential equation can be formulated:

$$dX_t = -\theta(X_t - \mu)dt + \sqrt{v(X_t)}dW_t, \ t \geq 0, \tag{1.15}$$

where $\theta > 0$, W_t is Brownian motion, μ is the mathematical expectation of f, and $\upsilon(X_t)$ is a non-negative function which is expressed as follows:

$$\upsilon(x) = \frac{2\theta}{f(x)} \int_l^x (\mu - y)f(y)dy, \ x \in (l, u). \tag{1.16}$$

Then the process X_t is mean-reverting and ergodic with invariant density f. If it is stationary, the autocorrelation function is

$$corr(X_{s+t}, X_s) = e^{-\theta t}, \ s, t \geq 0. \tag{1.17}$$

If W_t is multivariate processes where each coordinate is Brownian motion, dW_t is cross-correlated with a correlation coefficient matrix of C. Then the process X_t, constructed by Equation (1.15), also has the same correlation coefficient matrix.

Take wind power as an example; the wind speed may follow the Weibull distribution whose shape is determined by the parameters c and k:

$$f(x) = \frac{k}{c}(\frac{x}{c})^{k-1} exp\left[-(\frac{x}{c})^k\right]. \tag{1.18}$$

If f is a Weibull distribution as expressed in Equation (1.18), the corresponding items μ and $\upsilon(X_t)$ are as follows:

$$\mu = E(X) = c\Gamma\left(\frac{1}{k} + 1\right) \tag{1.19}$$

$$\begin{aligned}
\upsilon(x) &= \frac{2\theta}{f(x)} \int_l^x (\mu - y)f(y)dy = \frac{2\theta}{f(x)}\left[\mu F(x) - \int_l^x yf(y)dy\right] \\
&= \frac{2\theta}{f(x)}\left[c\Gamma\left(\frac{1}{k} + 1\right)\left(1 - exp\left[-(\frac{x}{c})^k\right]\right) - c\Gamma\left(\left(\frac{x}{c}\right)^k, \frac{1}{k} + 1\right)\right]
\end{aligned} \tag{1.20}$$

where F is the distribution function according to density f. $\Gamma(a)$ is gamma function and $\Gamma(x, a)$ is an incomplete gamma function

$$\Gamma(x, a) = \int_0^x y^{a-1}e^{-y}dy. \tag{1.21}$$

Thus, the wind speed of a singe wind farm can be simulated as follows:

$$\hat{v}_{it}^* = \hat{v}_{it-1}^* + dX_t. \tag{1.22}$$

How to simulate wind speed of multiple wind farms will be introduced in Chapter 6.

1.5 Stochastic Optimization

The stochastic programming problem is a framework for decision making under uncertainty. Given the probabilistic distribution information of uncertain factors, the goal is to find some policy that is feasible for all (or almost all) the possible instances and maximizes the expectation of objective function.

Generally, a deterministic mathematic programing problem is given by:

$$\min_{x \in X} \ g(x)$$
$$\text{s.t. } h(x) \leq 0 \tag{1.23}$$

where x represents decision variable, and $g(x)$ and $h(x)$ are the objective function and constraint condition, respectively. When random factor ξ is introduced, the deterministic programming problem becomes stochastic programming.

$$\min_{x \in X} \ g(x, \xi)$$
$$\text{s.t. } h(x, \xi) \leq 0 \tag{1.24}$$

Obviously, the random factor ξ exists in both the objective function and the constraint condition. Two common formulations are discussed in this section.

1.5.1 Two-Stage Stochastic Programming

The general formulation of the two-stage stochastic programming problem is given by:

$$\min_{x \in X} \ g(x) + E_\xi(Q(x, \xi))$$
$$\text{s.t. } a(x) \leq 0 \tag{1.25}$$

where $Q(x, \xi) = \left\{ \begin{array}{l} \min_{y \in Y} \ q(x, y(\xi), \xi) \\ \text{s.t. } h(x, y(\xi), \xi) \leq 0 \end{array} \right\}, \forall \xi \in \Omega$. In this formulation, at the first stage, "here and now" decision x is made with the goal of maximizing the expectation of the objective function before the realization of random factor ξ. Hence, the decision x does not depend on the realization of ξ. For the second stage, "wait and see" if decision $y(\xi)$ is made after knowing the actual realization of ξ. The optimal solution in the second-stage is considered as a recourse action pertaining to each realization of ξ, and $Q(x, \xi)$ is the recourse cost.

In order to solve the stochastic programming problem numerically, a common idea is assuming that random factor ξ has a finite number of possible realizations, called scenarios, and denoted by $\xi_1, ..., \xi_K$ with respective probability masses $p_1, ..., p_K$. Then, the stochastic problem can be transformed into a deterministic equivalent problem.

$$\min_{x \in X} \ g(x) + \sum_{k=1}^{K} p_k q(x, y_k, \xi_k)$$

$$\text{s.t. } a(x) \leq 0$$

$$h(x, y_k, \xi_k) \leq 0, \forall k$$

(1.26)

where y_k is the recourse action pertaining to the realization ξ_k.

1.5.2   Chance-constrained stochastic programming

Considering the situation that the decision x may not satisfy the constraint under all the possible realization of random factor ξ, a special principle is proposed that the decision x is permitted to satisfy the constraint with a minimum confidence level α, and the situation where the constraint is not satisfied will not be penalized. The general formulation is given by:

$$\min \ g(x)$$

$$\text{s.t. } P\{h(x, \xi) \leq 0\} \geq \alpha$$

(1.27)

When random factor ξ is described by a finite number of scenarios, $\xi_1, ..., \xi_K$. The deterministic equivalent formulation is given by:

$$\min_{x \in X} \ f(x)$$

$$s.t. \ g(x, \xi_k) \leq M(1\text{-}I_k), \forall k$$

$$\sum_{k=1}^{K} p_k I_k \geq \alpha$$

(1.28)

where I_k is a binary variable and M is a big positive constant. $I_k = 1$ represents the constraint pertained to the realization ξ_k that is satisfied. The second line of constraints ensures that the probability that the constraint is satisfied is higher than α.

1.6   Summary

In this chapter, we introduce several fundamental mathematics for power system uncertainty analysis, which will help us understand, analyze, model, and optimize the uncertainties brought about by high penetration of renewable energy. In addition to the individual uncertainty of a single object, there exist many dependencies among multiple objects such as multiple wind farms and PV station. How to model these dependencies is introduced in the next chapter.

2

Copula Theory and Dependent Probabilistic Sequence Operation

The integration of high penetration of renewable energy brings greater uncertainties for the operation of future power systems due to its intermittency and lack of predictability. The uncertainties brought by wide-scale renewable sources might have dependencies with each other because their outputs are mainly influenced by weather. Moreover, an analysis of such uncertainties with complex dependencies faces the curse of dimensionality. This challenges the power system uncertainty analysis in probabilistic forecasting, power system operation optimization, and power system planning. In this chapter, the principles of the copula function are introduced as a useful tool for the modeling and analysis of dependencies of renewable energy in this book. An extension of discrete convolution (DC), named dependent probabilistic sequence operation (DPSO), is proposed that considers the dependencies among stochastic variables using the copula function. An efficient DPSO calculation method based on a recursive stratagem is proposed to tackle the challenges of the curse of high dimensionality.

2.1 Introduction

The integration of high penetration of renewable energy presents greater uncertainties to power systems [1]. These uncertainties, i.e., outputs of wind power and photovoltaics (PV), electrical and heat demands, energy price, and charging behaviors of electrical vehicles (EVs), are coupled and result in complex dependencies based on relevant factors such as weather and time of the day. The modeling and analysis of high-dimensional dependent uncertainties are of great significance for the reliability and security of future power systems.

However, the uncertainties may not follow analytical distributions such as Gaussian distribution and Beta distribution, and their dependencies may be much more complex than linear. Most of the current methodologies such as Pearson correlation analysis and principal component analysis (PCA) cannot handle such dependencies well. The copula theory, as an effective way to model the nonlinear dependencies among various uncertainties, has been introduced into the modeling of stochastic dependence for power system uncertainty analysis [2].

The previous studies of copula-based uncertainty dependency mainly focus on modeling the dependency structures, and then sampling methods are applied to generate large numbers of samples of these uncertainties for basic operations (such as summation) to evaluate reliability, calculate PLF or formulate an optimization model. Even though high-performance computing can be efficiently applied for sampling [3], sampling-based calculation of high-dimensional uncertainties still needs to balance the stability of the calculation result and calculation time, i.e., exponentially growing samples are needed to obtain stable probabilistic distributions but result in a large computational burden. In addition, for many power system analyses, we focus more on the tail of the distributions such as the loss-of-load probability (LOLP). It is very hard for the sampling method to get a stable result on the tails of the distribution.

In the probabilistic analysis, convolution is a typical method to calculate the distribution of the aggregation (or subtraction) of two stochastic variables. Discrete convolution is used for its computerized implementation [4]. Discrete convolution is very efficient compared with other probabilistic analysis approaches such as Monte Carlo simulation. It thus has been widely applied in power system reliability analysis and probabilistic load flow calculations. A prerequisite of convolution is that the stochastic variables being convolved should be independent. This prerequisite holds for most of the cases in power system analysis, e.g., failure events of generators. However, in some situations, this independence may not hold, e.g., the power output of adjacent wind farms. Therefore, dependence must be specially handled before conducting discrete convolution to follow this strong prerequisite, which may increase the complexity and jeopardize the efficiency of discrete convolution.

This chapter first introduces the dependencies among stochastic variables and the copula theory. Then, a formulation of DC under the dependent circumstances using the copula function is proposed for DPSO. The dependent form of convolution is found to have an extra multiplier compared with traditional convolution. The proposed approach generalizes the discrete convolution that allows a direct calculation irrespective of the dependency between stochastic variables.

2.2   Dependencies and Copula Theory

Stochastic dependence refers to the behavior of a random variable that is affected by others. If one random variable has no impact on the probability distribution of the other, these two variables are regarded as independent variables; otherwise, they are regarded as stochastically dependent variables.

For example, the renewable energy outputs of two adjacent farms can be seen as two dependent probabilistic variables because both of them are affected by the weather of the area.

Modeling stochastic dependence is challenging because it is hard to find a multivariate analytical formula for variables with complex marginal distributions. For example, if one variable has a Weibull distribution and the other has a Gaussian distribution, their stochastic dependence can hardly be modeled by an analytical bivariate function. Though the concept of the correlation coefficient is widely used in probabilistic relationship studies, it only provides a way of measuring the stochastic dependence and does not contain all of its information, and therefore can hardly be used in modeling the stochastic dependence.

Copula theory provides an effective way of modeling stochastic dependence. Suppose x and y are random variables with a joint distribution F_{XY} and invertible cumulative distribution functions (CDF) F_X and F_Y, respectively. If we regard $F_X(x)$ and $F_Y(y)$ as random variables, then both of them follow the uniform distributions as below:

$$F_Y(x) \sim u(0,1), \; F_Y(y) \sim u(0,1). \tag{2.1}$$

In other words, $F_X(x)$ and $F_Y(y)$ transform x and y into uniform distributions, respectively.

In copula theory, F_{XY} can be written as:

$$F_{XY}(x,y) = C\left(F_X(x), F_Y(y)\right) \tag{2.2}$$

where the function C is named the copula function. It is a special kind of multivariate CDF that has uniform margins. If we set $u = F_X(x)$ and $v = F_Y(y)$, C can be expressed as:

$$C(u, v) = F_{XY}(F_X^{-1}(u), F_X^{-1}(v)). \tag{2.3}$$

Copula theory transforms the modeling of F_{XY} into the modeling of F_X, F_Y, and C, separately. It takes advantage of the fact that stochastic dependence is more easily recognized for uniform variables ($F_X(x)$ and $F_Y(y)$) than other arbitrarily distributed variables (x and y).

In fact, the above copula theory is named Sklar's theorem, which was proposed in 1959 to model the joint multivariate distribution function as a complex function of marginal distribution functions and a multivariate function showing its dependencies. Instead of two random variables, for any group of stochastic variables $(x_1, x_2, \cdots x_N)$, their joint cumulative distribution function can be expressed as:

$$F(x_1, x_2, \cdots x_N) = C(F_1(x_1), F_2(x_2), \cdots, F_N(x_N)) \tag{2.4}$$

where $F_i(x_i)$ is the marginal distribution of variable x_i; the function $C(\cdot)$ is called the copula function, the definition domain of which is $[0, 1]^N$.

Again, the core idea of Sklar's theorem is to decompose the modeling of the joint distribution function into separate models of the marginal distributions of each of these variables and their dependence structure (copula function) so that the copula function is much more easily approximated and formulated using well-defined analytical functions compared to the joint distribution function. Typical copula functions that are frequently used include Gaussian copula, t copula, Clayton copula, Gumbel copula, and Frank copula. The head and tail correlations of the former two are symmetrical and called the elliptical copula, whereas those of the latter three are asymmetrical and called the Archimedean copula.

In addition to the basic copula function, there are various types of copula-based modeling methods. The C-vine copula, which is suitable for high-dimensional dependencies, is proposed in [5] to model the uncertainties of the outputs of multiple PV stations. Dimensionality reduction, clustering, and parametric modeling techniques are applied for Vine copula construction in [6] and then Monte Carlo analysis is conducted to simulate the high-dimensional stochastic variables. Another innovative application of C-vine copula is to capture the uncertainties of residential electricity consumption and group the consumers through C-vine copula-based clustering [7]. Sun et al. apply a fuzzy copula to model wind speed output correlations and evaluate wind curtailment [8]. In power system analysis, the copula theory is used in the modeling of renewable energy outputs, load, and energy prices. A case study of multiple wind farms in Europe is conducted based on copula theory in [9], whereas the modeling of correlations among different wind farm outputs is used for probabilistic forecasting in [10]. By investigating and modeling the relational diagram of electrical load and real-time price, a probabilistic load forecasting method is proposed in [11]. The dependencies between wind power output and electricity price and between the prices of carbon emissions and the energy prices are modeled based on copula theory in [12], respectively. Lojowska et al. model the correlations among the time a vehicle leaves home, the time a vehicle arrives home, and the distance traveled for EV based on Copula in [13]. The dependency of wind and PV forecast error and the dependency of multi-area load forecast error are modeled in [14] and [15], respectively.

By taking the derivative of the joint cumulative distribution function, the joint probability density function (PDF) can be calculated as follows:

$$
\begin{aligned}
f(x_1, x_2, \cdots x_N) &= \frac{\partial^N F(x_1, x_2, \cdots x_N)}{\partial x_1 \partial x_2 \cdots \partial x_N} \\
&= c\left(F_1(x_1), F_2(x_2), \cdots F_N(x_N)\right) \prod_{n=1}^{N} f_n(x_n).
\end{aligned}
\tag{2.5}
$$

Equation (2.5) shows that only the individual marginal distributions and their copula density function are needed to calculate their joint PDF. The marginal distribution of each stochastic variable $f_n(x_n)$ can be obtained by using simple statistics or regression with certain analytical distributions.

Essentially, the copula-based modeling can be viewed as a multidimensional PDF regression problem. Thus, the selection of the copula function can be guided by using the χ^2 test. The test is used to quantify the significance of the assumed theoretical distribution fitting sampling distribution according to the difference between the frequency of statistical samples and the assumed theoretical probability [16]. A sum of squared errors can be calculated as follows:

$$\eta = \sum_{i=1}^{m} \frac{(f_i - np_i)^2}{np_i} \tag{2.6}$$

where n is the number of samples; m is the number of segmented regions of the copula PDF; f_i is the number of samples in the i-th region; and p_i is the theoretic probability of the i-th region of the copula function. Thus, the statistic η approximately follows the χ^2 distribution. The degree of freedom is $m - r - 1$, where r is the number of parameters in the copula functions. By choosing a significance level α, the dependent structure of the samples follows the copula function if $\eta < \chi_\alpha^2(k - r - 1)$ holds.

The typical copula functions can be parameterized by using Kendall's rank correlation coefficient τ, which measures the ordinal association between samples of two stochastic variables [17]. Kendall's rank correlation is not affected by the marginal distributions of the variables. For any two correlated stochastic variables with m samples, τ can be calculated by solving:

$$\tau = \frac{(\text{No.of concordant pairs}) - (\text{No. of discordant pairs})}{m(m-1)/2} \tag{2.7}$$

where the denominator is the number of all combinations. Then, we have $-1 \leq \tau \leq 1$.

After calculating the coefficient, the parameters of the typical copula function can be obtained. For example, for the Gaussian and t copulas, the parameter ρ can be calculated as follows:

$$\rho_n = \rho_t = \sin(\pi\tau/2) \tag{2.8}$$

where ρ_n and ρ_t are parameters to control the shape of the Gaussian copula and t copula, respectively [18]. The parameters for other copula functions can also be calculated according to τ. Thus, to model the dependence structures among multiple stochastic variables, we need to calculate Kendall's rank correlation coefficient of each pair of variables.

2.3   Dependent Probabilistic Sequence Operation

Assume that α and β have a joint probability distribution function (PDF) $f_{\alpha,\beta}(x,y)$ with marginal distributions $f_\alpha(x)$ and $f_\beta(x)$, respectively. The PDF of their sum $f_{\alpha+\beta}(a)$ can be calculated using $f_{\alpha,\beta}(x,y)$ as:

$$f_{\alpha+\beta}(a) = \int_{\underline{\alpha}}^{\overline{\alpha}} f_{\alpha,\beta}(x, a - x)dx, \ a \in [\underline{\alpha} + \underline{\beta}, \overline{\alpha} + \overline{\beta}] \tag{2.9}$$

where $[\underline{\alpha}, \overline{\alpha}]$ and $[\underline{\beta}, \overline{\beta}]$ define the domain of α and β, respectively. If α and β are independent, the $f_{\alpha,\beta}(x,y)$ equals the product of its marginal PDFs:

$$f_{\alpha,\beta}(x, y) = f_\alpha(x)f_\beta(y). \tag{2.10}$$

In this situation, Equation (2.9) becomes the convolution equation:

$$f_{\alpha+\beta}(a) = \int_{\underline{\alpha}}^{\overline{\alpha}} f_\alpha(x)f_\beta(a - x)dx, \ a \in [\underline{\alpha} + \underline{\beta}, \overline{\alpha} + \overline{\beta}]. \tag{2.11}$$

This convolution is an effective way of calculating the PDF of their aggregation. For computerized calculation, the discrete form of convolution is used:

$$P_{\alpha+\beta}(i) = \sum_{j=i_{\underline{\alpha}}}^{i_{\overline{\alpha}}} P_\alpha(j)P_\beta(i - j), \ i \in [i_{\underline{\alpha}} + i_{\underline{\beta}}, i_{\overline{\alpha}} + i_{\overline{\beta}}] \tag{2.12}$$

where $P_\alpha(i)$, $P_\beta(i)$ and $P_{\alpha+\beta}(i)$ are the discrete approximations of $f_\alpha(\cdot)$, $f_\beta(\cdot)$ and $f_{\alpha+\beta}(a)$ with a step of Δp, e.g.:

$$P_\alpha(i) = \int_{i\Delta p - \Delta p/2}^{i\Delta p + \Delta p/2} f_\alpha(x)dx \tag{2.13}$$

When variables are not independent, Equation (2.10) does not hold and convolution in Equations (2.12) and (2.13) can no longer be used.

According to Equation (2.5), the density function $f_{\alpha,\beta}(x,y)$ under dependent condition becomes:

$$f_{\alpha,\beta}(x,y) = \frac{F_{\alpha,\beta}(x,y)}{\partial x \, \partial y} = c\left(F_\alpha(x), F_\beta(y)\right) f_\alpha(x)f_\beta(y) \tag{2.14}$$

Equation (2.14) states that the copula PDF is the ratio of the joint PDF to the PDF if the variables were independent. Using (2.14), the dependent convolution becomes:

$$f_{\alpha+\beta}(a) = \int_{\underline{\alpha}}^{\overline{\alpha}} c\left(F_\alpha(x), F_\beta(a - x)\right) f_\alpha(x)f_\beta(a - x)dx. \tag{2.15}$$

To this regard, the DPSO can be formulated as:

$$P_{\alpha+\beta}(i) = \sum_{j=i_{\underline{\alpha}}}^{i_{\overline{\alpha}}} \iint_{j\Delta p - \Delta p/2, (i-j)\Delta p - \Delta p/2}^{j\Delta p + \Delta p/2, (i-j)\Delta p + \Delta p/2} f_{\alpha,\beta}(x, y)dxdy$$

$$= \sum_{j=i_{\underline{\alpha}}}^{i_{\overline{\alpha}}} \iint_{j\Delta p - \Delta p/2, (i-j)\Delta p - \Delta p/2}^{j\Delta p + \Delta p/2, (i-j)\Delta p + \Delta p/2} c\left(F_\alpha(x), F_\beta(y)\right) f_\alpha(x)f_\beta(y)dxdy. \tag{2.16}$$

If Δp is small enough, $c\left(F_\alpha(x), F_\beta(y)\right)$ can be seen as constant over the 2-dimensional interval of $[j\Delta p - \Delta p/2 , j\Delta p + \Delta p/2] \times [(i-j)\Delta p - \Delta p/2 , (i-j)\Delta p + \Delta p/2]$ and is noted as $c_{j\Delta p,(i-j)\Delta p}$. Equation (2.16) becomes:

$$
P_{\alpha+\beta}(i) = \sum_{j=i_{\underline{\alpha}}}^{i_{\overline{\alpha}}} c_{j\Delta p,(i-j)\Delta p} \int_{j\Delta p - \Delta p/2}^{j\Delta p + \Delta p/2} f_\alpha(x)dx \int_{(i-j)\Delta p - \Delta p/2}^{(i-j)\Delta p + \Delta p/2} f_\beta(y)dy
$$

$$
= \sum_{j=i_{\underline{\alpha}}}^{i_{\overline{\alpha}}} c_{j\Delta p,(i-j)\Delta p} P_\alpha(j)P_\beta(i-j).
$$

$$(2.17)$$

Comparing with the formulation of DC in Equation (2.12), the DPSO has a similar form of DC, except that there is an extra multiplier $c_{j\Delta p,(i-j)\Delta p}$ determined by the dependency between α and β. Note that the $F_\alpha(x)$ and $F_\beta(y)$ in $c_{j\Delta p,(i-j)\Delta p}$ can also be calculated using $P_\alpha(i)$ and $P_\beta(i)$, the complete formulation of DPSO would be:

$$
P_{\alpha+\beta}(i) = \sum_{j=i_{\underline{\alpha}}}^{i_{\overline{\alpha}}} c\left(\sum_{m=0}^{j} P_\alpha(m), \sum_{n=0}^{i-j} P_\beta(n)\right) P_\alpha(j)P_\beta(i-j) \quad (2.18)
$$

Equation 2.18 generalizes the discrete convolution approaches. When α and β are independent, $c_{j\Delta p,(i-j)\Delta p} \equiv 1$ and Equation (2.18) degenerates into Equation (2.12). When they are dependent, only an extra multiplier needs to be calculated for each discrete term of convolution. If the dependency structure of two variables can be expressed using a typical copula function, the copula-based multiplier can be calculated with high efficiency. It can be proved that $P_{\alpha+\beta}(i)$ is still a discrete probabilistic distribution function after adding this multiplier.

The proposed DPSO approach was tested in the generation reliability assessment of a modified IEEE RTS-79 system. Three wind farms with a nominal capacity of 170 MW each were added to the generation mix. The chronological outputs of these wind farms were extracted from the NREL Wind Integration Datasets, which are strongly correlated with linear correlation coefficients between 0.8 and 0.9. The results obtained from four different approaches are compared: 1) "MC" uses the Monte Carlo method and acts as a reference. The outputs of the three wind farms are synchronously sampled from their chronological outputs in order to maintain their inherent dependence. 2) "LHS" uses Latin hypercube sampling to capture the dependency of wind farms. The convergence limits of both MC and LHS are set to be 0.1%. 3) "DC" employs convolution to conduct the classic Capacity Outage Probability Table (COPT) approach for reliability assessment, however, the dependence of wind farms are not considered. 4) "DPSO" is similar to DC except that it uses DPSO to take into consideration the dependence of wind.

Figure 2.1 compares the PDF of the aggregated output of the three wind farms estimated by different approaches with its PDF obtained from MC. The results show that both dependent MC sampling and DPSO provide satisfactory results while ignoring the dependence would cause considerable errors.

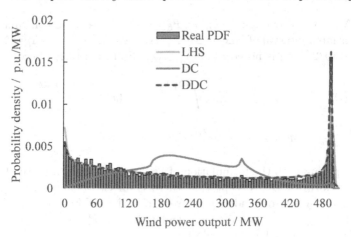

FIGURE 2.1 PDFs of the aggregated output estimated by different approaches.

Figure 2.2 compares the COPTs obtained from each of the approaches and Table 2.1 lists the results of LOLP and EENS. The results indicate that DPSO improves the calculation accuracy compared with DC by successfully modeling the dependency of wind power. Assuming independence in DC gives optimistic results as expected. The results of DPSO are close to LHS, indicating the effectiveness of this approach.

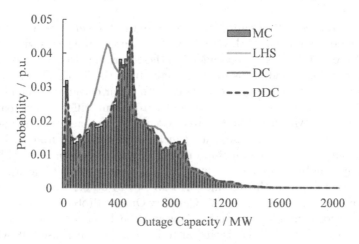

FIGURE 2.2 COPTs obtained from different approaches (each bar represents the sum of the outage probability within 20 MW capacity interval).

Table 2.1 also compares the computing time of each method, showing

TABLE 2.1 Reliability assessment results of different approaches

Approaches	LOLP / p.u.	EENS MWh/a	CPU Time / s
MC	3.50E-04	367.55	1051.96
LHS	3.54E-04	364.04	1326.8
DC	2.34E-04	231.82	3.07
DPSO	3.48E-04	364.94	4.5

that the DPSO significantly improves the calculation efficiency compared with Latin hypercube sampling. The superior calculation performance makes the proposed approach attractive in future power system uncertainty analysis.

2.4 High-Dimensional DPSO Computation

Further, the DPSO of two variables can be extended to N variables:

$$P_y(j) = \sum_{i_{x_1} + \ldots + i_{x_N} = j} \left(s_c(i_{x_1}, \ldots, i_{x_N}) \cdot \prod_{n=1}^{N} P_n(i_{x_N}) \right). \tag{2.19}$$

The traditional discrete convolution is denoted as follows:

$$y = x_1 \otimes x_2. \tag{2.20}$$

To distinguish between traditional discrete convolution and DPSO, the DPSO is represented as follows:

$$y = x_1 \overset{C}{\otimes} x_2. \tag{2.21}$$

Similarly, the DPSO for N stochastic variables $x_n (n = 1, 2, \ldots, N)$ is defined as follows:

$$y = \sum_{n=1}^{N} \overset{C}{\otimes} x_n. \tag{2.22}$$

It should be noted that the DPSO for multiple variables cannot be directly calculated by conducting DPSO between two variables successively according to Equation (2.19), as each variable is dependent on each remaining variable.

If the length of the discrete PDF of each variable is K, the complexity of the DPSO of two variables is K^2. The length of the discrete PDF is determined based on the variation area and the discretized step. A smaller discretized step will yield a higher computational burden but smaller calculation error. A balance between the computational burden and error should be achieved in DPSO. However, for multiple DPSOs, the computational complexity is K^N, making the proposed method unsuitable for high-dimensional DPSO. As the variables are coupled, the computation cannot be conducted in parallel. An exponential computational complexity of traditional DPSO limits its application in low (2 or 3) dimensional problems. In the next section, the strategies to make the calculation of high-dimensional DPSO more efficient is proposed.

This section introduces an algorithm for the proposed HD-DPSO to accelerate the computation. The algorithm can be divided into five stages, as shown in Algorithm 1. The first stage is to divide the variables into subgroups, where the variables in any two subgroups are independent. In this way, HD-DPSO can be implemented for each subgroup with lower dimensions than that of the original variables. The second stage is to handle the subgroups with the Gaussian distribution and Gaussian copula variables by direct expectation and deviation calculation. Then the third stage and fourth stage handle the remaining subgroups. The third stage is to conduct small-scale sampling according to their marginal distributions and dependencies, where the sampling will be used in the recursive dimension deflation process in the fourth stage. The fourth stage is to conduct recursive sample-guided DPSO for each subgroup. The fifth stage is to obtain the final result using the DPSO calculation result for each subgroup using the traditional discrete convolution. The details of the first four stages are introduced in the following.

Algorithm 1 Algorithm-HD-DPSO

Require: The marginal distribution of each variable; their dependencies, as described by Kendall's rank correlation coefficient.

  Grouping Stage: Divide the variables into subgroups.

  DPSO for Each Group:

  for $k = 1 : k$ **do**

    Aggregation Stage: Aggregate the variables of high-dimensional Gaussian distribution.

    Sampling Stage: Small-scale sampling according to their marginal distributions and dependencies.

    Recursive DPSO Stage:

    Set $j = 1$

    while $j < N$ **do**

      DPSO for two variables.

      Update their dependencies.

      $j = j + 1$.

    end while

  end for

  Summation Stage: Conduct traditional discrete convolution of the results obtained from the subgroups.

2.4.1 Grouping Stage

In this stage, the variables are divided into subgroups according to the rank correlation matrix. An illustration of the grouping is shown in Figure 2.3; five subgroups are obtained, and there are no dependencies among the subgroups.

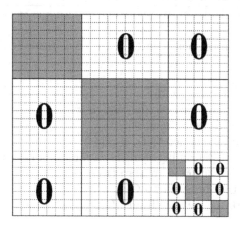

FIGURE 2.3 Illustration of grouping where each two variables in different groups are independent.

The grouping of variables depends on the problem to be analyzed. For example, if we analyze the output of several wind farms that are not so far from each other, the data are all highly dependent. All of them should be in the same group and there is no need for grouping. However, if we analyze the reliability of one system which takes the outage of conventional generations and uncertainty of wind power into consideration, there probably exist at least two groups. One is the conventional generations, the other is the wind power. The variable order influences the grouping. We suggest that the ordering of variables can be conducted from two aspects in power system analysis:

For power system analysis, we have prior knowledge of the dependency of different uncertainties in the power systems, such the uncertainties from the outage of conventional generations and wind power. Based on this prior knowledge, the variables can be primarily classified into different groups.

In addition, an iterative search can be implemented to obtain the optimal ordering or grouping results, as shown in Algorithm 3. For the variable set M, first, randomly select a variable denoted as x_1 and form a group G_i containing x_1. Then scan the rank correlations between the variables in G_i and M. If the correlation is nonzero, the corresponding variables should be grouped into G_i and deleted from M. The iteration is terminated when there is no variable added into G_i. Thus, the variables in the set G_i can be classified into a group. This subgroup satisfies the criterion that the variables in this group are independent of the variables in another group. Selecting another one in the remaining variables, another subgroup can be formed by repeating the process above. The grouping stage ends until all variables have been assigned into subgroups.

Algorithm 2 Algorithm-Variables-Grouping

Require: The variable set M, and Kendall's rank correlation coefficient matrix C.
  If the correlation coefficient in C is less than a small tolerance, it is set to zero.
  Set $i = 1$;
  while set M is not empty **do**
    Randomly select a variable denoted as x_1.
    Form a group G_i containing x_1.
    while no variable added to G_i do **do**
      for each correlation between the variables in G_i and M **do**
        if the correlation is nonzero **then**
          Add the variable into G_i.
          Delete the variable from M.
        end if
      end for
    end while
    Update their dependencies.
    $i = i + 1$.
  end while

2.4.2 Gaussian-Distribution-Based Aggregation Stage

Before handling the DPSO for each subgroup, we can first pick up the subgroups with Gaussian distributions for which the convolutions can be handled using analytical formulations other than numerical approaches. If the distributions of the variables and their dependencies fit Gaussian distributions and the Gaussian copula, respectively, then the distribution of their linear combination is also Gaussian distributed. Only the expectation and variance need to be calculated. Their expectation is the sum of expectations of individual variables. The variance of the sum of multiple variables can be calculated as follows:

$$D\left(\sum_{k=1}^{K} a_k x_k\right) = \sum_{k1=1}^{K}\sum_{k2=1}^{K} \rho_{x_{k1}x_{k2}} a_{k1}\sigma_{x_{k1}} a_{k2}\sigma_{x_{k2}} \qquad (2.23)$$

where σ_{x_k} denotes the standard deviation of the variable x_k; and $\rho_{x_{k1}x_{k2}}$ denotes the Pearson correlation between x_{k1} and x_{k2}.

In this way, with known expectation and variance, the distribution of the linear combination can be easily obtained without DPSO computation. It should be noted that in this stage we only handle the subgroups where all the variables follow Gaussian distribution and the Gaussian copula. Such special treatment is important for such an algorithm since in practical application the Gaussian distribution is widely used. Other subgroups are treated in the next stage. If there is no group in which the variables do not follow the Gaussian copula, this stage can be skipped and we can directly go to the next stage.

2.4.3 Small-Scale Sampling Stage

For the remaining subgroups that do not follow Gaussian distributions, DPSO will be conducted. Before implementing the convolution, a small scale-sampling that considers the correlation between variables is conducted for each of the subgroups. It should be noted that the sampling is not for calculating the distribution of the aggregation of variables, but for the re-modeling of copula functions in the next stage. The remodeling of copula functions only requires a small number of samples, which is significantly less than the number of samples that are required to obtain a stable distribution. The procedure, therefore, consumes much less computational time.

For a subgroup that has variables $x_1, x_2, ..., x_K$ with marginal distribution $F_1(x_1), F_2(x_2), \ldots, F_K(x_K)$ and dependency structure $C_K(u_1, u_2, ..., u_K)$, the sampling is calculated as:

$$x_1 = F_1^{-1}(\lambda_1), x_2 = F_2^{-1}(\lambda_2), \cdots, x_K = F_K^{-1}(\lambda_K) \qquad (2.24)$$

where $\lambda_1, \lambda_2, ..., \lambda_K$ is the sampling according to the copula function $C_K(u_1, u_2, ..., u_K)$:

$$[\lambda_1, \lambda_2, ..., \lambda_K] = \text{rnd}\,[C_K(u_1, u_2, ..., u_K)] \qquad (2.25)$$

where rnd $[\cdot]$ is the operation of sampling according to accumulated distribution function. Especially, if $C_K(u_1, u_2, ..., u_K)$ is Gaussian copula with the correlation coefficient matrix Σ, the sampling is calculated as:

$$x_1 = F_1^{-1}[\Phi(\xi_1)], x_2 = F_2^{-1}[\Phi(\xi_2)], \cdots, x_K = F_K^{-1}[\Phi(\xi_K)] \qquad (2.26)$$

where $\xi_1, \xi_2, ..., \xi_K$ is the sampling of multivariable Gaussian distribution:

$$[\xi_1, \xi_2, ..., \xi_K] = \text{rnd}\,[N\,(\mathbf{0}, \Sigma)] \qquad (2.27)$$

These samplings can be used in all of the following copula function updating processes.

2.4.4 Recursive Sample-Guided DPSO

For each subgroup that does not follow Gaussian distributions, a recursive sample-guided DPSO is proposed to tackle the "curse of dimensionality" originating from high-dimensional dependencies. The main idea of such a technique is to decompose the DPSO of multiple dependent variables into multiple stages, whereby in each stage, only the two dependent variables are convolved. The overall calculation can be largely reduced. More specifically, we first choose two dependent variables among all the K variables and calculate the DPSO of these two variables. The obtained new variable and the remaining K-2 variables form a new HD-DPSO problem with K-1 dimensions, one dimension lower than the number of dimensions of the original problem. Therefore, the overall DPSO calculation can be processed in a recursive manner following the above dimension deflation procedure. One problem remains: the dependency structure of the problem is changed during each dimension deflation such that the copula function model needs to be renewed in each step. For dependent variables $x_1, x_2, ..., x_K$ with copula function C_{K0}, the recursive DPSO calculation steps can be summarized as follows:

1. Calculate the DPSO of x_1 and x_2 using Equation (2.19), i.e., $x_{r1} = x_1 \overset{c_{1,2}}{\otimes} x_2$, where $c_{1,2}$ is part of the structure of C_{K0}.

2. Update the copula function among $x_{r1}, x_3, ..., x_K$; the new copula function is demoted to a C_{K-1}.

3. Regarding the DPSO calculation of $x_{r1}, x_3, ..., x_K$ and c_1 as a new problem, go back to step 1) to calculate the DPSO of x_{r1}, x_3, and then go back to step 2) to update the copula function to C_{K-2}. Repeat the procedure until $x_{r,K-2}$ and c_{K-2} are obtained.

4. The final result is $y = x_{r,K-2} \overset{c_{K-2}}{\otimes} x_K$.

The critical part of the above process is to update the copula function. Technically, the procedure can be conducted using the general copula function identification procedure described in Section 2.2. However, in a real case, there usually would be no realization of variables that can be used for modeling the copula function. Furthermore, the copula function might be distorted after the DPSO for some of the variables is conducted, i.e., the copula function between $x_{r1}, x_3, ..., x_K$ may not follow a well-defined copula function. We, therefore, propose a sampling-guided approach to update the copula function. As most of the high-dimensional Copula models use these two copulas, we illustrate in Figure 2.4 how the sampling-guided approach is conducted for these copula functions.

Each small box in Figure 2.4 denotes the dependency structure between two variables. c denotes the accordant copula function. When reducing the DPSO from five dimensions to four dimensions, $c'_{2,3}$ $c'_{2,4}$ and $c'_{2,5}$ need to be updated, and the others remain unchanged. When estimating $c'_{2,3}$, the sampling of x_1, x_2, and x_3 are used, then, $c'_{2,3}$ can be calculated based on the samples of $x_1 + x_2$ and x_3. By following the above approach, each element in the correlation matrix can be updated to obtain the updated copula function. In other words, the copula function updating process is guided by the sampling.

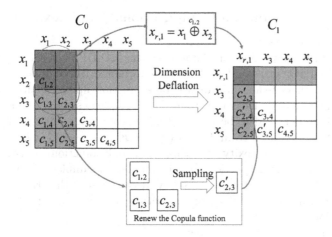

FIGURE 2.4 Dimension deflation and Copula function updating process.

The updated copula function may not be of the same type as the original copula function C_{K0}. In such a case, a vine copula can be used to link different copula types. As a lot of the high-dimensional copula models use the Gaussian copula or t copula, in such a special case, we also find that for these two copulas, the dependence structure after dimension deflation can still be modeled using the Gaussian copula or t copula. The dependency structure of such two copulas is fully determined by a correlation coefficient matrix. Therefore, updating the dependency structure is equivalent to updating the accordant elements of the matrix, which turns the copula updating procedure a very simple operation.

It should be noted that although the PDF of $x_1+x_2+...+x_K$ can also be directly estimated using the samplings mentioned above, the proposed method is different from the sampling method in two ways: 1) The accurate estimation of the PDF (especially tails of the PDF) using the sampling method requires a substantial number of samples. In the proposed method, only a small number of samples (about 100, which will be verified in the case study) are required to obtain a reliable and stable estimation of the correlation coefficient. 2) Most operations in the HD-DPSO approach are deterministic. In the sampling guided dimension reduction, the impact of the stochastic sampling on copula function identification is sufficiently small. Therefore the result of HD-DPSO is very robust against stochastic factors in the calculation. In contrast, the Monte Carlo-based sampling method, which is regarded as the only effective method for high-dimensional uncertainty analysis, involves the stochastic sampling process and therefore cannot usually obtain a stable result.

2.4.5   Discussions on Computational Complexity and Error

1) Complexity analysis

Using the above-mentioned approach, the HD-DPSO is decomposed into a series of DPSOs between two dependent variables. In the dimension deflation procedure, there is an additional copula function updating the operation in each reduction of dimension (e.g., in Figure 2.4, $c'_{2,3}$ should be identified, $c'_{2,4}$ and $c'_{2,5}$ do not need to be identified). The computational complexity of the HD-DPSO is reduced from K^N to $(N-1)K^2 + (N-1)Q$, where Q denotes the calculation complexity of fitting the copula function for two stochastic variables, which is proportional to the number of samples and accounts for a very small part of the total complexity. It can be seen that HD-DPSO makes the approach a polynomial time algorithm, which is preferable in terms of calculation tractability.

2) Computation error analysis

The computation error of the approach mainly comes from two sources: 1) the discretization error during the approximation; 2) the estimation error of the sampling-guided copula function updating process. The former error can be reduced by calculating the definite integral of the copula function instead of the copula density function; the latter can be controlled based on the number of samplings.

3) Potential for parallel computation

The proposed approach can be further extended to a distributed algorithm, whereby the dependent variables can be calculated as pairs, e.g., $x_1 \overset{c_{1,2}}{\otimes} x_2$, $x_3 \overset{c_{3,4}}{\otimes} x_4, \ldots, x_{K\text{-}1} \overset{c_{K-1,K}}{\otimes} x_K$. Then, the final result can be obtained by repeating the DPSO for the result of each pair. Such decomposition makes a parallel computation possible to further accelerate the calculation.

2.4.6   Case Study

This section introduces a typical application of the proposed method in reserve requirement evaluation with a high share of renewable energy. The numerical experiments are conducted using a modified IEEE RTS-79 system. This system contains 32 conventional generations with a total capacity of 3405 MW, and the peak load is 2850 MW [19]. Ten wind farms are added to the generation mix, of which the output time series is from NREL [20]. The capacity of each wind farm is 200 MW. The annual energy generated by each wind farm is 5% of the annual total energy demand, making the overall penetration of wind power 50% of the system load. The experiments are conducted using MATLAB® R2015a on a standard PC with an Intel® Core^{TM} i7-4770MQ CPU running at 2.40 GHz and with 8.0 GB of RAM.

Reserve requirements in a power system are quantified at different timescales, i.e., frequency regulation, load following, and forecast error. The balancing reserve is qualified based on the hourly net load forecast error. The load forecast errors are assumed as a Gaussian distribution, the standard deviations of which are 5% of the forecasted value. The wind power forecast error is modeled using the conditional forecast error [21]. The forecasted output of ten wind farms is shown in Figure 2.5. For simplicity, the short-term correlation of each two wind power forecast errors and load forecast errors is assumed to be r_{ws} and r_{ls}, respectively. The wind and load forecast errors are assumed to be independent. Both the dependencies among load forecast errors and wind power forecast errors are assumed to be Gaussian copula correlated.

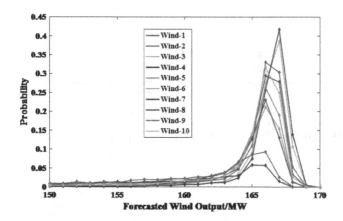

FIGURE 2.5 Distributions of the forecasted outputs of ten wind farms.

The distribution of "net errors" can be calculated as follows:

$$N_e = \sum_{n=1}^{N} \overset{c}{\otimes} L_{ne} - \sum_{M=1}^{M} \overset{c}{\otimes} W_{me} \qquad (2.28)$$

where L_{ne} and W_{me} denote the n-th wind and m-th load forecast error, respectively; N and M are the number of wind farms and load injection nodes, respectively; N_e denotes the system "net load error".

We set r_{ws} and r_{ls} as 0.7 and 0.3, respectively. Then, the distributions of the system "net errors" are obtained based on HD-DPSO and Monte Carlo (MC) method with 10^{12}, 10^8, and 10^6 samples, which are shown in Figure 2.6. The distribution calculated by HD-DPSO is smooth and coincided with the distribution obtained by MC method with 10^{12} samples. When the number of samples for the MC method decrease, the obtained distribution has some deviation compared with the distribution obtained with 10^{12} samples. It should be noted that the time consumed by HD-DPSO is 10.12 seconds which is comparable to the time consumed by the MC method with 10^8 samples. To cover the negative or the positive part of the "net errors", the up and down reserves are needed. By setting the confidence error to 90%, the up and down reserves can be determined from the distribution of "net errors", being 210 MW and 274 MW, respectively.

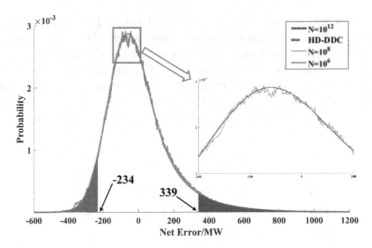

FIGURE 2.6 Distributions of the system "net errors" with Wind #1-#10.

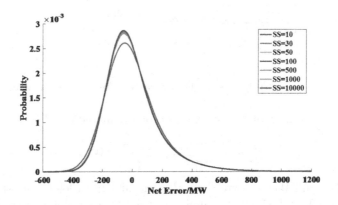

FIGURE 2.7 Distributions of the system "net errors" with different numbers of samples in HD-DPSO.

By varying the number of small-scale samples in the third stage of HD-DPSO, the distributions of the system's "net errors" are shown in Figure 2.7. Except for the distribution obtained with 10 samples, the remaining distributions are much similar. When the number of samples is larger than 100, the distributions are nearly the same. It shows that in our proposed HD-DPSO method, a very small number of samples (e.g., 500) are needed to obtain a stable distribution.

By varying r_{ws} from 0.3 to 0.9 with an interval of 0.1 and adding the wind farms one by one, the up and down reserve requirements are as shown in Figure 2.8 and Figure 2.9, respectively. The resolution of the reserve requirements is

1 MW and the results obtained by HD-DPSO and the MC method with 10^{12} samples are identical in this resolution. It shows that our proposed method has high accuracy.

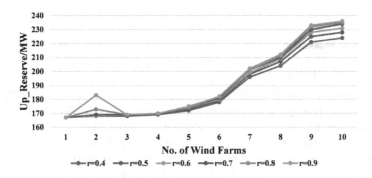

FIGURE 2.8 Up reserve requirements with different penetration wind power values and correlations.

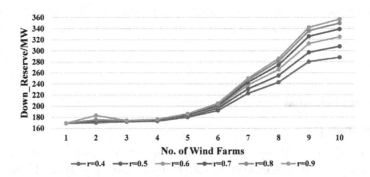

FIGURE 2.9 Down reserve requirements with different penetration wind power values and correlations.

It can be seen that both the up and down reserve requirements increase with the increase of wind power penetration and their correlations. When the wind power penetration is low (the number of wind farms is smaller than 5), the reserve requirement remains at a relatively lower level. However, when the penetration of wind power increases (the number of wind farms is larger than 6), the reserve requirement increases rapidly. As the forecast error of wind power is asymmetric, the requirements for up and down reserves are not identical. It can be observed that there is a spike in the reserve requirements when there are 2 wind farms compared with the reserve requirements of 3 wind farms. It is very possible that there exists complementarity between the third wind farm and the first two wind farms. However, the overall trend for reserve requirements is to increase with the penetrations of wind power increases.

2.5 Summary

This chapter proposes an efficient approach to power system uncertainty analysis with high-dimensional dependencies, called HD-DPSO. Three measures including grouping, a Gaussian distribution-based aggregation stage, and a recursive sample-guided DPSO are implemented to ease the computational burden. The potential applications of the proposed method in future power systems with high penetration of renewable energy are summarized from the perspective of probabilistic forecasting, power system operation, and power system planning. Illustrative examples of reserve requirement evaluation and capacity credit assessment are proposed to verify our proposed method. In future power systems with high penetration of renewable energy, the types and size of data generated in power systems will increase dramatically and push the power system sector into the big data area. The proposed method can effectively model and calculate the variables with high-dimensional dependencies.

Bibliography

[1] Y. Wang, N. Zhang, C. Kang, M. Miao, R. Shi, and Q. Xia. An efficient approach to power system uncertainty analysis with high-dimensional dependencies. *IEEE Transactions on Power Systems*, 33(3):2984–2994, May 2018.

[2] G. Papaefthymiou and D. Kurowicka. Using copulas for modeling stochastic dependence in power system uncertainty analysis. *IEEE Transactions on Power Systems*, 24(1):40–49, Feb 2009.

[3] I. Konstantelos, G. Jamgotchian, S. H. Tindemans, P. Duchesne, S. Cole, C. Merckx, G. Strbac, and P. Panciatici. Implementation of a massively parallel dynamic security assessment platform for large-scale grids. *IEEE Transactions on Smart Grid*, 8(3):1417–1426, May 2017.

[4] N. Zhang, C. Kang, C. Singh, and Q. Xia. Copula based dependent discrete convolution for power system uncertainty analysis. *IEEE Transactions on Power Systems*, 31(6):5204–5205, Nov 2016.

[5] W. Wu, K. Wang, B. Han, G. Li, X. Jiang, and M. L. Crow. A versatile probability model of photovoltaic generation using pair copula construction. *IEEE Transactions on Sustainable Energy*, 6(4):1337–1345, Oct 2015.

[6] M. Sun, I. Konstantelos, S. Tindemans, and G. Strbac. Evaluating composite approaches to modelling high-dimensional stochastic variables in power systems. *2016 Power Systems Computation Conference (PSCC)*, pages 1–8, June 2016.

[7] M. Sun, I. Konstantelos, and G. Strbac. C-vine copula mixture model for clustering of residential electrical load pattern data. *IEEE Transactions on Power Systems*, 32(3):2382–2393, May 2017.

[8] C. Sun, Z. Bie, M. Xie, and J. Jiang. Fuzzy copula model for wind speed correlation and its application in wind curtailment evaluation. *Renewable Energy*, 93:68–76, 2016.

[9] S. Hagspiel, A. Papaemannouil, M. Schmid, and G. Andersson. Copula-based modeling of stochastic wind power in Europe and implications for the Swiss power grid. *Applied Energy*, 96(August 2012):33–44, 2012.

[10] G. D'Amico, F. Petroni, and F. Prattico. Wind speed prediction for wind farm applications by extreme value theory and copulas. *Journal of Wind Engineering & Industrial Aerodynamics*, 145:229–236, 2015.

[11] Y. He, R. Liu, H. Li, S. Wang, and X. Lu. Short-term power load probability density forecasting method using kernel-based support vector quantile regression and copula theory. *Applied Energy*, 185:254–266, 2017.

[12] V. Marimoutou and M. Soury. Energy markets and CO2 emissions: Analysis by stochastic copula autoregressive model. *Energy*, 88:417–429, 2015.

[13] A. Lojowska, D. Kurowicka, G. Papaefthymiou, and V. Lou. Stochastic Modeling of Power Demand Due to EVs Using Copula. *IEEE Transactions on Power Systems*, 27(4):1960–1968, 2012.

[14] J. Vasilj, P. Sarajcev, and D. Jakus. Estimating future balancing power requirements in wind-PV power system. *Renewable Energy*, 99:369–378, 2016.

[15] J. Cheng, N. Zhang, Y. Wang, C. Kang, W. Zhu, M. Luo, and H. Que. Evaluating the spatial correlations of multi-area load forecasting errors. *International Conference on Probabilistic Methods Applied to Power Systems*, 2016.

[16] F. Yates. Contingency tables involving small numbers and the χ^2 test. *Supplement to the Journal of the Royal Statistical Society*, 1(2):217–235, 1934.

[17] D. Cai, D. Shi, and J. Chen. Probabilistic load flow computation using copula and Latin hypercube sampling. *IET Generation Transmission & Distribution*, 8(9):1539–1549, 2014.

[18] Rodríguez-Lallena, A José, úbeda Flores, and Manuel. Distribution functions of multivariate copulas. *Statistics & Probability Letters*, 64(1):41–50, 2003.

[19] Probability Methods Subcommittee. IEEE Reliability Test System. *IEEE Transactions on Power Apparatus & Systems*, PAS-98(6):2047–2054, 1979.

[20] M. Milligan, E. Ela, D. Lew, D. Corbus, Y. Wan, and B. Hodge. Assessment of simulated wind data requirements for wind integration studies. *IEEE Transactions on Sustainable Energy*, 3(4):620–626, 2012.

[21] N. Zhang, C. Kang, Q. Xia, and J. Liang. Modeling conditional forecast error for wind power in generation scheduling. *IEEE Transactions on Power Systems*, 29(3):1316–1324, 2014.

Part II

Uncertainty Modeling and Analytics

3

Long-Term Uncertainty of Renewable Energy Generation

Modeling renewable energy stochastic behavior is the basis of the planning and scheduling studies of renewable energy integration. The output of adjacent renewable energy farms has strong spatial and temporal correlations, which brings challenges in characterizing the stochastic generation behavior. This chapter provides a methodology to model the wind power generation considering the spatiotemporal correlation of wind speed using copula theory and to model the PV power generation by separating the determined and stochastic parts of PV output time series. Empirical studies using realistic wind and solar power measurements are presented to validate the effectiveness of the proposed method. This study provides new insights into the modeling of long-term renewable energy generation.

3.1 Overview

Modeling long-term uncertainty model of renewable energy is usually used in power system planning studies and power system reliability assessments. The long-term uncertainty refers to the distribution of renewable energy output over a long time period. Here, the uncertainty denotes the dispersity of renewable energy output.

The long-term PDF of wind power output is one basic model used in many studies. A variety of kinds of models have been proposed, e.g., beta distribution [1], gamma distribution [2], or an analytical formula transformed from a Weibull distribution of wind speed [3]. However, the outputs of wind farms are found to have evident spatiotemporal correlations due to the meteorological nature of wind [4]. The above-mentioned models are based on a single wind farm and, thus, may not be applicable in multiple wind farm models.

The spatiotemporal correlations of wind power have significant impacts on the overall uncertainty, and in turn, impacts on the power system. From the spatial point of view, the weak correlation would make clustered wind power more reliable in generating less disturbance to the power system, e.g., a higher capacity credit and less variance in power flow [5]. From the temporal point of view, the dispersion of wind farms would result in a smaller fluctuation, which is widely known as the "smoothing effect" [6]. Such an effect would bring down the requirement of flexibility [7], which would further reduce the system operation cost [8], [9]. This temporal correlation would also affect the operation of energy storage [10]. Several approaches based on stochastic sampling have also been proposed. The most common method is to generate a multi-dimensional Gaussian time series and convert each marginal distribution into the assigned distributions [8]. Autoregressive-moving average (ARMA) models [11] or a Markov chain [12] could be used to perform the temporal correlations. The three aspects could also be simultaneously considered using a stochastic differential equation [13]. Several studies have used copula theory to model the spatial-temporal correlation of wind power, PV output and load demand [14, 15].

The long-term uncertainty in PV power generation is mainly affected by changes in season and weather, as well as the alternation between day and night. Many researchers have studied the random characteristics of PV outputs using the clearness index which is denoted by K_t and defined as the ratio of the total solar radiation incident on the horizontal plane to the astronomical irradiation. Hollands et $al.$ [16] provided a PDF for hourly K_t from a meteorological perspective based on the hourly mean value of K_t of each month. Graham et $al.$ [17] used K_t to describe the uncertainty in PV outputs. Chen et $al.$ [18] employed a statistical method to produce a $1km \times 1km$ spatial distribution map of the average value of K_t for China for each month of the year. Tina et $al.$ [19] established a K_t-based power generation model for PV panels. Jin et $al.$ [20] provided a functional relationship between PV output and solar radiation intensity and obtained a PDF for hourly PV power outputs based on a PDF for K_t. Wang et $al.$ [21] and Fang et $al.$ [22] examined the operating reserves and credible capacities of PV grid-connected systems based on K_t-based power generation models for PV power plants. Karim et $al.$ [23] modeled solar radiation through wavelet transformation and curve fitting. Liang et $al.$ [24] noted that the intensity of solar radiation that reaches the ground could be approximated to follow a beta distribution within a short period of time (hours) and provided PDFs for short-term PV outputs.

In this chapter, the long-term uncertainties of both wind and PV generation are analyzed. For wind power, we first analyze the long-term distribution of wind speed and wind power output and how the wind power outputs are correlated. The wind power outputs are modeled using Copulas to capture their long-term spatial and temporal correlations. An empirical analysis is carried out using historical measurements, and the proposed approach is further applied to the evaluation of a planned 10-GW cluster of wind farms in China. For PV output, we first divide the PV output into a deterministic part and a random part. For the deterministic part, the effects of uncertain factors, such as shade, blocking by clouds and changes in weather and temperature, are ignored, and this part is modeled using the global solar radiation intensity model. For the random part, the focus is placed on the effects of the aforementioned factors, and a concept of "PV output shedding factor" is proposed to describe the effects of uncertain factors on PV outputs.

3.2 Wind Power Long-Term Uncertainty Characteristics

3.2.1 Power Generation Model of a Wind Turbine

A wind turbine converts the wind's kinetic energy into electricity when the wind turns its blades. The output characteristic curve for most wind turbines is similar. Generally, the output characteristic curve for a wind turbine can be approximately linearized into four sections and formulated by the following equation,

$$P^w = \begin{cases} 0 & x < x_{cut_{in}} \\ k \cdot (x - x_{start}) & x_{start} \leq x < x_{rate} \\ P_{rate} & x_{rate} \leq x < x_{cut_{off}} \\ 0 & x_{cut_{off}} \leq x \end{cases} \tag{3.1}$$

where $x_{cut_{in}}$, x_{rate} and $x_{cut_{off}}$ are the cut-in, rated and cut-off wind speeds, respectively. When the wind speed surpasses the cut-in wind speed, the output of wind turbines begins to increase linearly. When the wind speed reaches the rated wind speed, the output of wind turbines is at the maximum. If the wind speed continues to increase after surpassing the rated wind speed, the output of wind turbines keeps unchanged. If the wind speed surpasses the cut-off wind speed, wind turbine will be turned off.

3.2.2 Probabilistic Distribution of Wind Power

The long-term uncertainty in wind power generation is attributed to the variation and uncertain nature of wind speed. Extensive research finds out that wind speed well follows a Weibull distribution:

$$f_{W(c,k)}(x) = \frac{k}{c} \left(\frac{x}{c}\right)^{k-1} \exp\left[-\left(\frac{x}{c}\right)^k\right] \tag{3.2}$$

where c and k are the scale and shape parameters of the Weibull distribution, respectively. The mean wind speed, x, and its standard deviation, σ, can be calculated using Equations (3.3) and (3.4), respectively:

$$\bar{x} = c\Gamma\left(1 + 1/k\right) \tag{3.3}$$

$$\frac{\sigma}{\bar{x}} = \sqrt{\left[\Gamma\left(1 + \frac{2}{k}\right)/\Gamma^2\left(1 + \frac{1}{k}\right)\right] - 1}. \tag{3.4}$$

In general, the standard deviation of wind speed determines k, and the average value of wind speed is positively correlated to c. Thus, parameters c and k can be estimated based on the mean wind speed $\bar{x}$ and its standard deviation σ. First, k can be estimated using Equation (3.5). On this basis, c can be estimated based on the Equation (3.3).

$$k = (\sigma/\bar{x})^{-1.086}. \tag{3.5}$$

Figure 3.1 illustrates the fitting of a wind farm. The mean wind speed is 9.4391 m/s with a standard deviation of 4.5925 m/s. The parameters of the Weibull distribution are as follows: $c = 10.53$ and $k = 2.315$. The fitting results demonstrate that the probability distribution of wind speeds is in good agreement with the Weibull distribution.

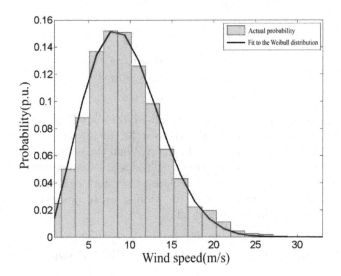

FIGURE 3.1 The probability distribution of wind speeds and its fit to the Weibull distribution.

For a wind turbine generator system, the curve depicting the relationship between its output and wind speed is referred to as the characteristic curve. The characteristic curve of a wind turbine is divided by its cut-in, rated and cut-out wind speeds into four sections. The probability distribution of the output of a wind turbine generator system is determined collectively by the probability distribution of wind speeds and the characteristic curve. In addition, the PDF of the wind power output is also affected by the reliability of the wind turbine generator systems and the wake effect. Generally, the PDF of the wind power output can be fitted to a beta distribution:

$$f_{B(a,b)}(x) = \frac{\Gamma(a)\Gamma(b)}{\Gamma(a+b)} x^{a-1}(1-x)^{b-1}. \tag{3.6}$$

Figure 3.2 shows the analysis of the probability distribution of the output of the wind power farm corresponding to the same anemometer tower like the one in Figure 3.1, as well as its fit to the beta distribution. Because of the saturation effect of the characteristic standard power curve of a wind turbine, the probabilities corresponding to the zero and full outputs of the wind power farm are significantly higher than those corresponding to the other output stages. The fit to the beta distribution ($a = 0.3767$, $b = 0.3817$) is not ideal. Therefore, numerical discrete distribution is usually used to model its output instead of analytical distribution.

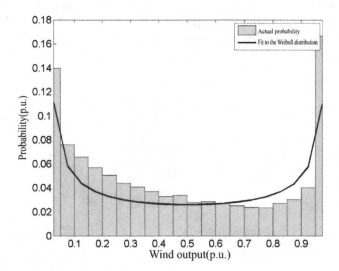

FIGURE 3.2 The probability density of the output of the wind power farm and its fit to a Beta distribution.

3.2.3 Spatio-Temporal Correlations of Wind Power Output

Using copula theory to model the spatiotemporal correlations of wind power from historical measurements follows the two steps below.

1. Model the marginal distribution of the output of a single wind farm. A variety of techniques can be used in this step, i.e., regressing using a beta distribution or recording the distribution function numerically.

2. Model the spatiotemporal correlations using copula. The historical wind power outputs are then transformed into a uniform distribution by the marginal distribution obtained in the first step. This is used as a basic guide to identify the copula function. A classical fitting procedure can be utilized in choosing the appropriate copula function and parameters. Usually, the commonly used copulas are competent for such modeling, i.e., the t copula and Gaussian copula in the elliptical copula category. Especially, the parameters of these copulas have a fixed relationship with Kendall's rank correlation coefficient and are easy to estimate [25].

The time series involved in the copula modeling can be the synchronized wind power outputs from different sites (spatial correlation) or the same site with different time lags (temporal correlation) or a mixture of the above (spatiotemporal correlation). Figure 3.3 illustrates the modeling of two wind farms located 100 km apart using scatter plots. The left figure shows the joint distribution of the outputs, while the right one shows its copula obtained. According to the work of Zhang [26], the copula within the wind power could be well fitted by the t copula.

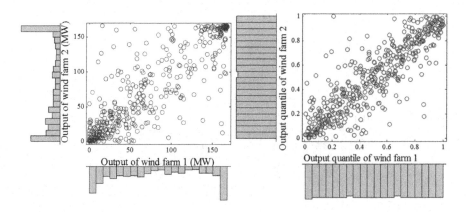

FIGURE 3.3 The joint distribution of the outputs of two wind farms (left) and their copulas (right).

3.2.4 Empirical Study

This section carries out the empirical study using real-world wind power data. The purpose of this section is to demonstrate for clustered wind farms to what extent the outputs are correlated and how this correlation affects the overall uncertainty.

1) Data sources and basic settings

The wind power and the corresponding forecast data are from National

Renewable Energy Laboratory (NREL) [27], which contains a consistent set of wind data of more than 30,000 sites over the United States, with a ten-minute time resolution from 2004 to 2006. As shown in Figure 3.4, we choose a series of sites along the East Coast of the United States. At these sites, the wind speed is not an actual measurement but is actually calculated using the mesoscale model in MASS v.6.8 with a 2km^2 resolution. Assuming 170-MW-rated wind power for each site, the wind speeds are transformed into wind power using IEC standard wind turbine curves. The day-ahead forecast is performed using a forecast tool named SynForecast that is based on a statistical model. The details of these datasets can be found in [28]. Though the wind data chosen in this case study is somewhat simulated, it is derived from real meteorology measurements and contains similar uncertainty characteristics to actual wind farms.

FIGURE 3.4 Sites chosen for the measurement of wind power and the forecast power.

2) The spatio-temporal correlations

Since the modeling of the marginal distribution of single wind farms has been widely discussed, only the spatiotemporal correlations are investigated. The t copula is used in this chapter to capture the correlations around the extreme value. The key parameter of the t copula is its correlation coefficient matrix. Its elements are in the range from -1 to 1. The values can be determined by Kendall's rank correlation coefficient of the input data [25]. It should be noted that this correlation coefficient is not exactly the same as the well-known linear correlation coefficient since t copula represents a nonlinear dependence.

Figure 3.5 shows the spatial correlation coefficients of each two-site-pair for both wind power outputs and day-ahead forecast errors. These coefficients are plotted with the accordant distances between the sites. It is clear from the results that the spatial correlations of both the outputs and forecast errors decrease as the wind farms are located farther away from each other. Especially, the wind farms within a distance of 50 km would have such a strong correlation (approximately 0.9) that the smoothing effect could be ignored and could be considered integrally as one single wind farm. Beyond this distance, the wind farms should be considered separately. The spatial correlation of the forecast errors is weaker than that of the outputs for the same site pair. This suggests that the smoothing effect of forecast errors would be more evident, which could benefit the system in terms of declining the overall forecast errors.

The relationship between the correlation coefficient and the distance is regressed using an exponential function. The damping ratio of the correlation coefficient for the outputs and forecast errors are 0.0018 and $0.0035 \mathrm{km}^{-1}$, respectively. Such a relationship could be used empirically to estimate the spatial correlations of planned wind farms based on their locations.

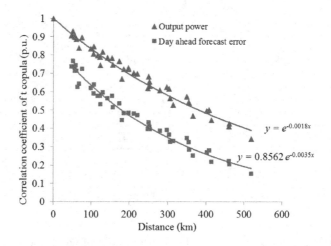

FIGURE 3.5 Spatial correlation coefficients of t copula as a function of the distance between wind farms.

Figure 3.6 shows the temporal correlation coefficients of the wind power outputs for both single sites and the aggregation of all sites. Similarly, the coefficients are plotted with the time lags increases, and the damping ratio of the first 20 hours is regressed using an exponential function. As expected, the temporal correlation coefficient of the aggregate output damps slower than that of the single site, which reveals in part the smoothing effect of clustered wind power.

Figure 3.7 shows the joint spatiotemporal correlation coefficients. The results show that the outputs of wind farms with adjacent locations and time periods have relatively strong correlations. This result suggests that the similarities of the outputs might be used as redundant information to cross-check the short-term forecast, which might contribute to its accuracy. However, the effectiveness of such a connection is limited within a certain range of space and time. For instance, taking the benchmark of the correlation coefficients as 0.5, the limit would be 400 km in space and four hours in time.

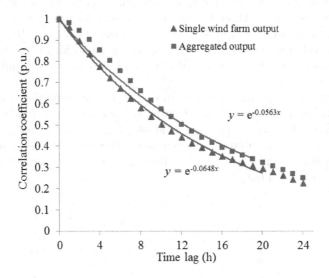

FIGURE 3.6 Temporal correlation coefficient of t copula as a function of the time lag.

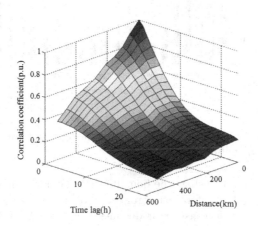

FIGURE 3.7 Surface diagram of the spatio-temporal correlation coefficient of t copula.

3) The overall uncertainty characteristic

The purpose of this subsection is to evaluate the impact of the spatiotemporal correlation on the overall uncertainty of clustered wind power. Three common indices are chosen, namely, the aggregate output, day-ahead forecast error, and the hourly variation. For each index, statistics are performed at four different scales of the area along the coast of the United States and are of lengths of 2 km (single wind farm), 20 km, 150 km, and 500 km.

The results of the three indices are shown in Figures 3.8, 3.9, and 3.10. The results confirm the argument given in Subsection 3.1 that the wind farms within a 20-km area are hardly affected by the smoothing effect because the distributions of the three indices of the wind power from 2 km and 20 km are very close.

For the aggregate output shown in Figure 3.8, the wind power outputs are more concentrated when the wind farms are geographically widely dispersed, i.e., in a 500-km area. The most evident changes are the reduction in the probabilities of the minimum output and maximum output. This suggests that the dispersion of wind farms would largely reduce the risks caused by extreme wind farm outputs.

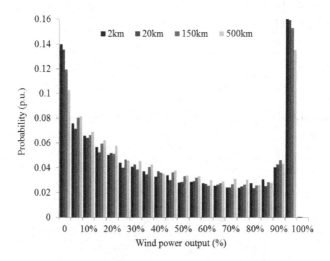

FIGURE 3.8 Probabilistic distribution of the aggregate outputs for different geographically distributed wind farms.

Figures 3.9 and 3.10 show the benefits that the dispersion brings, in terms of reducing the relative forecast errors and variations. For the 500-km case, the relative average forecast error declines from 13.88% to 9.74%, while the relative average hourly variation declines from 6.59% to 4.18%. Such benefits will shrink when the dispersed area is 150 km or less.

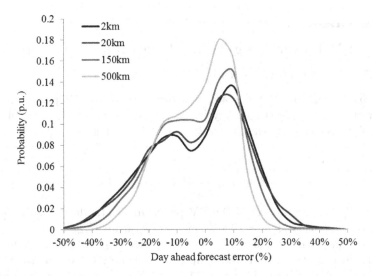

FIGURE 3.9 Probabilistic distribution of the day-ahead forecast errors for different geographically distributed wind farms.

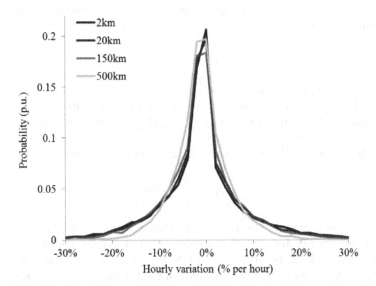

FIGURE 3.10 Probabilistic distribution of the hourly variations for different geographically distributed wind farms.

3.3 PV Power Long-Term Uncertainty Characteristic

3.3.1 PV Output Model

The basic principle of PV power generation is to use PV panels to convert solar energy into electric energy based on the PV effect. The actual output, P_t^v, of a PV panel at time t can be calculated using the following equation [29]:

$$P_t^v = P_{stc}^v \frac{I_{r,t}}{I_{stc}} \left[1 + \alpha_T \left(T_t - T_{stc}\right)\right] \tag{3.7}$$

where P_{stc}^v is the output of the PV panel under standard conditions (solar radiation intensity, $I_{stc} = 1000\text{W/m}^2$; temperature, $T_{stc} = 25°\text{C}$); α_T is the power temperature coefficient of the PV panel; $I_{r,t}$ is the solar radiation intensity at time t; and T_t is the temperature of the PV panel at time t.

As demonstrated in Equation (3.7), $I_{r,t}$ and T_t are two factors affecting P_t^v. $I_{r,t}$ is affected by a myriad of external factors, including the position of the Sun, shade, blocking by clouds and changes in weather. The effects of T_t are related to α_T. Here, Equation (3.7) is transformed to separate the deterministic and random factors. The output of a PV power station when the effects of blocking by the cloud and temperature are not taken into consideration, $P_{c,t}^v$, is defined as follows:

$$P_{c,t}^v = P_{stc}^v \frac{I_t}{I_{stc}} \tag{3.8}$$

where I_t is the maximum solar radiation intensity when no shade is considered, which is only related to the latitude, longitude and altitude of the location and has a certain variation pattern. Thus, $P_{c,t}^v$ only contains the deterministic part of the PV output.

The PV power shedding factor, η_t, is defined as follows:

$$\eta_t = \frac{P_{c,t}^v - P_t^v}{P_{c,t}^v} \tag{3.9}$$

where η_t characterizes the relative difference between the P_t^v and $P_{c,t}^v$ of a PV power station under the action of shade, blocking by clouds, and changes in weather and temperature, all of which are of relatively high uncertainty. Thus, η_t characterizes the random part of the PV output. Based on Equations (3.7)-(3.9), we have:

$$\eta_t = \frac{I_t - I_{r,t} \left[1 + \alpha_T \left(T_t - T_{stc}\right)\right]}{I_t} \tag{3.10}$$

Based on Equations (3.8) and (3.10), $P_{c,t}^v$ determines the outer envelope of the PV output. However, P_t^v will be reduced by various random factors.

η_t describes the reduction effect of shade, blocking by clouds, and changes in weather and temperature on the PV output on the basis of its outer envelope. The physical meaning of η_t is the relative difference between the P_t^v and $P_{c,t}^v$ of a PV power station. Naturally, the value of η_t is less than or equal to 1.

In the following sections, P_t^v and $P_{c,t}^v$ are modeled separately.

3.3.2   Unshaded Solar Irradiation Model

Based on the aforementioned analysis, $P_{c,t}^v$ is determined by I_t. Because of the regular rotation and revolution of the Earth, It is an analytical function with respect to time and geographic location and is also referred to as the global solar radiation intensity model. Its derivation is as follows:

The non-shade solar irradiance at ground level I_{0t} equals to the product of solar irradiance I_0 and the sine of the solar zenith angle. For any point on earth, the exo-atmospheric solar irradiance I_0 is determined, which is only related to the relative positions of the sun and Earth [30]; its calculation formula is as follows:

$$I_0 = A_0 \left(1 + 0.033 \cos \left(\frac{2\pi (G + 10)}{365} \right) \right) \tag{3.11}$$

where A_0 is a solar constant representing the total solar irradiance entering Earth's atmosphere per unit area, measured on the plane normal to the incident light outside Earth's atmosphere; it equals approximately $1367W/m^2$. G is the day sequence number, beginning from January 1st of each year.

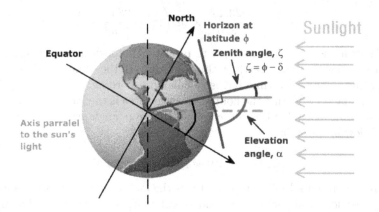

FIGURE 3.11 Solar zenith angle diagram.

The solar zenith angle α can be calculated by:

$$\sin \alpha = \sin \phi \sin \delta + \cos \phi \cos \delta \cos \omega \tag{3.12}$$

where ϕ is the latitude of the area (positive at north and negative at south); δ represents the declination angle, which is the angle between the plane of Earth's equator and the line connecting the sun and the center of Earth; and ω is the hour angle.

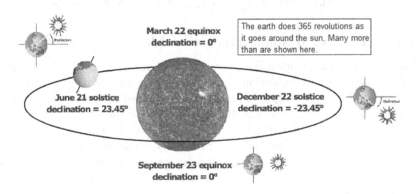

FIGURE 3.12 Solar declination diagram.

Since Earth rotates around the sun, the declination angle changes constantly throughout the year and can be calculated by the following equation. The declination angle is positive when the sun is directly in the northern hemisphere and negative when the sun is in the southern hemisphere:

$$\delta = 2\pi \times \frac{23.45°}{360°} \times \sin\left(2\pi \times \frac{284 + G}{365}\right). \tag{3.13}$$

Hour angle is a quantity used to measure the position of the sun relative to a certain place on Earth within one day. If the time is not calculated according to the local time but is calculated according to the standard time, the main factor affecting the hour angle of the area is the local longitude. Using Beijing time (GMT+8) as an example, the hour angle can be calculated by the following formula:

$$\omega = (t - 12) \times 15° + (\psi - 120°). \tag{3.14}$$

where t represents Beijing time and ψ is the longitude of the solar panel location (east longitude is positive, and west longitude is negative). The midday hour angle is 0, the morning hour angle is negative, and the afternoon hour angle is positive.

The solar zenith angle at a certain area can be obtained by combining Equations (3.12)-(3.14). When the effects of random factors of atmospheric attenuation such as scattering effects of the atmosphere on the sunlight and cloud cover are not considered, the instantaneous solar irradiance on any point on earth I_{t0} is:

$$I_{t0} = I_0 \sin \alpha. \tag{3.15}$$

where I_{t0} is the aforementioned maximum direct solar irradiance which can be received on Earth.

However, sunlight is weakened to some degree when it passes through the atmosphere; upon reaching the ground surface, solar radiation is divided into direct and scattered radiation. The sum of these two radiations is the total solar radiation reaching the ground surface. In terms of the global average, the total solar radiation reaching the ground surface is approximately 45% of the solar radiation above the atmosphere. Direct radiation refers to the solar radiation reaching Earth's surface in the form of parallel light; scattered radiation is the phenomenon of sunlight redistribution in all directions when passing through the atmosphere following a certain pattern. Generally, solar radiation is mainly direct radiation on sunny days and scattered radiation on cloudy days.

Considering the influential atmospheric factors, the solar irradiance reaching the ground surface is calculated as follows. The atmospheric transparency coefficient must be calculated first. The atmospheric transparency coefficient refers to the percentage of solar radiation allowed to pass into Earth's atmosphere and is expressed as the ratio of the flow of solar radiation in the direction of the zenith to the ground to the flow at the upper atmosphere. Some studies have proposed the following empirical formula for the transparency coefficient of direct solar radiation under sunny conditions [31]:

$$\tau_b = 0.56 \left(e^{-0.56M_h} + e^{-0.095M_h} \right) \tag{3.16}$$

where M_h is atmospheric mass. Considering local altitude conditions, the atmospheric mass can be calculated by the following equation:

$$M_h = ((1229 + (614 sin\alpha)^2)^{\frac{1}{2}} - 614 sin\alpha \times \left(\frac{288 - 0.0065h}{288} \right)^{5.256} \tag{3.17}$$

where h is the local altitude.

Under sunny conditions, the atmospheric transparency coefficients of scattered radiation and direct radiation have the following relationship:

$$\tau_d = 0.271 - 0.274\tau_b. \tag{3.18}$$

Based on the definitions of direct solar radiation and atmospheric transparency coefficient, it is very easy to obtain the calculation formula for direct solar irradiance at a certain place:

$$I_b = I_0 \tau_b \sin \alpha. \tag{3.19}$$

The mechanism of scattered radiation is more complex; it is related to a variety of weather conditions. According to an empirical equation [31], one can obtain the following formula for scattered solar irradiance:

$$I_d = \frac{1}{2} \times \sin \alpha \times \frac{1 - \tau_d}{1 - 1.4 \ln (\tau_d / M_h)}. \qquad (3.20)$$

Equation (3.21) shows the empirical equation for scattered solar radiation intensity, I_d:

$$I_d = \frac{1}{2} \times \sin \alpha \times \frac{1 - \tau_d}{1 - 1.4 \ln (\tau_d / M_h)} \times k \qquad (3.21)$$

where k is a parameter related to the atmospheric quality [32]. When the atmospheric quality is relatively low, the value of k falls in the range of $[0.60, 0.70]$. When the atmospheric quality is normal, the value of k falls in the range of $[0.71, 0.80]$. When the atmospheric quality is relatively high, the value of k falls in the range of $[0.81, 0.90]$.

In summary, the total solar irradiance after considering air mass and direct and scattered radiation is:

$$I_t = I_b + I_d. \qquad (3.22)$$

The above equations constitute the solar irradiance model. It is evident from the above model that the local non-shade solar irradiance at ground level can be obtained when the local geographic information (latitude and longitude, elevation), time information (day of the year and time of the day) and air mass information are given.

Based on Equations (3.11)-(3.22), the I_t at any point on the Earth's surface at any time can be simulated when the shedding of cloud is not considered. By substituting I_t into Equation (3.8), $P_{c,t}^v$ can be determined.

We use PV output data measured at eight observation sites throughout the year of 2012 obtained from the website of NREL to validate the model [18]. These PV output data were collected at intervals of one hour. Figure 3.13 shows the geographic locations of the eight observation sites.

The distances between the selected eight observation sites, which are located in areas of various landforms, vary from 4 to over 1,200 km. Here, the observation data collected at the observation site #1 are used as an example. Figure 3.14 shows the output data for all the hours (8,784 h) of the entire year of 2012 (a leap year) measured at observation site #1 (the top plot shows the $P_{c,t}^v$ values calculated using the model described in this section; the bottom plot shows the Pt data). The P_t curve is similar to the $P_{c,t}^v$ curve in terms of shape, the P_t^v curve contains a notable random fluctuation. In this chapter, this random fluctuation is described using η_t, which is defined in Subsection 3.3.1. In the following section, η_t is separated from P_t based on $P_{c,t}^v$, to study its random characteristics.

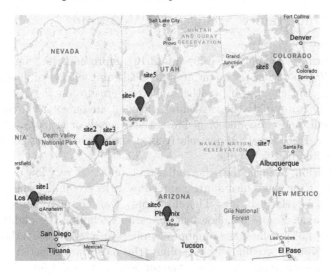

FIGURE 3.13 Locations of the PV observation sites.

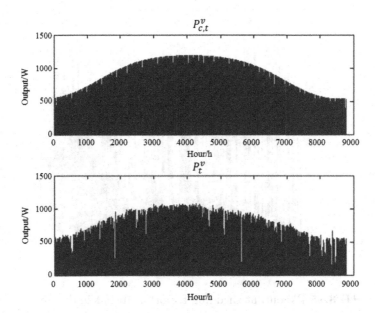

FIGURE 3.14 Comparison of the deterministic output and actual output of the PV station at site #1.

3.3.3   Uncertainty Analysis of PV Output

1) Overview

The term η_t in Equations (3.9) and (3.10)) represents the relative difference between P_t and $P_{c,t}^v$ and quantifies the weakening effects of shade, blocking by clouds, and changes in weather and temperature on the PV output. On a relatively sunny day, the weakening effect of air on solar radiation is relatively small, and the value of η_t is relatively small or even zero (completely unshaded). In contrast, under relatively poor meteorological conditions, the value of η_t is relatively large.

In addition, because the solar radiation intensity is very low or even zero at night and at the times near sunrise and sunset, the credibility of the calculated value of η_t is relatively low. Therefore, only the daytime values of η_t are of research value. Figure 3.4 shows the η_t series excluding the nighttime points (a total of 4,371 points; the number of points selected for each day is related to the season: approximately 14 points (the most) are selected for summer, whereas approximately 10 points (the fewest) are selected for winter). In this section, the random characteristics of η_t are empirically analyzed from three perspectives, namely, probability distribution, temporal characteristics, and spatial characteristics.

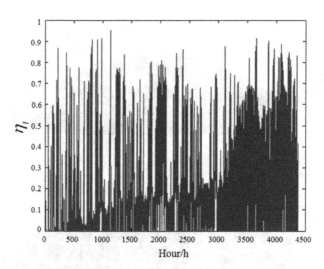

FIGURE 3.15 PV output shedding factor of site #1 in daytime periods.

2) Probability distribution of η_t

Figure 3.16 shows the probability distribution of η_t determined based on the observation data measured at observation site #1 to site #8. As demonstrated in Figure 3.16, as η_t increases, its probability decreases nearly monotonically. The probability for η_t to fall in the interval of $[0, 0.1]$ is the highest. There is a relatively even probability distribution of η_t in the interval of $[0.2, 0.7]$. The probability of η_t to be near 1 is very low. This suggests that the probability of occurrence of sunny weather is relatively high at the observation location, whereas the probability for sunshine to be completely shed is relatively low.

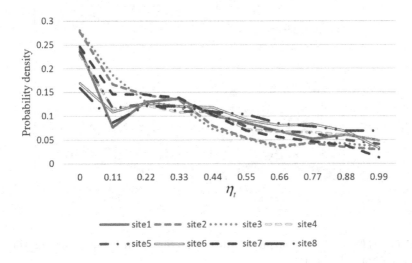

FIGURE 3.16 Probabilistic distribution factors of the PV output shedding factor of sites #1 to #8.

3) Temporal characteristics of η_t

Figure 3.17 shows the scatter plots of η_t at observation site #1 at different times of each day in 2012. The four plots are produced based on the η_t data at 10:00, 12:00, 14:00 and 16:00 of each day.

As demonstrated in Figure 3.17, η_t displays some temporal patterns:

1. A comparison of the distribution of scattering points in the plots finds that η_t displays certain daily characteristics - the probability distribution of η_t is not completely consistent at various times of the day.

2. A comparison of the distribution of scattering points of different seasons in the same plot finds that changes in η_t display notable seasonal (monthly) characteristics - the probability distribution of η_t at the same time of each day changes to a certain extent as the season (month) changes.

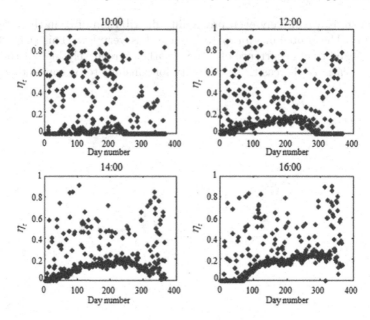

FIGURE 3.17 PV output shedding factor of different hours in the day throughout the year.

Next, the fluctuating characteristics of η_t are analyzed by calculating its autocorrelation coefficient. Because only the values of η_t during the day can be obtained in practice, data discontinuity is encountered when calculating the fluctuations in η_t. Thus, data preprocessing is performed via the following steps:

1. Because the nighttime PV output is zero, the nighttime values of η_t are of no practical meaning. However, to retain its time-series information, nighttime points must be taken into consideration when calculating the autocorrelation coefficient of η_t. Thus, the nighttime values of η_t are uniformly set to zero.

2. The values of η_t at the times near sunrise and sunset may be distorted. This should be ignored when calculating the time-series correlation coefficient.

3. Neither the time series obtained after a time-delay operation nor the original time series should contain nighttime points. This is because changes in η_t at nighttime points are fixed (always zero), and the pattern of change in η_t during the day is the area of focus. If the values of η_t at nighttime points are considered, some of the daytime time-series characteristics of η_t will be concealed.

Figure 3.18 shows the values of the autocorrelation coefficient of η_t at observation site #1 with a phase difference from 0 to 120 h and calculated using the above-mentioned process.

Clearly, as the time delay increases, the autocorrelation coefficient of η_t changes periodically and gradually tends to approach zero. Based on this result, it can be inferred that the η_t series should contain a periodic component that changes intermittently on a daily basis. Specifically, the value of η_t at the same time of the day tends to be consistent, or it changes or fluctuates in a relatively stable manner.

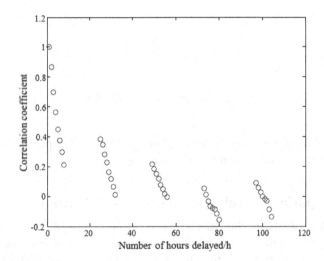

FIGURE 3.18 Autocorrelation coefficient of the PV output shedding factor of site #1.

Figure 3.19 shows the values of the autocorrelation coefficient of the η_t series, excluding the daily characteristics, and their fit to an exponential function. The exponential decay coefficient of the autocorrelation coefficient of the η_t series is -0.0692. Based on the calculation, η_t shows a strong autocorrelation within 24 h and nearly no autocorrelation at a time interval of over three days. This is in agreement with the actual weather factor affecting the PV power generation. PV outputs are affected by weather processes. η_t displays certain continuity within a day. However, a weathering process often lasts for fewer than three days.

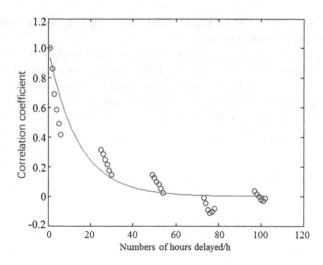

FIGURE 3.19 Autocorrelation coefficient of the PV output shelter factor of site #1 with the day characteristics removed.

3.3.4   Spatial Correlation between PV Outputs

1) Overview of the spatial correlation between PV outputs

Observation data show that the outputs of PV power stations geographically close to one another are highly similar. The extent of similarity between the outputs of geographically close PV power stations can be described by using the spatial correlation between their outputs. The spatial correlation between PV output refers to the extent of similarity between the PV output series in different areas. Solar radiation intensity, as well as random factors (e.g., changes in weather and shedding by clouds), vary between different areas. These differences increase as the distance between two areas increases. In the zonal direction, solar radiation intensity gradually decreases with an increasing latitude. In the meridional direction, the phase difference in a PV output series between two areas gradually increases with an increasing time difference. These factors will all affect the spatial correlation between PV outputs.

Research on the randomness of wind power generation finds out that the correlation between the outputs of wind power farms significantly affects the total wind output. The weaker the correlation is, the stronger the smoothing effect of the total output is, and thus, the lower the standby, ramp rate and transmission channel capacity requirements for the system are, which have a significant impact on the safety and economic efficiency of system operation [14]. The random characteristics of PV outputs are similar to those of wind outputs. Therefore, studying the spatial correlation between PV power stations in different areas is of vital importance to the understanding of the overall randomness of PV power generation.

PV outputs are determined by radiation intensity. Therefore, the correlation between PV outputs is affected by a myriad of factors, including geographic location, changes in weather and blocked by clouds. In this chapter, the correlation between PV outputs and the mechanism by which it is affected are studied from deterministic-factor and random-factor perspectives.

2) Spatial correlation between values of $P_{c,t}^v$

Latitude and longitude are the main factors affecting the spatial correlation between $P_{c,t}^v$ values. Longitude affects the time difference between PV outputs, whereas latitude affects the heights of the outer envelopes of PV outputs, i.e., the magnitudes of PV outputs.

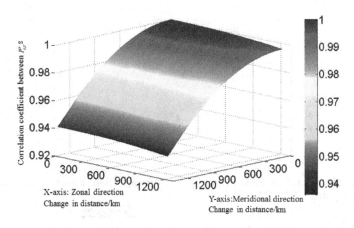

FIGURE 3.20 Correlation of the PV output at different latitude and longitude.

A total of 15×15 sites (including observation site #1) are evenly selected within the area extending from observation site #1 in the zonal and meridional directions, each by 1,200 km (i.e., 15° in the meridional direction (one time zone) and 11° in the zonal direction). Figure 3.20 shows the correlation coefficient between the annual output series at each of the sites (the X-axis represents the change in distance corresponding to a change in latitude when the longitude remains the same; the Y-axis represents the change in distance corresponding to a change in longitude when the latitude remains the same; and the Z-axis represents the correlation coefficient between the $P_{c,t}^v$ at a certain site). As demonstrated in Figure 3.20, as the zonal and meridional distances increase, the correlation coefficient between the annual output series at the two sites gradually decreases. In addition, the correlation coefficient decreases faster with an increasing meridional distance than an increasing zonal distance. This suggests that the spatial correlation between PV outputs is more

significantly affected by a time difference (mainly affected by longitude) than
solar radiation intensity (mainly affected by latitude).

3) Spatial correlation between values of η_t

The effects of longitude and latitude on η_t display no notable pattern.
This is because η_t mainly describes meteorological factors, such as blocking
by clouds. These meteorological factors are directly related to the distance
between different locations. Figure 3.21 shows the spatial correlations among
the η_t values for the eight selected observation sites. The closest and farthest
distances between any two observation sites are 4 and 1,338 km, respectively.

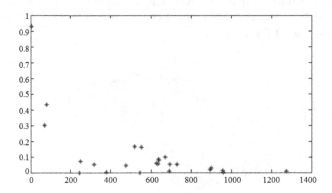

FIGURE 3.21 Correlation of the PV output shelter factor in different observation sites.

As demonstrated in Figure 3.21, there is a strong correlation between the
η_t values for two relatively close observation sites. However, this correlation decreases rapidly as the distance between the observation sites increases. When
the distance is 4 km, the correlation coefficient is 0.9267. When the distance
increases to 79 km, the correlation coefficient decreases to 0.4323. The correlation coefficient between the η_t values for two observation sites (7 and 8)
that are 69 km apart is only 0.3027, which may be attributed to the difference
in altitude between these two sites (1,000 m). When the distance increases
to over 200 km, the correlation coefficient decreases to below 0.2. For random variables, when their correlation coefficient is below 0.2, modeling them
as independent variables will not generate very large errors. Therefore, when
the distance between two PV power stations is over 200 km, the weather factors affecting their outputs can be considered independent of one another. 4)
Spatial correlation between the P_t values of PV power stations

In the previous sections, the spatial correlation between PV outputs is analyzed from the perspectives of deterministic and random factors affecting PV
power generation. As demonstrated in Figure 3.21, when the distance between
two locations exceeds 80 km, the correlation coefficient between the η_t values
(mainly representing the weather factor) is very small. As shown in Figure

3.20, the correlation coefficient between the $P_{c,t}^v$ values (mainly representing the changes in solar radiation that occur as day and night alternate) of two PV power stations 1,200 km apart in the meridional or zonal direction can still reach over 0.92. Therefore, what is the spatial correlation between the P_t values of PV power stations?

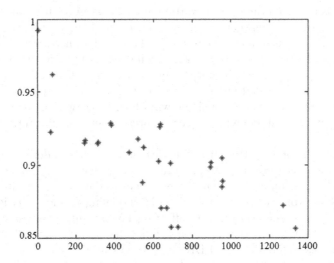

FIGURE 3.22 Correlation of the PV outputs in different observation sites.

Figure 3.22 shows the spatial correlations among the values of P_t at the eight observation sites. As demonstrated in Figure 3.22, the correlation coefficient between the P_t series at any two observation sites is relatively large (over 0.85). Based on the previous analysis, it can be inferred that the spatial correlation between the P_t values of PV power stations is mainly affected by their $P_{c,t}^v$ values and relatively insignificantly affected by their η_t values. The results show that there remains a strong correlation between the outputs of PV power stations distributed over a wide area, whereas the long-time-scale smoothing effect of their outputs is relatively weak, which represents a significant difference between PV and wind outputs from the perspective of long-term characteristics. This is also the cause of the "duck curve" challenge facing high-proportion PV grid-connected power systems that are currently attracting substantial attention. In high-proportion PV-connected power systems, the long-time-scale smoothing effect of PV outputs is not significant. As a result, these systems will face relatively high peak-load regulation and ramp pressure.

3.4 Summary

This chapter analyzes the long-term characteristics of renewable energy. The spatiotemporal correlation of renewable energy is given special attention. For wind power, copula theory is used to model the marginal distribution and the spatiotemporal correlation in separate fashions. The empirical study based on historical measurements shows that the correlation between wind power outputs decreases as the connection of wind farms becomes weaker in terms of both space and time. The outputs of wind farms within a 20-km area show a high degree of correlation, and the variations in their outputs are so similar that they can be treated as a single wind farm. The spatial smoothing effect is not evident until the wind power is spread over a significantly wide area, i.e., 500 km.

In this chapter, a PV output is decomposed into a deterministic part, $P_{c,t}^v$, and an uncertain part to analyze its uncertainty. $P_{c,t}^v$ is affected by the changes in solar radiation intensity that occur as day and night alternate, and its analytic expression can be obtained by modeling I_b and I_d. η_t is proposed to describe the uncertain factors affecting PV output. The results obtained based on P_t data show that η_t displays a notable periodic, as well as daily and seasonal, characteristic. The PDF for η_t values in different areas is relatively stable and can be fitted and modeled using a typical probability distribution or nonparametric approach. In addition, the spatial correlation between PV outputs changes significantly as longitude and latitude change. Moreover, there is also a strong spatial correlation between the values of η_t for any two locations. However, this correlation decreases rapidly as the distance between the two locations increases. Overall, there is a relatively strong spatiotemporal correlation between PV outputs, suggesting a relatively weak smoothing effect.

The findings and methodology in this chapter are useful for both planning and operation studies involving large-scale wind power and PV power clusters, i.e., a probabilistic load flow, generation and transmission expansion in systems with a high penetration of renewable energy.

Bibliography

[1] H. Louie. Characterizing and modeling aggregate wind plant power output in large systems. *Power and Energy Society General Meeting*, pages 1–8, 2010.

[2] H. Louie. Evaluation of probabilistic models of wind plant power output

characteristics. *IEEE International Conference on Probabilistic Methods Applied to Power Systems*, pages 442–447, 2010.

[3] X. Liu and W. Xu. Economic load dispatch constrained by wind power availability: A here-and-now approach. *IEEE Transactions on Sustainable Energy*, 1(1):2–9, 2010.

[4] Y. Wan, M. Milligan, and B. Parsons. Output power correlation between adjacent wind power plants. *Journal of Solar Energy Engineering*, 125(4):551–555, 2003.

[5] J. M. Morales, L. Baringo, A. J. Conejo, and R. Minguez. Probabilistic power flow with correlated wind sources. *IET Generation Transmission and Distribution*, 4(5):641–651, 2010.

[6] C. Yang, M. U. Gang, L. Yu, and G. Yan. Spatiotemporal distribution characteristic of wind power fluctuation. *Power System Technology*, 35(2):110–114, 2011.

[7] H. Holttinen. Impact of hourly wind power variations on the system operation in the Nordic countries. *Wind Energy*, 8(2):197–218, 2005.

[8] D. Villanueva, A. Feijoo, and J. L. Pazos. Simulation of correlated wind speed data for economic dispatch evaluation. *IEEE Transactions on Sustainable Energy*, 3(1):142–149, 2012.

[9] L. Xie, Y. Gu, X. Zhu, and M. G. Genton. Power system economic dispatch with spatio-temporal wind forecasts. pages 1–6, 2011.

[10] H. Bludszuweit and J. A. Dominguez-Navarro. A probabilistic method for energy storage sizing based on wind power forecast uncertainty. *IEEE Transactions on Power Systems*, 26(3):1651–1658, 2011.

[11] A. Lojowska, D. Kurowicka, G Papaefthymiou, and L. Van, der Sluis. Advantages of ARMA-GARCH wind speed time series modeling. *IEEE International Conference on Probabilistic Methods Applied To Power Systems*, pages 83–88, 2010.

[12] G. Papaefthymiou and B. Klockl. MCMC for Wind Power Simulation. *IEEE Transactions on Energy Conversion*, 23(1):234–240, 2008.

[13] N. Zhang, C. Kang, C. Duan, X. Tang, J. Huang, Z. Lu, W. Wang, and J. Qi. Simulation methodology of multiple wind farms operation considering wind speed correlation. *International Journal of Power and Energy Systems*, 30(30):264–173, 2010.

[14] G. Papaefthymiou and P. Pinson. Modeling of spatial dependence in wind power forecast uncertainty. *International Conference on Probabilistic Methods Applied to Power Systems*, pages 1–9, 2008.

[15] H. V. Haghi, M. T. Bina, M. A. Golkar, and S. M. Moghaddas-Tafreshi. Using copulas for analysis of large datasets in renewable distributed generation: PV and wind power integration in Iran. *Renewable Energy*, 35(9):1991–2000, 2010.

[16] K. Hollands and R. G. Huget. A probability density function for the clearness index, with applications. *Solar Energy*, 30(3):195–209, 1983.

[17] V. A. Graham and K. Hollands. A method to generate synthetic hourly solar radiation globally. *Solar Energy*, 44(6):333–341, 1990.

[18] Z. Chen. Exploring the monthly clearness index models in china. *Journal of Nanjing Institute of Meteorology*, 28(5):649–656, 2005.

[19] G. Tina, S. Gagliano, and S. Raiti. Hybrid solar/wind power system probabilistic modelling for long-term performance assessment. *Solar Energy*, 80(5):578–588, 2006.

[20] J. Peng, A. Xin, and X. Jia. An economic operation model for isolated microgrid based on sequence operation theory. *Proceedings of the CSEE*, 32(25):52–59, 2012.

[21] H. Wang and X. Bai. Short-term operating reserve assessment for grid-connected photovoltaic system. *Automation of Electric Power Systems*, 37(5):55–60, 2013.

[22] X. Fang, Q. Guo, D. Zhang, and S. Liang. Capacity credit evaluation of grid-connected photovoltaic generation considering weather uncertainty. *Automation of Electric Power Systems*, 26(9):31–35, 2012.

[23] S. Karim, B. Singh, R. Razali, and N. Yahya. Data compression technique for modeling of global solar radiation. *IEEE International Conference on Control System, Computing and Engineering*, pages 348–352, 2012.

[24] S. Liang, X. Hu, D. Zhang, H. Wang, and H. Zhang. Probabilistic models based evaluation method for capacity credit of photovoltaic generation. *Automation of Electric Power Systems*, 36(13):32–37, 2012.

[25] E. W. Frees and E. A. Valdez. Understanding relationships using copulas. *North American Actuarial Journal*, 2(1):1–25, 1998.

[26] N. Zhang and C. Kang. Dependent probabilistic sequence operations for wind power output analyses. *Journal of Tsinghua University*, 52(5):704–709, 2012.

[27] National Renewable Energy Laboratory (NREL). Wind integration datasets. http://www.nrel.gov/wind/integrationdatasets/.

[28] M. Brower. Development of Eastern Regional Wind Resource and Wind Plant Output Datasets: March 3, 2008-March 31, 2010. *NREL Subcontract Report NREL/sr*, 2009.

[29] L. Xu, X. Ruan, C. Mao, B. Zhang, and Y. Luo. An improved optimal sizing method for wind-solar-battery hybrid power system. *Proceedings of the CSEE*, 4(3):774–785, 2013.

[30] S. Zhong-Xian, J. Zhou, and Y. Pan. Some output power issues of fixed PV panel. *Mechanical and Electrical Engineering Magazine*, 2008.

[31] Frank Kreith and Jan F Kreider. Principles of solar engineering. *Washington, DC, Hemisphere Publishing Corp., 1978. 790 p.*, 1978.

[32] J. Yang, Z. Liu, and B. Meng. Resources calculation of solar radiation based on MATLAB. *Energy Engineering*, 2011.

4

Short-Term Renewable Energy Output Forecasting

The uncertainty and variability of weather is a significant factor leading to unpredictable renewable generation. Therefore, it is essential to design a robust and adaptive forecasting method. This chapter proposes a generalized bad data identification method for the raw data of NWP aiming to improve the performance of wind power output forecast. The K-means algorithm acts as the NWP clustering solution. The BIC is employed to identify the bad data. Finally, the neural network is applied as the main forecasting engine.

4.1 Overview

To guarantee a reliable system operation, the ISO has to schedule sufficient spinning reserve for wind power. It is acknowledged by most of the researchers that a more accurate wind power forecast could help in reducing the cost of supplying spinning reserve without losing the reliability of power system operation [1].

The WPO denoted as P_t^w, is affected by varieties of factors, such as wind speed, wind direction, temperature, turbine type, turbine position, terrain roughness, air density and wake effect [2]. Numerous approaches to forecasting wind power in a single site have been developed, and generally, they can be classified into the following three categories: model-driven approaches, data-driven approaches, and ensemble approaches.

The solutions in the first category mainly focus on the physical model analysis [3] and reveal the influence of the relevant factors on P_t^w from the physical point of view. These solutions are capable of establishing a specialized wind power forecast model for a specialized wind farm. Meanwhile, they do not require a large amount of historical data to train the model. They can perform very well in the long-term wind power forecast as well [4]. However, these approaches need abundant meteorological knowledge and detailed physical characteristics of wind turbines in order to achieve an accurate model. The accuracy of the modeling is critical to the forecast performance.

To overcome the drawbacks of the model-driven solutions, many data-driven mathematical statistics approaches or artificial intelligence approaches are proposed [5–8]. The data used for model training in these approaches are NWP-related indices such as wind speed, wind direction, and air temperature. Several famous statistical methods are like PM [9], ARMA [10], ARIMA [11], GP [12], KF [13], et al. The artificial intelligence methods include NN [14], SVM [15], evolutionary algorithms [16], WA [17], et al. The approaches in this category are self-adaptive to types of turbine types, diverse geographical terrains, and different wind farm locations, avoiding the relatively weak influence of other factors, for example, air density. However, they all require a large amount of historical data as well as a sufficient training process. The model cannot guarantee a reliable estimation for the long look ahead of time and are therefore usually used for short-term wind power forecast.

Later, the research moves forward to the combination of model-driven and data-driven solutions [18]. The ensemble approaches are not only able to take advantages of every single approach but also to some extent avoid some drawbacks by selecting a reasonable model combination. Comparing to the approaches of the other two categories, the ensemble approaches should theoretically give relatively better results [19]. Several studies have proved its validity [20]. However, as discussed in [20], although all of the above works contribute a lot to the development of wind power forecast area, the short-term wind power forecast is still a young area and needs more contributions from different aspects.

All the solutions mentioned above only focus on the improvement of the forecast engine itself. This chapter follows up the author's idea in [1] and tries to improve the wind power forecasting accuracy in a different manner.

Among all the factors that influent the accuracy of the wind power forecast, the WSF, denoted as $\hat{s}_t$ and derived from NWP is acknowledged as the most significant [1]. However, the $\hat{s}_t$ usually comes from a third party like the National Weather Service Center, and its accuracy is still not high enough to support the next-stage wind power forecast. There are already a few works on the wind speed data adjustment [21]. However, the author studies NWP errors point by point in the scale of per hour, while these NWP errors are more likely to follow certain patterns that last for a longer time, e.g., 12 hours. In [1], the authors observe the NWP waveform and manually select four major types of errors, which can be categorized as miss-peak faults and wrong-peak faults. However, the type of error is obviously not supposed to be limited to four and a generalized data mining based solution should be studied.

4.2 Short-Term Forecasting Framework

4.2.1 Dataset and Definitions

To better present the idea of this chapter, several definitions are given first. In wind power forecasting, three types of datasets are usually used: the historic P_t^w, the historic $\hat{s}_t$, and the future $\hat{s}_t$.

The error rate of WPF, denoted as $\hat{P}_t^w$, is usually defined as the ratio of the absolute forecast error and the total installed wind power capacity shown in Equation (4.1):

$$\varepsilon_r = \frac{\left| \hat{P}_t^w - P_t^w \right|}{Cap^w} \times 100\% \tag{4.1}$$

where ε_r is the error rate, and Cap^w is the installed wind farm capacity.

An AWPF is defined as a forecast with an error rate greater than an assigned threshold value, while other forecasts with error rates less than the threshold value are regarded as an NWPF. For each forecast, besides the forecasted wind power, a corresponding data *Struct* is generated, which includes the past 12-hour's $\hat{s}_t$, the past 24-hour's P_t^w, as well as the future 12-hour's $\hat{s}_t$. Equation (4.2) gives a clearer illustration of the *Struct*. In addition, each *Struct* contains a binary attribute, i.e., AWPF or NWPF.

$$\begin{aligned} Struct_t &= \{\hat{s}_{t-1}, \cdots, \hat{s}_{t-12}, \hat{s}_t, \cdots, \hat{s}_{t+11}, P_{t-1}^w, \cdots, P_{t-12}^w\} \\ t &= 1, 2, \cdots, 12 \end{aligned} \tag{4.2}$$

4.2.2 Proposed Methodology

In previous work [1], an observation on $\hat{s}_t$ with a 24-hour horizon has been implemented and four obvious types of error patterns have been extracted as shown in Figure 4.1. These errors may happen because of the abnormal temperature, air pressure, air density, or even some other special events. No matter what reason, once the $\hat{s}_t$ has a large deviation, the accuracy of the wind power forecast would be spoiled.

Instead of identifying the above four patterns, we try to propose a general way of finding out the errors in the NWP raw data using the data mining technique and embed this method into the WFO procedure.

The overall flowchart of this methodology is illustrated in Figure 4.2. For every wind power forecast point, the system firstly generates a corresponding *struct* as the preprocessing. Then, the data of the *struct* will run through an abnormal detection engine and is identified as AWPF or NWPF. The NWPFs are directly sent to the wind power forecast engine to obtain the WFO results, while AWPFs are sent to the data adjustment engine based on their types of error pattern to calculate the estimated $\hat{P}_t^w$ deviation. This

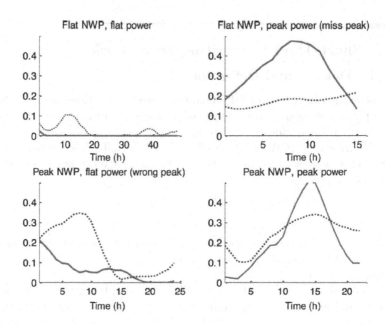

FIGURE 4.1 Four different types of error patterns. (Red solid line is wind power curve; blue dotted line is regularized NWP curve).

estimated deviation will be used to compensate the result directly from the wind power forecast engine.

To implement the framework, each module in Figure 4.2 is required to be well trained before the real application. The next section will explain each module in detail.

4.3   Improving Forecasting Using Adjustment of MWP

4.3.1   Wind Power Forecast Engine

The wind power forecast engine forms the basis of the whole forecasting system. As shown in Figure 4.3, the ANN is adapted to run as the forecast engine. The inputs are the same as those in Figure 4.2, i.e. the future 12-hour's $\hat{s}_t$, past 24-hour's P_t^w and the past 12-hour's $\hat{s}_t$. The output of the forecast engine is the next 1-hour's $\hat{P}_t^w$. In this chapter, the forecasting process is run in a rolling base, hour by hour, as shown in Figure 4.4.

To improve the efficiency and performance of the ANN, data preprocessing is implemented. In detail, wind speed usually ranges from 0 to 25 m/s for a wind turbine, therefore the input speeds are normalized to [0, 1] by using a

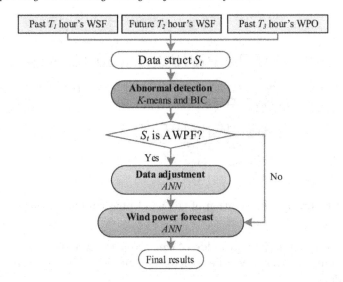

FIGURE 4.2 The flowchart of the short-term wind power forecasting system.

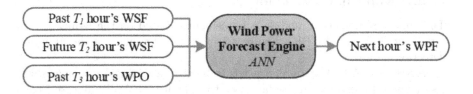

FIGURE 4.3 Framework of the wind power forecast engine.

linear transformation shown as Equation (4.3). Moreover, the wind power is normalized to [0, 1] as well by taking the wind farm capacity as the base value, shown as Equation (4.4),

$$\hat{s}_t = \frac{\hat{s}_{t,m}}{25} \tag{4.3}$$

$$P_t^w = \frac{P_{t,m}^w}{Cap^w} \tag{4.4}$$

where $\hat{s}_{t,m}$ is the original wind speed forecast value, and $P_{t,m}^w$ is the original wind power output.

Empirically, one hidden layer with MLP is a nonlinear function that is proved to be able to approximate any normal function to an arbitrary degree of accuracy, therefore it is sufficient for a 12-hour wind power forecast. Moreover, model selection techniques are usually required to specify the number of neurons for each layer. However, since the NN design is not the core work in this chapter, the network is finalized experimentally with three neurons as an input layer, four neurons as a hidden layer, and one neuron for the output

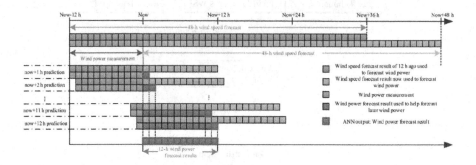

FIGURE 4.4 The illustration of the wind power forecast process.

layer. The input layer and hidden layer use a log-sigmoid transfer function, and the output layer uses a linear transfer function. The forecaster is trained with BP.

4.3.2 Abnormal Detection

The abnormal detection module is implemented based on the K-means clustering method and the BIC. Figure 4.5 gives the flowchart of the abnormal detection module.

Firstly, the AWPF dataset S needs to be generated for K-means learning. As shown in Figure 4.5, all the data used to generate the AWPF datasets come from the historic P_t^w and $\hat{P}_t^w$. The WPF engine is run for each historic time point and gets the wind power forecast of that time. Moreover, the forecasted P_t^w is compared with the corresponding P_t^w of that time point. Based on the definition of the AWPF, the forecast result whose error rate is higher than a threshold, i.e. 30%, will be selected, with its corresponding *struct*. An example is shown in Figure 4.6.

Next, a feature vector that can represent the raw AWPF dataset is extracted for each forecast. The feature vector should possess the capability to facilitate the clustering algorithm to distinguish the AWPF effectively and efficiently. Usually, those features associated with the same type of abnormal forecast are supposed to be close to each other. In this chapter, several features are extracted and listed as follows.

The features extracted from the real $\hat{s}_t$ include:

1. $\hat{s}_t$ value at time t:

$$\hat{s}_t, \quad \hat{s}_t \in Struct_t \tag{4.5}$$

2. $\hat{s}_t$ peak-vally difference

$$\begin{aligned} d_t^s = \max(\hat{s}_t | \hat{s}_t \in Struct_t) \\ - \min(\hat{s}_t | \hat{s}_t \in Struct_t) \end{aligned} \tag{4.6}$$

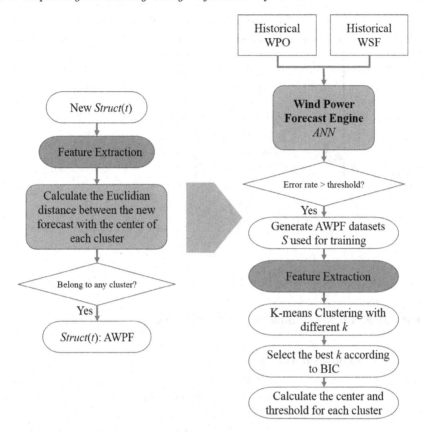

FIGURE 4.5 Framework of the abnormal detection module.

where d_t^s is the peak-valley difference of $\hat{s}_t$; $\max(\cdot)$ and $\min(\cdot)$ returns the maximum and minimum values, respectively.

3. $\hat{s}_t$ average value

$$\mu_t^s = \frac{\sum\limits_{i=t-12}^{t+11} \hat{s}_i}{num(\hat{s}_{t-12}, ..., \hat{s}_t, ..., \hat{s}_{t+11})} \qquad (4.7)$$

4. Maximum $\hat{s}_t$ ramp up

$$RU_t^s = \max\left(\hat{s}_{i+1} - \hat{s}_i \,|\, i = t-12, ..., t+10\right) \qquad (4.8)$$

5. Maximum $\hat{s}_t$ ramp down

$$RD_t^s = \max\left(\hat{s}_i - \hat{s}_{i+1} \,|\, i = t-12, ..., t+10\right) \qquad (4.9)$$

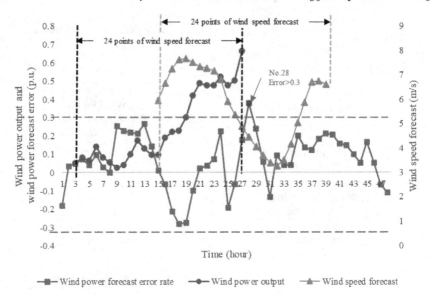

FIGURE 4.6 An example of the AWPF dataset.

6. $\hat{s}_t$ Standard deviation

$$\sigma_t^s = \sqrt{\frac{\sum\limits_{i=t-12}^{t+11}(\hat{s}_i - \mu_t^s)^2}{num(\hat{s}_{t-12}, ..., \hat{s}_t, ..., \hat{s}_{t+11})}} \tag{4.10}$$

The features extracted from the real P_t^w include:

1. P_t^w value at time $t - 1$

$$P_{t-1}^w, \ P_{t-1}^w \in Struct_t \tag{4.11}$$

2. P_t^w peak-valley difference

$$d_t^w = \max(P_{t-1}^w | P_{t-1}^w \in Struct_t) - \\ \min(P_{t-1}^w | P_{t-1}^w \in Struct_t) \tag{4.12}$$

3. P_t^w average value

$$\mu_t^w == \frac{\sum\limits_{i=t-12}^{t-1} P_i^w}{num(P_{t-12}^w, P_{t-11}^w, ..., P_{t-1}^w)} \tag{4.13}$$

4. Maximum P_t^w ramp up

$$RU_t^w = \max\left(P_{i+1}^w - P_i^w \,|\, i = t-12, ..., t-2\right) \tag{4.14}$$

5. Maximum P_t^w ramp down

$$RD_t{}^w = \max\left(P_i^w - P_{i+1}^w \,|\, i = t - 12, ..., t - 2\right) \quad (4.15)$$

6. P_t^w Standard deviation

$$\sigma_t^w = \sqrt{\frac{\sum\limits_{i=t-12}^{t-1} (P_i^w - \mu_t^w)^2}{num(P_{t-12}^w, P_{t-11}^w, ..., P_{t-1}^w)}} \quad (4.16)$$

Furthermore, the K-means algorithm is applied here to implement clustering based on the feature vectors generated in the previous step.

The number of clusters is denoted as k, which will be learnt by the algorithms itself based on the BIC. The center of each cluster k is denoted by $B_k = (B_k^1, B_k^2, \cdots, B_k^m)$, where m is the feature number. The threshold radius of the kth cluster is denoted as e_k, which is the maximum Euclidean distance between set elements and the set center, as in Equation (4.17). In Equation (4.17), $F_{k,q}^j$ is the value of feature j for the qth point in the kth cluster.

The threshold radius of each cluster will be utilized as the rules to identify the error events in the real application, as Equation (4.18), where $I(\cdot)$ is the judge function of every new input *Struct*, and is a factor ($[0, 1]$) to adjust the threshold value.

$$e_k = \max\left(\sqrt{\sum_{j=1}^{m} (F_{k,q}^j - B_k^j)^2}, q = 1, 2, ..., \text{points in Set k}\right) \quad (4.17)$$

$$I(Struct_t) = \begin{cases} Features\ of\ Struct_t \in k, & \sqrt{\sum\limits_{j=1}^{m} (F_t^j - B_k^j)^2} \leqslant \alpha e_k \\ Features\ of\ Struct_t \notin k, & otherwise \end{cases}$$

$$(4.18)$$

The BIC for each k and B_k will be calculated as Equation (4.19) [22]:

$$BIC = \log \prod_q \frac{N_{B(q)}}{N} \frac{1}{\sqrt{2\pi}\sigma_{B(q)}} \exp(-\frac{\|x_q - m_{B(q)}\|}{2\sigma_{B(q)}^2}) - \frac{p}{2} \log N \quad (4.19)$$

In Equation (4.19), N denotes the number of points, $B(q)$ denotes the cluster to which x_q is assigned, $N_{B(q)}$ denotes the number of points assigned to $B(q)$, $m_{B(q)}$ denotes the center vector of $B(q)$, $\sigma_{B(q)}$ denotes the standard deviation of points in $B(q)$ with respect to $m_{B(q)}$, and p denotes the number of parameters in the statistical model, which equals to $k \cdot m$, i.e. the multiplication of number of clusters and the feature space dimension. The first term of Equation (4.19) is the log likelihood of the k-means fit, the higher of which reflects a more precise clustering. The second term reflects the complexity of the model which increases with k. As the model becomes more complex, i.e. k increases, both the first and the second terms in Equation (4.19) will increase. The resulting BIC will reach a peak point at some k. This very k will be selected and its corresponding clustering result will be used.

The complete procedure of the algorithm is presented in the pseudo-code below:

Algorithm 3 Algorithm-K-means

Require: A set K of candidate k.

  for $k \in K$ **do**

    Make initial guesses for the means $m_1, m_2, \cdots, m_k$.

    Set the counts $n_1, n_2, \cdots, n_k$ to zero.

    while uninterrupted **do**

      Acquire the next example, x.

      if m_q is closest to x **then**

        Increment n_q.

        Replace m_q by $m_q + 1/n_q(x - m_q)$.

      end if

    end while

    Calculate BIC (k) according to Equation (4.21).

  end for

  Select the k with max BIC (k), and return corresponding clustering results.

The number of clusters, as well as the center and threshold of each cluster, is calculated as a basis of the bad data identification. The abnormal detection engine will evaluate each wind power forecast whether it belongs to some type of AWPF by calculating the Euclidian distance between the *Struct* of the forecast and the center of each cluster, and moreover, compare the distance with the threshold.

4.3.3 Data Adjustment Engine

Once a *Struct* is labeled as the AWPF, it will be sent to the data adjustment engine first. The data adjustment engine is also constructed using the neural network. The procedure is shown in Figure 4.7. Different from the wind power forecast engine, the input of the data adjustment engine is the features extracted from the abnormal detection engine. The used network structure of ANN here is identical with the one in the wind power forecast engine, an input layer with 3 neurons, a hidden layer with 4 neurons, and an output layer with one neuron.

Since the neural network naturally has the over-fitting problem, which is known as the bias-variance dilemma, feature and model selection are usually required. A proper design should aim at reducing the ratio of the complexity to the number of training sets as low as possible. Many related works exist

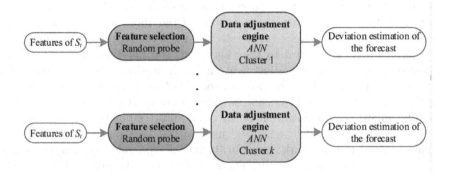

FIGURE 4.7 The framework of data adjustment.

[23–25]. In this chapter, the Random Probe [26] is applied because of its two desirable properties, i.e. it does not rely on any assumption about the probability distribution of the variables, and it guarantees a user-defined upper bound on the risk of keeping a variable, although it is irrelevant. In addition, it is easy to implement, computationally efficient, and has been proved successful in a variety of areas.

In detail, a set of realizations of random, irrelevant features, defined as "probe features", is firstly generated by randomly shuffling the components of the vectors of candidate features. This will be applied later to discriminate the irrelevant and relevant features. The generated probe features, together with the candidate variables, are ranked in the order of decreasing relevance by orthogonal forward regression. The probability for a probe feature to rank better than a candidate feature is estimated, and the threshold rank is determined such that the probability for an irrelevant variable to be selected is smaller than a threshold chosen by the model designer.

Ranking is performed by orthogonal forward regression, based on Gram-Schmidt orthogonalization, which provides a list of the candidate variables, ranked in order of decreasing relevance to the quantity to be predicted.

The next step is the definition of the threshold rank such that all variables that are ranked below the threshold are discarded. To that end, n_p realizations of the probe feature are generated by randomly shuffling the components of each vector of candidate variables, so that they have the same probability distributions as the original candidate feature. The resulting random features are appended to the set of vectors of the candidate features and ranked with the latter. The probability that a probe ranks better (i.e. is more relevant) than a candidate feature is estimated as follows: the estimated probability that the probe rank is smaller than or equal to a given rank r is the ratio n_{rp}/n_p, where n_{rp} is the number of realizations of the random probe whose rank is smaller than or equal to r. During the ranking process, when a rank r is reached such that $n_{rp}/n_p > \delta$ (where δ is the risk chosen by the designer as described above), the procedure is terminated and the threshold rank r_0 is set equal to $r - 1$. A general presentation of the above method put into the perspective of alternative variable selection methods is provided in [25].

Before moving forward to the validation, one question needs to be answered. Why is the data adjustment engine trained separately for different types of error patterns instead of all together? Actually, if it is trained together, there is no need to cluster the error patterns at all. Then the solution will be almost the same as what is described in [21], a point-by-point forecast. As the author claimed, the method used in [21] performs very well for the datasets used in their work. However, it does not provide the same contribution in this work experimentally. Theoretically, since the reasons for different types of abnormal forecast vary from each other, there is a high possibility that certain features may have completely contrary influences on the error rates. Taking Figure 4.1 as an example, the wind speed ramp-up rate for 3rd and 4th figure are close to each other, however, these happened in different patterns, and it can be seen that the wind power of these two ramp-up durations are smaller and larger than the expected value, respectively.

4.4   Case Study

In this section, one case study is carried out using the data of one wind farm from 'Global Energy Forecasting Competition 2012 - Wind Forecasting' to validate the proposed method. The dataset contains the wind power measurement data and NWP data including wind speed and direction which updates the 48-hour-ahead forecast every 12 hours for more than one year. The time interval is one hour, and in this chapter, the target is to forecast the future 12-hour P_t^w when the future NWP data is available. The data from 01/07/2009 to 12/12/2010 (530 days) are adopted in this chapter to conduct the case study. Because the $\hat{s}_t$ data is updated every 12 hours, there would be a total of 1060 sets and each set contains the 12-hour actual wind power measurement and 48-hour $\hat{s}_t$ results. In this chapter, 200 sets are used for training the wind power forecast engine, 800 sets are used for training the data adjustment engine, and the rest are used for algorithm evaluation. Parameters of the proposed method are shown in Table 4.1.

TABLE 4.1 Parameters of the data adjustment forecasting approach

Name	Value
Threshold Error Rate of Error Event	0.3
α	1

4.4.1 Indices for Evaluating the Prediction Accuracy

This chapter will use the index of RMSE, denoted as ε_{RMS}, to study the prediction accuracy, which is calculated as Equation (4.20)

$$\varepsilon_{RMS} = \sqrt{\frac{\sum\limits_{t=1}^{N}\left(P_t^w - \hat{P}_t^w\right)^2}{N}} \qquad (4.20)$$

where P_t^w is the actual wind power output, and N is the total forecast point number.

4.4.2 Wind Power Forecast Engine

The detailed algorithm has been explained in Subsection 4.3.1. The results analysis for the historical data is given in Figure 4.8. It is the error rate distribution for different forecast horizons from 1 hour ahead to 12 hours ahead. In the figure, it can be seen that more than 50% of forecast error rates are within [-0.05, 0.05] for the 1-hour~4-hour prediction horizon. The closer the forecasting time point is to the current time point, the higher the accuracy is. Therefore, the wind power forecast engine is run in a rolling way hour by hour.

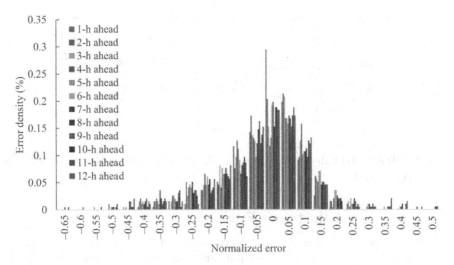

FIGURE 4.8 Forecast error distribution for the final prediction results with each prediction horizon from 1 hour ahead to 12 hours ahead.

4.4.3 Abnormal Detection

As in the discussion in Subsection 4.3.2, the datasets used for abnormal detection are generated by the wind power forecast engine. Moreover, the extracted features of each dataset are calculated. Finally, putting these features into the k-means clustering algorithm, based on the BIC, a total of 6 clusters are generated as shown in Figure 4.9 (only the centers are shown in the figure). The threshold value to be used as the rules identify the error event is given in Figure 4.10.

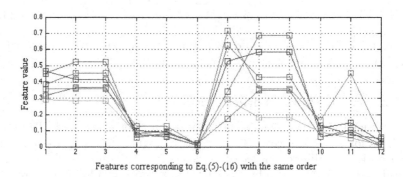

FIGURE 4.9 Using the K-means clustering method to classify AWPFs into 6 clusters.

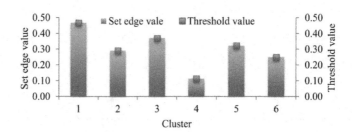

FIGURE 4.10 Edge value of each AWPF cluster and the corresponding threshold value.

4.4.4   Data Adjustment Engine

The data adjustment engine is trained based on the features selected in the abnormal detection stage. The outputs are error rates. After the training, once a forecast is defined as an AWPF, the potential error rate can be calculated by using the data adjustment engine. Then the final results can be adjusted by using Equation (4.21)

$$\hat{P}_t^{w,a} = \frac{\hat{P}_t^w}{1 + \varepsilon_r} \qquad (4.21)$$

where $\hat{P}_t^{w,a}$ is the adjusted wind power forecast.

4.4.5 Results Analysis

The wind power forecast results of the test datasets are shown in Figure 4.11. Both the forecast results by the wind power forecast engine and the results by the adjustment engine are compared with the measured wind power. Figure 4.11b-d are three extracted segments from Figure 4.11a, and the results show that AWPFs can be identified effectively. The wind power forecast with adjustment has higher accuracy compared with that without adjustment.

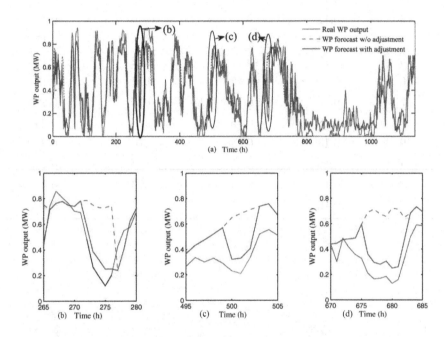

FIGURE 4.11 Comparison between the forecasted wind power by the WPF engine and the one after adjustment with the actual measured wind power.

The RMSE results are shown in Figure 4.12. The blue curve is the RMSE of preliminary wind power prediction derived from the base forecaster; and the red curve is the RMSE of the final wind power prediction after adjustment. It shows that the prediction accuracy is higher for larger prediction horizons. Detailed results of comparison of different methods are shown in Table 4.2.

TABLE 4.2 RMSE comparison of different approaches

Time Horizon (hour)	WPF w/o Adjustment	WPF with Adjustment	SVM with Adjusted NWP	Persistence Model
1	0.09	0.06	0.06	0.07
2	0.11	0.07	0.08	0.12
3	0.12	0.09	0.09	0.14
4	0.14	0.10	0.10	0.17
5	0.15	0.12	0.12	0.18
6	0.16	0.12	0.13	0.19
7	0.16	0.13	0.16	0.22
8	0.18	0.15	0.16	0.24
9	0.18	0.14	0.16	0.23
10	0.18	0.15	0.14	0.22
11	0.17	0.14	0.13	0.21
12	0.16	0.14	0.13	0.20

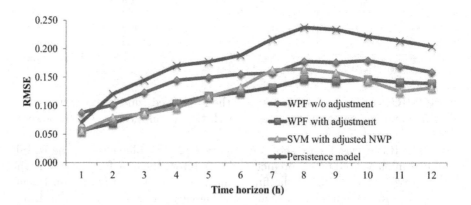

FIGURE 4.12 The RMSE of the preliminary wind power forecast result, the final wind power forecast result after adjustment, the persistence model forecast result, and the SVM forecast result with adjusted NWP.

4.5 Summary

The main contribution of this chapter is to introduce the NWP data adjustment to short-term wind power forecasting. It is implemented by identifying and clustering the errors of NWP using data mining techniques, and then the abnormal raw NWP data will be adjusted before it is sent to the wind power forecast engine. The simulation results demonstrate that the proposed approach can effectively reduce the total WFO error.

However, there are still challenges remaining for the proposed work. First, the physical explanation and reason for different error patterns are still unclear. A purely data-driven solution without model explanation is not solid enough to guarantee the generality of the proposed algorithm. Second, the threshold to select the AWPF is experimental, which obviously needs a more sophisticated adaptive solution. This will guarantee the practical application of the proposed method. Finally, there are too many parameters for the currently proposed method; compared with the improvement in accuracy, it seems still far away from being in a real application. Since this work is an attempt to do data adjustment for wind power forecasting, more efforts should be expected in future work.

Bibliography

[1] G. Qu, J. Mei, and D. He. Short-term wind power forecasting based on numerical weather prediction adjustment. *Industrial Informatics (IN-DIN), 2013 11th IEEE International Conference on*, pages 453–457, July 2013.

[2] L. Li, Y. Wang, and Y. Liu. Impact of wake effect on wind power prediction. *Renewable Power Generation Conference (RPG 2013), 2nd IET*, pages 1–4, Sept 2013.

[3] D. Carvalho, A. Rocha, M. Gmez-Gesteira, and C. Santos. A sensitivity study of the WRF model in wind simulation for an area of high wind energy. *Environmental Modelling and Software*, 33(0):23–34, 2012.

[4] L. Mei, S. Luan, C. Jiang, H. Liu, and Y. Zhang. A review on the forecasting of wind speed and generated power. *Renewable and Sustainable Energy Reviews*, 13(4):915–920, 2009.

[5] M. Khalid and A.V. Savkin. Closure to discussion on "a method for short-term wind power prediction with multiple observation points". *IEEE Transactions on Power Systems*, 28(2):1898–1899, May 2013.

[6] J. Tastu, P. Pinson, P. J. Trombe, and H. Madsen. Probabilistic forecasts of wind power generation accounting for geographically dispersed
information. *IEEE Transactions on Smart Grid*, 5(1):480–489, Jan 2014.

[7] C. Wan, Z. Xu, P. Pinson, Z. Dong, and K. Wong. Probabilistic forecasting of wind power generation using extreme learning machine. *IEEE
Transactions on Power Systems*, 29(3):1033–1044, May 2014.

[8] P. J. Trombe, P. Pinson, and H. Madsen. A general probabilistic forecasting framework for offshore wind power fluctuations. *Energies*, 5(3):621–
657, 2012.

[9] H. Bludszuweit, J.A. Dominguez-Navarro, and A. Llombart. Statistical
analysis of wind power forecast error. *IEEE Transactions on Power Systems*, 23(3):983–991, Aug 2008.

[10] S. Rajagopalan and S. Santoso. Wind power forecasting and error analysis
using the autoregressive moving average modeling. *IEEE Power Energy
Society General Meeting*, pages 1–6, July 2009.

[11] P. Chen, T. Pedersen, B. Bak-Jensen, and Z. Chen. ARIMA-based time
series model of stochastic wind power generation. *IEEE Transactions on
Power Systems*, 25(2):667–676, May 2010.

[12] N. Chen, Z. Qian, I.T. Nabney, and X. Meng. Wind power forecasts using
Gaussian processes and numerical weather prediction. *IEEE Transactions
on Power Systems*, 29(2):656–665, March 2014.

[13] P. Louka, G. Galanis, N. Siebert, G. Kariniotakis, P. Katsafados,
I. Pytharoulis, and G. Kallos. Improvements in wind speed forecasts for
wind power prediction purposes using Kalman filtering. *Journal of Wind
Engineering and Industrial Aerodynamics*, 96(12):2348 – 2362, 2008.

[14] H. Quan, D. Srinivasan, and A. Khosravi. Short-term load and wind
power forecasting using neural network-based prediction intervals. *IEEE
Transactions on Neural Networks and Learning Systems*, 25(2):303–315,
Feb 2014.

[15] J. Zeng and W. Qiao. Short-term wind power prediction using a wavelet
support vector machine. *IEEE Transactions on Sustainable Energy*,
3(2):255–264, April 2012.

[16] E.E. Elattar. Short term wind power prediction using evolutionary optimized local support vector regression. *Innovative Smart Grid Technologies (ISGT Europe), 2011 2nd IEEE PES International Conference and
Exhibition on*, pages 1–7, Dec 2011.

[17] J. Shi, Y. Liu, Y. Yang, and W. J. Lee. Short-term wind power prediction

based on wavelet transform-support vector machine and statistic characteristics analysis. *Industrial and Commercial Power Systems Technical Conference (I CPS), 2011 IEEE*, pages 1–7, May 2011.

[18] N. Amjady, F. Keynia, and H. Zareipour. Wind power prediction by a new forecast engine composed of modified hybrid neural network and enhanced particle swarm optimization. *IEEE Transactions on Sustainable Energy*, 2(3):265–276, July 2011.

[19] S. Fan, J. R. Liao, R. Yokoyama, L. Chen, and W. J. Lee. Forecasting the wind generation using a two-stage network based on meteorological information. *IEEE Transactions on Energy Conversion*, 24(2):474–482, June 2009.

[20] A. Costa, A. Crespo, J. Navarro, G. Lizcano, H. Madsen, and E. Feitosa. A review on the young history of the wind power short-term prediction. *Renewable and Sustainable Energy Reviews*, 12(6):1725 – 1744, 2008.

[21] G. Sideratos and N.D. Hatziargyriou. An advanced statistical method for wind power forecasting. *IEEE Transactions on Power Systems*, 22(1):258–265, Feb 2007.

[22] D. Pelleg, A. W. Moore, et al. X-means: Extending k-means with efficient estimation of the number of clusters. *ICML*, pages 727–734, 2000.

[23] N. Amjady and F. Keynia. Day-ahead price forecasting of electricity markets by mutual information technique and cascaded neuro-evolutionary algorithm. *IEEE Transactions on Power Systems*, 24(1):306–318, Feb 2009.

[24] I. Drezga and S. Rahman. Short-term load forecasting with local ANN predictors. *IEEE Transactions on Power Systems*, 14(3):844–850, Aug 1999.

[25] I. Drezga and S. Rahman. Input variable selection for ANN-based short-term load forecasting. *IEEE Transactions on Power Systems*, 13(4):1238–1244, Nov 1998.

[26] H. Stoppiglia, G. Dreyfus, R. Dubois, and Y. Oussar. Ranking a random feature for variable and feature selection. *Journal of Machine Learning Research*, 3:1399–1414, Mar 2003.

5

Short-Term Uncertainty of Renewable Energy Generation

The integration of renewable energy requires additional operating reserves to cope with the uncertainty of power system operation. Previous research shows that the uncertainty of the renewable energy forecast varies with the level of its output. Therefore, allocating reserves dynamically according to the specific distribution of the renewable energy forecast would benefit system scheduling. In this chapter, a statistical model is presented to formulate the conditional distribution of forecast error for multiple renewable energy farms using copula theory. The proposed model is tested using a set of synchronous data of renewable energy and its day-ahead forecast. The results show that scheduling reserves dynamically according to the modeled conditional forecast error reduces the probability of reserve deficiency while maintaining the same level of operating costs.

5.1 Overview

The short-term uncertainty of renewable energy generation refers to the deviation of renewable energy output compared with its forecasts. In power system operation, renewable energy is usually forecasted a day ahead or several hours ahead. Due to the stochastic nature of renewable energy, its output cannot be forecasted perfectly, and the forecast error should be compensated for by operational reserves during power system operation. Recent research shows that the uncertainty of renewable energy forecasts varies with its output level. Therefore, modeling its conditional forecast error to dynamically allocate reserves according to the specific distribution of renewable energy forecast is necessary because the integration of renewable energy requires additional operating reserves to cope with the uncertainty in power system operation.

For wind power, the amount of reserve needed largely relies on the uncertainty of its forecast error. Typically, the forecast error is modeled using a Gaussian distribution. Other distribution types have been proposed over the years, including the beta distribution [1], Weibull distribution, and hyperbolic distribution [2]. More thorough studies have found that the uncertainty of a wind power forecast varies with its output level [3–6]. Lange [3] proposes a conditional forecast error model based on the wind speed forecast error and the wind turbine power curve. Pinson and Kariniotakis [4] calculate the conditional prediction intervals for different wind power forecasts using a fuzzy inference model. Menemenlis *et al.* [5] use gamma-like functions to approximate the forecast error distributions for various levels of forecasted wind power. Zhang *et al.* [6] propose another universal analytical formulation for the same purpose. Such detailed modeling of forecast error makes it possible to dynamically adjust reserves for different wind power forecast levels rather than use a fixed reserve rate, such as that used to compensate for loads. It is estimated that up to half of the extra cost due to the inaccuracy of wind power forecast could be saved if the forecast error was more strategically considered [7]. Furthermore, there are several stochastic forecast techniques that also model the short-term uncertainty of renewable energy from a forecast perspective [8].

Currently, there is very limited domestic and foreign research on the short-term uncertainty of photovoltaic (PV) output. Compared with the uncertainty modeling of wind power output [9–11], the uncertainty modeling of PV power generation is still in an early phase. In some studies, the forecast error of PV output is viewed as a normal distribution, and a method is adopted that superimposes the forecast of PV points on a normal distribution to constitute the PV probabilistic forecasting [12]. A framework of stochastic differential equations was used to carry out probabilistic forecasting for solar radiation in [13]. The probabilistic distribution of the hourly clearness index was first predicted through Bayesian autoregressive time series models, and then the probabilistic distribution of PV output was estimated using a Monte Carlo simulation to carry out random sampling on the clearness index in [14]. On the basis of analyzing the relationships between various factors that affect PV output, the dynamic Bayesian network theory was used to construct a forecasting model for the short-term probability of PV power generation to give the probabilistic distribution of PV output at the next moment under the condition of known current factors in [15].

This chapter contributes to the technique of modeling the conditional forecast error of multiple wind farms and PV stations by generation scheduling. The advantage of this approach is that it requires only the historical point forecast records and can obtain a conditional forecast error estimation without improving forecasting skills. The method also considers the spatial correlations observed in the forecast errors between adjacent wind farms or PV stations [16], which will help geographically allocate the reserve capacity to guarantee the delivery of reserve power under congested transmission conditions.

In this chapter, a statistical model is presented to formulate the conditional distribution of forecast error for multiple renewable energy farms using copula theory. The proposed model is tested using a set of synchronous data of renewable energy and its day-ahead forecast. The results show that scheduling reserves dynamically according to the modeled conditional forecast error reduces the probability of reserve deficiency while maintaining the same level of operating costs.

5.2 Wind Power Short-Term Uncertainty Modeling

5.2.1 Modeling Conditional Error for a Single Wind Farm

Due to the stochastic nature of wind, the wind power forecast and the actual power output do not have a fixed relationship and thus can be regarded as a pair of random variables with stochastic dependence, where there would be a change of the conditional distribution of one variable when the other variable is altered. The joint distribution of wind power output and its forecast can be modeled using the copula function. Suppose x is the actual output, while y is the forecasted power. Their joint probability density function (PDF) f_{XY} can be rewritten as:

$$f_{XY}(x,y) = \frac{F_{XY}(x,y)}{\partial x\, \partial y} = c\left(F_X(x), F_Y(y)\right) f_X(x) f_Y(y) \qquad (5.1)$$

where f_X and f_Y are the marginal PDFs of x and y, respectively, and c is the copula density function. Given the point forecast $y=\hat{p}$, the conditional probability density function (CPDF) of the actual power output $f_{X|Y}$ can be expressed as:

$$f_{X|Y}(x|y=\hat{p}) = \frac{f_{XY}(x,\hat{p})}{f_Y(\hat{p})} = c\left(F_X(x), F_Y(\hat{p})\right) f_X(x) \qquad (5.2)$$

Assuming the forecast error $e = x - y$, the corresponding conditional distribution of the forecast error $f_{E|Y}$ is:

$$f_{E|Y}(e|y=\hat{p}) = \frac{f_{XY}(x,\hat{p})}{f_Y(\hat{p})} = c\left(F_X(e+\hat{p}), F_Y(\hat{p})\right) f_X(e+\hat{p}) \qquad (5.3)$$

The domain of e is $[-\hat{p}, \max(x) - \hat{p}]$.

Equation (5.3) provides a method for quantifying the conditional forecast error of wind power using the point forecasts. It shows that the distribution of the forecast consists of two parts: the distribution of actual power output as the base function and the copula density function as a variant multiplier. The calculation of conditional forecast error is therefore transformed into the calculation of these two functions.

5.2.2 Modeling Conditional Errors for Multiple Wind Farms

Equation (5.3) can be expanded to model multiple wind farms. Assuming $y_1=\hat{p}_1, y_2=\hat{p}_2,...,y_N=\hat{p}_N$ are the point forecasts of wind farms from #1 to #N, and e_1, e_2,...,e_N represent the forecast errors, the joint CPDF of e_1, e_2,...,e_N conditional to $y_1=\hat{p}_1$, $y_2=\hat{p}_2$,...,$y_N=\hat{p}_N$, namely, $f_{E_1,...,E_1|Y_1,...,Y_N}$, can be formulated as:

$$
\begin{aligned}
&f_{E_1,...,E_1|Y_1,...,Y_N}(e_1,...,e_N|y_1=\hat{p}_1,...,y_N=\hat{p}_N) \\
&=\frac{f_{X_1,...,X_N,Y_1,...,Y_N}(e_1+P_1,...,e_N+P_N,P_1,...,P_N)}{f_{Y_1,...,Y_N}(P_1,...,P_N)} \\
&=\frac{c_{2N}\left(F_{X_1}(e_1+P_1),...,F_{X_2}(e_N+P_N),F_{Y_1}(P_1),...,F_{Y_N}(P_N)\right)}{c_N\left(F_{Y_1}(P_1),...,F_{Y_N}(P_N)\right)} \cdot \prod_{i=1}^{N} f_{X_i}(e_i+P_i)
\end{aligned}
$$

$$(5.4)$$

where $f_{Y_1,...,Y_N}$ denotes the joint PDF of the point forecasts among all the wind farms, and $f_{X_1,...,X_N,Y_1,...,Y_N}$ denotes the joint PDF for all the actual outputs together with their forecasts. c_N and c_{2N} are the copula density functions that correspond with $f_{Y_1,...,Y_N}$ and $f_{X_1,...,X_N,Y_1,...,Y_N}$, respectively. $F_{Y_1},F_{Y_2},...,F_{Y_N}$ are the CDFs of the forecast power for each wind farm.

The marginal CPDF of each single wind farm $f_{E_i|Y_1,...,Y_N}$ can be formulated as:

$$
\begin{aligned}
&f_{E_i|Y_1,...,Y_N}(e_i|y_1=P_1,...,y_N=P_N) \\
&=\frac{c_{i,N}\left(F_{X_i}(e_i+P_i),F_{Y_1}(P_1),...,F_{Y_N}(P_N)\right)}{c_N\left(F_{Y_1}(P_1),...,F_{Y_N}(P_N)\right)} f_{X_i}(e_i+P_i),\ i \in [1,N]
\end{aligned}
$$
$$(5.5)$$

where $c_{i,N}$ is the copula density function that involves only the dependence between the actual output of a single wind farm i and the forecast for all the wind farms.

Equations (5.4) and (5.5) not only model the relationship between the forecast error and the point forecast of each wind farm but also take into consideration the correlations among all the wind farms. The marginal CPDFs for each forecast error presented in Equation (5.5) may also have correlations with each other.

5.2.3   Standard Modeling Procedure

According to the above derivation, the procedure of modeling the conditional forecast error of wind power using its point forecasts can be summarized as follows:

1. Prepare historical measurements of the wind power outputs $p_{i,t}$ and the corresponding record of their point forecasts $\hat{p}_{i,t}$, where $i \in [1,N]$ denotes the number of wind farms, and $t \in [1,T]$ denotes the time scale of the historical measurements, e.g., hourly resolution for a year.

2. For each single wind farm, model the unconditional CDF of the wind power output and its forecast power, namely, F_{X_i} and F_{Y_i}. Such modeling has been widely studied, and various forms of models are currently available, e.g., using a beta distribution, a mixed normal distribution or nonparametric distributions [1], [2].

3. Transform the $p_{i,t}$ and $\hat{p}_{i,t}$ into uniform distributed records using their own CDFs, respectively:

$$u_{i,t} = F_{X_i}(p_{i,t}), \hat{u}_{i,t} = F_{X_i}(\hat{p}_{i,t})\ i \in [1,N], t \in [1,T]. \qquad (5.6)$$

4. Find the best copula (c_{2N} in Equation(5.4)) for $p_{i,t}$ and $\hat{p}_{i,t}$ by fitting the joint PDF of all of the $u_{i,t}$ and $\hat{u}_{i,t}$, $i \in [1,N], t \in [1,T]$. The modeling can follow a standard copula function identification procedure: choosing several typical forms of copula and identifying the key parameters using the rank correlation coefficient between all of the $u_{i,t}$ and $\hat{u}_{i,t}$ values and selecting the model with the best performance by comparing the fitness indices. A detailed technique for copula function regression and its parameter identification can be found in [17]. Once the c_{2N} is formulated, the c_N in Equation (5.4) and the $c_{i,N}$ in Equation (5.5) can also be obtained since they present a part of the stochastic dependence of c_{2N}.

5. Calculate the CPDF of the forecast error for any output level using Equations (5.3) to (5.5).

5.2.4 Discussion

The advantage of the proposed model is that it employs only two simple functions, namely the unconditional distribution of the wind power output and the copula function, to formulate the uncertainty of forecast error overall output levels. According to previous studies [5] and [6], the CPDF of the forecast error can also be modeled using a series of distributions, in which each distribution model is used for a certain level of wind power output. However, such approaches use a number of models to capture the forecast uncertainty for only a finite number of output levels. The distribution parameters for certain output levels are difficult to determine using historical measurements because of the low occurrence probability of these output levels. In contrast, the functions in the proposed model are not specific to any output level. Thus, all of the historical data can be used to give the best estimate of the parameters. The proposed model also takes into account the possible correlations among the forecast errors of different wind farms. Therefore, it can provide a comprehensive estimate of the forecast error for each single wind farm using the forecasting from multiple wind farms. The proposed CPDF model can be calculated efficiently in a discrete form using the sequence operation theory [18].

Although the proposed model seems similar to that proposed by Bessa [19],

TABLE 5.1 Fitting testing of stochastic dependence modeling using copula functions

Type of Copula	Gaussian	t	Clayton	Frank	Gumbel
Statistics	42.94	47.03	237.81	65.12	39.75
Benchmark			$59.30\left(\chi^2_{0.05}(43)\right)$		

it has significant differences. The approach proposed in this chapter models the stochastic dependence between forecasted power and actual output rather than the dependence between forecasted wind speed and actual output because the proposed model aims to provide a statistical method to estimate the forecast error that utilizes the results of point forecasts instead of a probabilistic forecasting method. In this way, the forecasted wind power and the actual output can be linked by an analytical, closed-form function.

5.2.5 Empirical Analysis: The U.S. East Coast

1) Stochastic dependence modeling

The wind power output and forecast data used in this chapter are identical as those used in Subsection 3.2.4.

Figure 5.1 demonstrates the probabilistic relationship between wind power output and its forecast for wind farm 1 (the northeastern most site) using the scatter plot of each pair of data. The marginal distributions of both the actual output and the point forecast are also plotted by each coordinate axis. Figure 5.2 shows the same data as Figure 5.1 after a uniform transformation (step 3 in Subsection 5.2.3). Five typical copula functions were chosen to fit the stochastic dependence. As shown in Table 5.1, the test shows that the Gumbel copula, Gaussian copula, and t copula are valid, and the Gumbel copula has the best performance. The Gaussian copula is more flexible for multivariate modeling than the Gumbel copula, and thus, this chapter uses the Gaussian copula in the case study. Figure 5.3 shows the identification results for the correlation coefficient in c_{2N}, which clearly shows the spatial correlation of different wind farm outputs, their forecasts and the correlation between them.

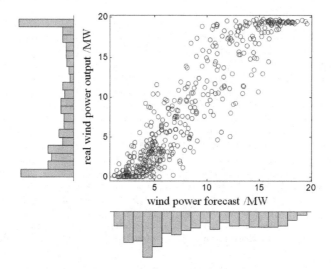

FIGURE 5.1 Scatter plot of the joint distribution of the actual wind power output and its forecast for wind farm 1

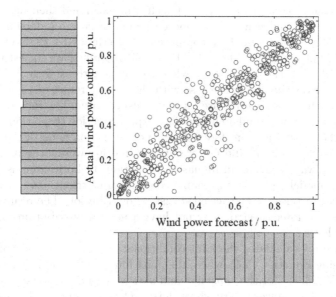

FIGURE 5.2 Scatter plot of the joint distribution of the actual wind power output and its forecast after CDF transformation for wind farm 1

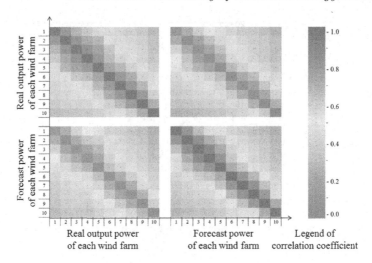

FIGURE 5.3 Copula identification: the correlation coefficients of the Gaussian copula for the actual outputs and forecasted power for all the wind farms

2) Conditional forecast error estimation

The conditional forecast error of each wind farm and their aggregate output were estimated using the proposed model. Figure 5.4 demonstrates the performance of the model by comparing the estimated CPDF with the actual distribution for wind farm 1. The left histogram shows the estimated and actual CPDF of the wind power when the point forecast is 5.30 MW, while the right histogram shows the case with the point forecast of 14.12 MW. The CPDFs of the forecast error are estimated using both a bivariate copula function (Equation (5.3)) and a multivariate copula function (Equation (5.5)). The results show that the point forecast may not be the mathematical expectation of all possible realized wind power outputs; in other words, the point forecast does not always provide an unbiased estimation. In both circumstances, the proposed models provide reasonable conditional distributions of the forecast errors that are close to the actual statistical distribution. The results also suggest that the forecast error may not have a normal distribution for all point forecast levels.

Figure 5.5 calculates the joint conditional forecast error for wind farms 1 and 2 using Equation (5.4), where the point forecast of wind farms 1 and 2 are 4.99 MW (24.95% of the capacity) and 9.81 MW (49.3% of the capacity), respectively. The figures clearly show the correlation of the conditional forecast error between these two wind farms. Such correlations can be easily understood in that the forecast errors of an adjacent wind farm are always driven by the same factor of weather pattern which the forecast tool is unaware of. Therefore, the forecast errors will have inherent interconnectedness. When a

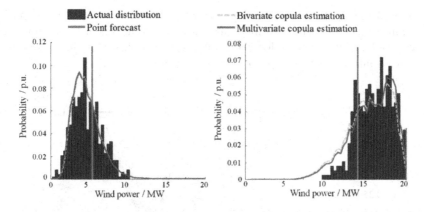

FIGURE 5.4 Conditional distribution estimation of wind farm 1 when the point forecast is 5.30 MW (left) and 14.12 MW (right)

large forecast error occurs on a wind farm, there is a high probability that adjacent wind farms also have large deviations.

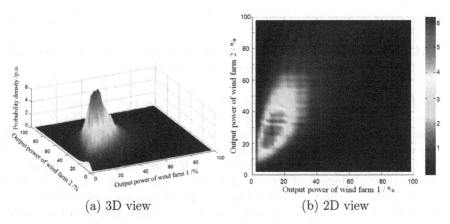

(a) 3D view   (b) 2D view

FIGURE 5.5 The joint distribution of the outputs of two wind farms (left) and their copulas (right)

Figure 5.6 demonstrates the estimation of conditional forecast errors for the aggregate output of all wind farms. A four-day period of the point forecast and the actual output are plotted with confidence intervals for the different occurrence probabilities. This figure shows that the uncertainty of the forecast varies notably with the output level. The uncertainty of very low or very high point forecasts is small while the uncertainty of medium point forecasts is large. Another interesting characteristic is that the point forecast always overestimates the actual wind power when the forecast is low while it underestimates the actual wind power when the forecast is high. In certain circumstances (30 h to 36 h in Figure 5.6), the point forecast has an evident positive bias with very little uncertainty, which implies that this forecast will overestimate the actual wind power in all circumstances.

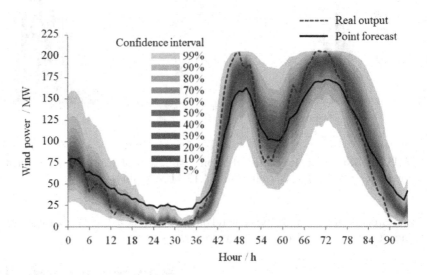

FIGURE 5.6 Four-Day period of point forecasts, actual output and confidence interval for the different occurrence probabilities

Figure 5.7 shows the overall distribution of wind power output and partitions the distribution by different point forecast levels using different colors. The conditional distribution of wind power output is differentiated as the point forecast increases from 0% to 100% with a step of 10%. The outer contour of the plot forms the marginal distribution of the wind power output. Figure 5.8 shows the distribution of the forecast error conditional to different levels of point forecast, in which the outer contour forms the marginal distribution of the forecast error. These two figures verify the observation from Figure 5.6: the forecast error has high bias and low variance when the forecast is very low or high, while it has low bias and high variance when the forecast is medium.

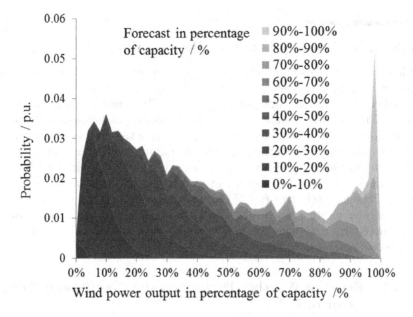

FIGURE 5.7 Cumulative plot of the conditional distribution of the wind power output for different forecast levels

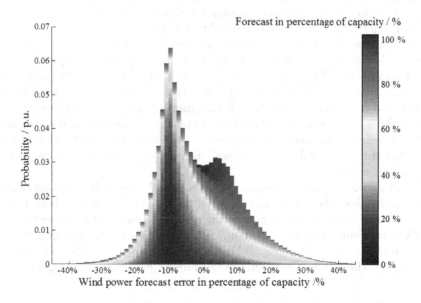

FIGURE 5.8 Cumulative plot of the conditional distribution of forecast error for different forecast levels

5.3   PV Power Short-Term Uncertainty Modeling

Photovoltaic (PV) output also has strong uncertainty. Different from wind power output, PV power generation is more significantly affected by the weather. The stability and predictability of PV power generation vary under different weather conditions, causing significant randomness in PV output. Therefore, in the present chapter, we first analyze the effect of weather factors on the conditional PV output forecast error and then establish a probabilistic distribution model applicable to the probability of error for the PV output conditions, achieving a more accurate estimation of the probabilistic distribution of the forecast error of PV output by establishing copula-based models for different weather types. On this basis, an accuracy evaluation index for PV short-term uncertainty modeling is proposed, and the effectiveness of the proposed method is verified by empirical analysis.

5.3.1   Effect of Weather Factors on the Conditional Forecast Error of PV

This section addresses the classification method for different weather types. Since it is very difficult to model uncertainty of PV power generation output for different weather conditions using a single model, most current research on forecasting methods involves first clustering the PV output according to the weather conditions and then adopting different models or the same model with different parameters to model different weather conditions for forecasting. The weather types are often classified by integrating the conditions of multiple types of meteorological factors such as cloudiness, temperature, and solar radiation. The standards of the China Meteorological Administration divide weather conditions into 33 types. Given that excessive classifications will lead to overly complicated classification algorithms, and the number of samples for each weather type will be too few, most of the PV power generation forecasting methods merge these 33 specialized weather types into fewer representative typical weather types, such as sunny, cloudy, rain shower, or heavy rain [20].

Clustering algorithms commonly used in PV forecasting include the K-means method, the self-organizing map (SOM) method, among others. Herein, the K-means method is adopted to divide the PV output data using the daily PV output curve (the curve is composed of 24 points, and each hour is a point) as an object to be clustered. Taking the Euclidean distance between the two curves as the distance between the two objects in the K-means algorithm:

$$d = \sqrt{\sum_{t=1}^{24} \left(P_t^{v(1)} - P_t^{v(2)}\right)^2}.$$

In the formula: $P_t^{v(1)}$ and $P_t^{v(2)}$ are the PV output on the $i - th$ hour of two different days, respectively; and d is the Euclidean distance between the two curves. An output curve is selected as the initial cluster center from each type of generalized typical weather type (sunny, cloudy, rain shower, moderate rain, and heavy rain) described in [20].

To analyze the difference in the PV output forecast error for different weather conditions, an artificial neural network model (widely used in PV forecasting) is adopted to conduct virtual forecasting and PV output verification. The verification data is taken from the Solar Power Forecasting section in the Second Global Energy Forecasting Competition in 2014 held by the Institute of Electrical and Electronics Engineers (IEEE) Working Group on Energy Forecasting [21]-[22]. The forecasted target PV power plant is located somewhere in the southern hemisphere, and the PV panels are fixed. The given data includes the hourly historical output data for the PV panels and the historical output values correspond to 12 weather-related factors. When forecasting, it is necessary to estimate the hourly PV output on the basis of the given forecast values of various relevant factors.

In this chapter, a forecasting model for the PV day-ahead output points based on the back propagation (BP) artificial neural network is established. Through repeated testing, the solar radiation outside the Earth's atmosphere, the solar radiation on Earth's surface, and temperature are selected as the neural network input variables, and the output variable is the PV forecast output. Each amount is the data from 6:00 to 21:00 of the day to be forecasted, for a total of 16 hours. That is, there are a total of $3 \times 16 = 48$ input variables and 16 output variables. The neural network contains one hidden layer and 16 hidden neurons. The training data set for the neural network includes a total of 365 days, and the testing dataset includes a total of 30 days. The Levenberg-Marquardt method embedded in MATLAB®is used for training, the sigmoid function is adopted from the input layer to the hidden layer, and the linear function is utilized from the hidden layer to the output layer. See [23] for details on more research on forecasting methods for PV points based on an artificial neural network.

In the aforementioned forecasting competition, the data from May 2012 to May 2013 was used as examples for analyzing the effect of weather types on the PV output forecast error. First, the PV output forecast value for each day was obtained based on the algorithm for PV point output point forecasting. Then, the K-means method was used to carry out classification on the weather type for each day based on the given surface radiation, and a scatter diagram was made from the forecast values and actual values for PV power generation output for different weather types, as shown in Figure 5.9; wherein, the output forecast values and actual values were processed by per-unit value normalization. The abscissa of each subgraph represents the PV output forecast values, and the ordinate represents the actual PV output values.

Figure 5.9 shows that the degree of correlation between the PV output forecast values and actual values on sunny days is very high, and the scatter

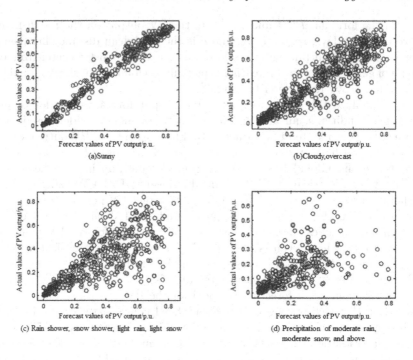

FIGURE 5.9 Scatter plot of forecasted power and actual output for different weather types

diagram almost becomes a straight line; certain differences between the forecast values and actual values appear for cloudy weather, but the degree of correlation between the two is still comparatively high; for weather with rain showers and a small amount of precipitation, the degree of correlation between the output forecast values and actual values decreases significantly as the output increases; and in weather with a large amount of precipitation, there is a certain correlation between the forecast values and the actual values only when the PV output is very small. These results illustrate that the PV output forecast is more accurate under sunny conditions, but the forecast uncertainty of PV output is higher when the weather conditions are more complicated. Using a quantitative indicator—the Kendall rank correlation coefficient between the PV output forecast values and actual values—to describe this difference, the results for the four weather types are 0.8997, 0.7964, 0.6983, and 0.6463, and the differences are obvious. Therefore, in using copula theory to model the conditional PV power generation output forecast error, it is necessary to consider the effect of different types of weather on the forecast error distribution. Specifically, before the conditional forecast error is determined, all the historical PV data (including output and weather-related factors) should first be classified according to weather type, and then a conditional forecasting model for each type of weather can be established.

5.3.2 Standard Modeling Procedure

In adopting the copula function for the conditional wind power output forecast error to model the conditional PV power generation output forecast error, the difference from the traditional wind power output forecast error is that we achieve a more accurate forecast error estimation for the probabilistic distribution of the PV output by establishing copula-based models for different weather types. To summarize, the estimation process for conditional forecast error distribution of the weather-affected PV power generation output is as follows.

1. Use a clustering algorithm such as K-means to divide the historical PV power generation data into several categories according to weather type (assuming a total of N categories). The actual output values and the point forecast values should be included in the PV historical data. Wherein, the point forecast values can be obtained by virtual forecasting using a neural network algorithm or other deterministic forecasting methods. Record the actual values of historical output for each type of weather as $\alpha_{k,t}$, and record the historical point forecast values as $\beta_{k,t}$, wherein, $k \in [1, N]$ represents the weather type, $t \in [1, T]$ represents the time period, and T is the total number of time periods in the historical data.

2. Separately add up the historical actual output values $\alpha_{k,t}$ and the historical point forecast values $\beta_{k,t}$ for each weather type to obtain the marginal distribution of the actual values F_{Xk} and the marginal distribution of the forecast values F_{Yk} of the PV output.

3. Obtain the Kendall rank correlation coefficient between the historical actual output values $\alpha_{k,t}$ and the historical point forecast values $\beta_{k,t}$ for each weather type, identify the parameters for each type of copula function according to the method described in Section 5.1, and derive the copula function with the best fitness.

4. Use a neural network algorithm or other forecasting methods to obtain the point forecast values of the PV output in the future, $\hat{p}$.

5. Use the modeled copula function to calculate the conditional probabilistic distribution of the forecast error under known point forecast $\hat{p}$ conditions using Equation (5.3).

5.3.3 Accuracy Analysis

In this chapter, the evaluation method utilized in probabilistic forecasting is adopted to evaluate the accuracy of the proposed method for estimating the forecast error probabilistic distribution of the PV output. Since the result obtained in probabilistic forecasting is not one numerical value but a probabilistic distribution, it is impossible to utilize the usual error evaluation indicators for evaluation. The primary evaluation dimensions of probabilistic forecasting include the forecasting calibration (which evaluates the approximate degree of the shape of the forecasted probabilistic distribution with the actual probabilistic distribution) and sharpness (which evaluates the concentration degree of the forecast probabilistic distribution), among others [24]. Four methods are adopted separately to carry out the evaluation, including the quantile scoring method, the marginal calibration method, the central probability interval index, and the continuously ranked probability score index. They are described as follows:

1. Quantile scoring method

   The organizing committee of the 2014 Global Energy Forecasting Competition adopted the quantile scoring method to evaluate the probabilistic forecasting result; the method's principles are as follows:

   For each time period to be forecasted within the time range, assuming that the output of the $q-th$ quantile provided by the competitor is $P_q^v (q = 1, 2, ...99)$, and assuming that the actual value of PV output is P^v, the score S calculated through the pinball loss function is as shown in the equation:

   $$S(P_q^v, P^v) = \begin{cases} (1 - q/100) \cdot (P_q^v - P^v), & P^v < P_q^v \\ (q/100) \cdot (P^v - P_q^v), & P^v \geq P_q^v. \end{cases} \qquad (5.7)$$

   The final total score, namely, the mean value of the sum of all the quantile scores for S is:

   $$S_{total} = \frac{1}{99} \sum_{q=1}^{99} S(P_q^v, P^v). \qquad (5.8)$$

   Equation (5.7) shows that the S metric is the error between the forecast value and the actual value; therefore, a lower final score S_{total} indicates a more accurate forecasting result.

   The definitions of Equation (5.7) and Equation (5.8) show that the quantile scoring method can comprehensively account for the main features of the aforementioned probabilistic forecasting. In terms of calibration, if the output of a certain quantile P_q^v is farther away from the actual value P^v, then the score S_{total} is higher; if the forecasted output interval fails to cover the actual value P^v, then the distance between P^v and the majority of the P_q^v will be farther, and the corresponding score for S_{total} will be higher. From the sharpness perspective, if the forecasted output interval is too wide, although it can contain the actual value P^v, due to the presence of many P_q^v in the forecasted output interval that is farther away from the actual value P^v, the score for S_{total} will still not be reduced. In summary, the quantile scoring method can comprehensively assess probabilistic forecasting.

2. Marginal calibration

The marginal calibration method graph compares the differences between the forecasted probabilistic distribution and the actual probabilistic distribution through mapping; this type of graph is called the marginal calibration plot. The horizontal axis of this graph is the value of the random variable to be forecasted (or its per-unit value), and the vertical axis is the difference between the forecast quantile number and the actual quantile number corresponding to each value. The closer the curve is drawn to 0, the better the calibration of the probabilistic forecasting result. Since it is impossible to precisely know the probabilistic distribution of the actual quantity to be forecasted, historical statistical values are often used to calculate the actual quantile numbers in the assessment.

3. Central probability interval

The central probability interval represents the mean width between the centers of the prediction intervals corresponding to a certain degree of confidence, which is used to measure the extent of dispersion in the quantity to be forecasted. The central probability interval reflects the sharpness of the probabilistic forecasting. Under ordinary circumstances, when the mean width of the prediction interval is smaller, the amount of information carried by the probabilistic forecasting is larger, and the sharpness of the probabilistic forecasting is greater.

4. Continuous ranked probability score

The definition of the continuous ranked probability score is given by the equation:

$$\mathrm{CRPS}(F, P) = \begin{cases} \displaystyle\int_{-\infty}^{\infty} \{F(y) - 1\}^2 dy, & y \geq P \\ \displaystyle\int_{-\infty}^{\infty} (F(y))^2 dy, & y < P. \end{cases} \tag{5.9}$$

In the equation, P is the actual observed value, and F is the forecasted probabilistic distribution. This index can comprehensively evaluate the calibration and sharpness of the probabilistic forecasting.

Refer to [25] for additional introduction and discussion related to probabilistic forecasting evaluation methods.

5.3.4 Empirical Analysis

The PV output data and the data on related factors from the Solar Power Forecasting section in the Second Global Energy Forecasting Competition were used to verify the effectiveness of the proposed conditional forecast error distribution estimation method for PV power generation output.

First, the PV output data was divided into four categories corresponding to the four categories of weather types described in the previous text through the K-means algorithm. In each weather type, the Gaussian copula was used to establish a conditional forecast error model, and the conditional probabilistic distribution obtained from the copula was compared with the conditional probabilistic distribution obtained from actual statistics, as shown in Figure 5.10.

Each subgraph in Figure 5.10 gives the conditional probabilistic distribution of the actual PV output for different weather types when the point forecast is 0.3. In each subgraph of Figure 5.10, the vertical straight line represents the point forecast value of 0.3, the curve represents the fitting result of the copula function, and the histogram represents the statistics of the actual output corresponding to when the point forecast is 0.3. The copula fitting results are close to the actual statistics, thereby verifying the effectiveness of the conditional forecast error distribution estimation method proposed above.

Figure 5.10 shows that although the given point forecast value is the same, there are very large differences in the conditional probabilistic distribution of the actual output for different weather types. The conditional probabilistic distribution on sunny days is more concentrated, but the conditional probabilistic distribution for overcast and rainy weather is more dispersed. The point forecast values on sunny days are very close to the expectations of the conditional probabilistic distribution, but the point forecast values on rainy days deviate widely from the expectations of the conditional probabilistic distribution, and the situation for cloudy weather is between these two results. These results illustrate that the degree of correlation between the forecast values and the actual values is high on sunny days, and the forecast error is small, and the degree of correlation between the forecast values and the actual values is low on rainy days, and the forecast error is larger. These results are consistent with the modeling process analysis results in this chapter. Figure 5.10 also shows that the conditional probabilistic distributions are not all normally distributed. In reality, the conditional probabilistic distribution is closer to a normal distribution only when the weather is clearer.

Figure 5.11 gives the conditional probabilistic distribution of the PV output corresponding to the cloudy, overcast weather type when the point forecast

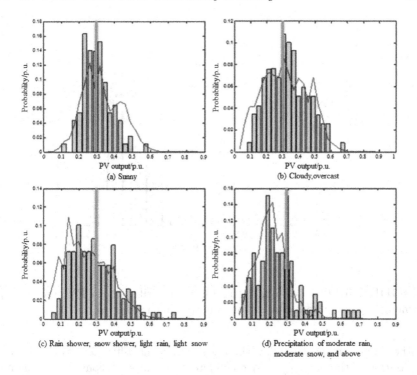

FIGURE 5.10 Conditional probabilistic distribution of PV power output under different weather types when point forecast is 0.3 p.u.

is 0.7. Comparing Figure 5.11 and Figure 5.10(b), when the point forecast differs for the same weather type, there may exist a very large difference in the probabilistic distribution of the actual output, and this is precisely the reason that a conditional forecast error model must be established.

Seven days were selected from the aforementioned results. The PV output point forecast, actual output values, and the forecasted probabilistic distribution within this time period were plotted on the same graph, as shown in Figure 5.12. The method for seeking the confidence interval of the different degrees of confidence in the graph is similar to the box plot process. For example, in selecting the PV output values corresponding to the cumulative probabilities of 47.5% and 52.5% from the forecasted output cumulative probabilistic distribution, between these two PV output values is a confidence interval corresponding to 5% confidence. In Figure 5.12, the weather during the first two days and on the seventh day was sunny; the third and fourth days had rain showers and light rain, respectively; and the fifth and sixth days were cloudy. Figure 5.12 shows that the confidence interval is relatively "narrow" when the weather is sunny, indicating that the accuracy of probabilistic forecasting is higher for sunny days than for cloudy or overcast rainy weather. On the

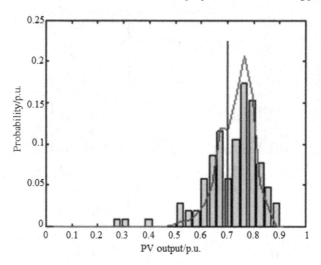

FIGURE 5.11 Conditional probabilistic distribution of PV power output when the point forecast is 0.7 and the weather is overcast

other hand, on the same day, there are significant differences in the error distribution of the point forecast for different output situations. When the point forecast is comparatively small, the error is also relatively small, but when the point forecast at noon is comparatively large, its error fluctuation range is also comparatively large.

The PV output actual values and the relevant weather factors from April 1, 2012, to May 31, 2013, in the aforementioned forecasting competition were used as training data to forecast the PV output from June 1 to June 30, 2013. First, the neural network algorithm was used to obtain the point forecast of the PV output, and then the conditional forecast error distribution estimation method for PV power generation output proposed in this chapter was used to convert the point forecast into a probabilistic distribution.

The proposed conditional forecast error distribution estimation method for PV power generation output was compared separately with the probabilistic forecasting method based on a normal distribution and with the empirical distribution method. The probabilistic forecasting method based on a normal distribution assumes that the actual values of PV power generation output obey an expected normal distribution using a point forecast and that the size of the variance is proportional to the point forecast values, taking the variance sizes corresponding to the values with the best accuracy after making multiple attempts. The empirical distribution method does not make any assumptions about the probabilistic distribution function and form of the model, but it obtains the probabilistic distribution model of the variables to be forecasted based on statistical calculations from historical values. Since there is a need

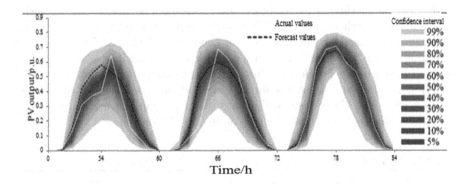

FIGURE 5.12 Seven-day period of point forecast, actual output, and confidence interval for the different occurrence probabilities

to calculate the conditional probabilistic distribution and sample points completely equal to the point forecast values may not exist in the historical data, it is often necessary to perform statistics within a small interval with a point forecast value as the center when generating an empirical probabilistic distribution. The window width of the small interval taken in this work was 2% of the PV panel capacity.

The four evaluation methods of quantile scoring, marginal calibration, central probability interval, and the continuous ranked probability score introduced in Subsection 5.3.3 of this chapter were adopted to compare the several aforementioned forecasting methods. Table 5.2 is the result of comparing the various methods corresponding to quantile scoring and continuous ranked probability score. Table 5.3 is a comparison of the mean widths of the 50% and the 90% prediction intervals for the various methods. Figure 5.13 shows the marginal calibration plot corresponding to the normal distribution method and the proposed method for PV output when the point forecast is 0.7; F_{fcast} represents the cumulative distribution function of the forecasted output when the point forecast is 0.7, and F_{obs} represents the empirical distribution function of the output when the point forecast is 0.7. Since it was necessary to adopt historical data as an estimate for the "actual probabilistic distribution", the beginning and end of the curve corresponding to the empirical distribution method is $y = 0$. Only the normal distribution method was compared with the proposed method.

The analysis and evaluation of the forecast results show that in Figure 5.13, from a calibration perspective, the red line is closer than the blue line to the ordinate scale at 0, indicating that the proposed method is closer to the actual probabilistic distribution than to a normal distribution. In terms of sharpness, Table 5.3 indicates that the mean width of the forecast interval for the proposed method is comparatively small, and due to the normal distribution method, the sharpness of the method is close to the empirical distribution method. The computational results of quantile scoring and the continuous ranked probability score are shown in Table 5.2. Under the conditions of the known PV output point forecast, regardless of whether the quantile scoring evaluation or the continuous ranked probability score evaluation is adopted, the scores for the method proposed in this chapter are the lowest. The above results illustrate that the proposed method can more robustly carry out probabilistic forecasting for PV output.

TABLE 5.2 Evaluation result of the probabilistic forecasting

Forecast Error Estimation Method	Quantile Scoring	Continuous Ranked Probability Score
Normal Distribution	0.0139	0.72
Empirical Distribution	0.0130	0.61
The Method in This Chapter	0.0128	0.58

TABLE 5.3 Average width of the central prediction intervals

Forecast Error Estimation Method	Mean Width of the 50% Prediction Interval	Mean Width of the 90% Prediction Interval
Normal Distribution	0.18	0.39
Empirical Distribution	0.15	0.30
The Method in This Chapter	0.14	0.31

5.4   Summary

This chapter proposes a statistical model to calculate the conditional forecast error for renewable energy generation. The model uses copula theory to build a joint dependence structure between the actual renewable energy output and its forecast. The proposed modeling technique can improve generation scheduling by more accurate scheduling of dynamic reserves for renewable energy. The

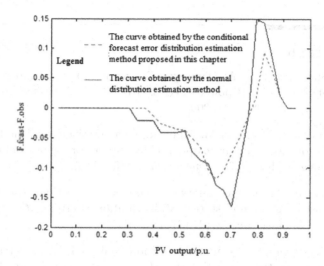

FIGURE 5.13 Marginal calibration plot for a Normal Distribution and the proposed method

PDF of the error deviating from its point forecast is further obtained from the copula-based model. Empirical analysis using historical measurements of renewable energy demonstrates that the distributions of renewable energy forecast errors under various point forecast levels have large differences. For wind power, the forecast error has high bias and low variance when the point forecast is very low or high, while it has low bias and high variance when the point forecast is medium. The forecast errors of PV output are distinct for different weather types. The distribution of renewable energy forecast errors also has a spatial correlation between adjacent renewable energy generation locations. The empirical analysis also implies that the proposed model is effective in capturing such stochastic features.

The proposed model can be applied to a power system in which only the renewable energy point forecasts are available. It is not restricted to the scenario-based UC model and is widely applicable. For instance, a chance-constrained UC model can directly utilize the proposed CPDF model for the forecast error. For the robust UC model, the conditional confidence interval of each renewable energy farm can also be obtained from the CPDF in the proposed model.

Bibliography

[1] H. Bludszuweit, J. A. Dominguez-Navarro, and A. Llombart. Statistical analysis of wind power forecast error. *IEEE Transactions on Power Systems*, 23(3):983–991, 2008.

[2] B. Hodge, E. Ela, and M. Milligan. Characterizing and modeling wind power forecast errors from operational systems for use in wind integration planning studies. *Wind Engineering*, 36(5), 2012.

[3] M. Lange. On the uncertainty of wind power predictions—analysis of the forecast accuracy and statistical distribution of errors. *Journal of Solar Energy Engineering*, 127(2):177–184, 2005.

[4] P. Pinson and G. Kariniotakis. Conditional prediction intervals of wind power generation. *IEEE Transactions on Power Systems*, 25(4):1845–1856, 2010.

[5] N. Menemenlis, M. Huneault, and A. Robitaille. Computation of dynamic operating balancing reserve for wind power integration for the time-horizon 1–48 hours. *IEEE Transactions on Sustainable Energy*, 3(4):692–702, 2012.

[6] Z. Zhang, Y Sun, W. Gao, J. Lin, and L. Cheng. A versatile probability distribution model for wind power forecast errors and its application in economic dispatch. *IEEE Transactions on Power Systems*, 28(3):3114–3125, 2013.

[7] P. Pinson, C. Chevallier, and G. N. Kariniotakis. Trading wind generation from short-term probabilistic forecasts of wind power. *IEEE Transactions on Power Systems*, 22(3):1148–1156, 2007.

[8] A. Costa, A. Crespo, J. Navarro, G. Lizcano, H. Madsen, and E. Feitosa. A review on the young history of the wind power short-term prediction. *Renewable and Sustainable Energy Reviews*, 12(6):1725–1744, 2008.

[9] Y. Ming, S. Zhu, X. Han, and H. Wang. Joint probability density forecast for wind farm output in multi-time-interval. *Automation of Electric Power Systems*, 2013.

[10] Z. Feng, L. Jin, L. Jian, and Z. Zhang. Probabilistic wind power forecasting based on multi-state space and hybrid Markov chain models. *Automation of Electric Power Systems*, 36(6):29–33+84, 2012.

[11] C. Wang, Z. Lu, Y. Qiao, Y. Min, and S. Zhou. Short-term wind power forecast based on non-parametric regression model. *Automation of Electric Power Systems*, 34(16):78–82, 2010.

[12] Z. Ziadi, M. Oshiro, T. Senjyu, A. Yona, N. Urasaki, T. Funabashi, and C. H. Kim. Optimal voltage control using inverters interfaced with pv systems considering forecast error in a distribution system. *IEEE Transactions on Sustainable Energy*, 5(2):682–690, 2014.

[13] E. B. Iversen, J. M. Morales, J. K. Møller, and H. Madsen. Probabilistic forecasts of solar irradiance using stochastic differential equations. *Environmetrics*, 25(3):152–164, 2014.

[14] A. Bracale, P. Caramia, G. Carpinelli, A. R. D. Fazio, and G. Ferruzzi. A Bayesian method for short-term probabilistic forecasting of photovoltaic generation in smart grid operation and control. *Energies*, 6(2):733–747, 2013.

[15] L. Dong, W. Zhou, P. Zhang, G. Liu, and W. Li. Short-term photovoltaic output forecast based on dynamic Bayesian network theory. *Proceedings of the CSEE*, 33:38–45, 2013.

[16] J. Tastu, P. Pinson, E. Kotwa, H. Madsen, and H. A. Nielsen. Spatio-temporal analysis and modeling of short-term wind power forecast errors. *Wind Energy*, 14(1):43–60, 2011.

[17] E. Bouyé, V. Durrleman, A. Nikeghbali, G. Riboulet, and T. Roncalli. Copulas for finance - a reading guide and some applications. *Social Science Electronic Publishing*, 2000.

[18] C. Kang, Q. Xia, and N. Xiang. Sequence operation theory and its application in power system reliability evaluation. *Automation of Electric Power Systems*, 78(2):101–109, 2002.

[19] R. J. Bessa, J. Mendes, V. Miranda, A. Botterud, J. Wang, and Z. Zhou. Quantile-copula density forecast for wind power uncertainty modeling. pages 1–8, 2011.

[20] F. Wang, Z. Mi, Z. Zhen, G. Yang, and H. Zhou. A classified forecasting approach of power generation for photovoltaic plants based on weather condition pattern recognition. *Proceedings of the CSEE*, 33(34):75–82, 2013.

[21] T. Hong, P. Pinson, and S. Fan. Global energy forecasting competition 2012. *International Journal of Forecasting*, 30(2):357–363, 2014.

[22] Global energy forecasting competition 2014 [EB/OL].[2014-08-15]. https://www.crowdanalytix.com/contests/.

[23] S. Wang and N. Zhang. Short-term output power forecast of photovoltaic based on a grey and neural network hybrid model. *Automation of Electric Power Systems*, 36(19):37–41, 2012.

[24] P. Pinson. Estimation of the uncertainty in wind power forecasting. *Engineering Sciences*, 2007.

[25] T. Gneiting, F. Balabdaoui, and A. E. Raftery. Probabilistic forecasts, calibration and sharpness. *Journal of the Royal Statistical Society*, 69(2):243–268, 2007.

6

Renewable Energy Output Simulation

In the uncertainty analysis of renewable energy, one important task is to regenerate the renewable energy output according to its stochastic characteristics. This is because the realization of "possible future" is the basic input of renewable energy system optimizations. However, simulating the output of wind power and PV is not as easy as Monte Carlo sampling since it is a complex stochastic process with certain time sequential patterns. Furthermore, the outputs of multiple wind farms/PV stations show a strong dependency, which complicates the simulation. This chapter presents methodologies for simulating the generation outputs of multiple wind farms and PV stations according to given statistical indices. Especially, the methodologies are able to well address the dependency issue.

6.1 Overview

Output time series for wind power and PV act as the foundation of power system planning and operation, considering their stochastic behavior. In most cases, however, it is difficult or even impossible to get real output curves from projected wind farms or PV stations. This increases the necessity for accurate techniques for renewable energy operation simulation.

Wind farm simulation models can be divided into two groups: the first group disregards the time sequencing of the wind power curve and the second group simulates time sequencing data. The models in the first group use a wind power distribution function to generate a wind power curve. Stochastic approaches are used to sample every point of the wind power curve. The models in the second group consider not only a distribution function of wind power but also the intensity of wind power variation. Approaches appropriate for time series are often employed in these models. In most power curve simulation models, wind speed is simulated first, and then, the wind power curve is obtained using the wind turbine output characteristic function. Alternately, some models directly sample wind power according to its statistical characteristics[1].

Sun, Y. et al. [2] sampled wind speed from its Weibull distribution function to study a stochastic model for dynamic economic dispatch considering unit commitment. In [3], several wind speed series were obtained using a similar technique and transformed into correlated series to simulate multiple wind farms. Models ignoring time sequencing often cause inaccuracy in simulating the wind stochastic profile. Kris [4] studied a case to test the differences between models with or without time sequencing and concluded that considerable error is possible when time sequencing is ignored. Martin and Carlin [5] also stress the importance of using time sequencing wind speeds and demonstrate a strong correlation of wind speed velocities between consecutive hours. Billinton [6] and Rajesh [7] simulated wind speed using a time-series model including auto-regressive and moving average parameters. Wind speed autocorrelation was performed by its recursive coefficient in the model. The wind turbine output characteristic was modeled to calculate wind farm output. Wu [8] used a similar technique to generate hourly wind speed according to mean wind speed and its standard deviation from statistical data. Models in [8] considered wind power generation random forced outages and weather effects [9]. Doquet and Michel [10] used the stochastic differential equation proposed in [11] to sample single wind farm output curves with a marginal beta-type distribution.

Currently, some studies at home and abroad use the concept of a clearness index to study the stochastic properties of PV power output. The clearness index is defined as the ratio of the horizontal global irradiance to the corresponding available atmospheric irradiance. Given a known hourly mean clearness index value for each month and from the perspective of meteorology, Hollands, K. G. and Huget, R. G. [12] presented an hourly clearness index probability density function. Tina, G. M. *et al.* [13] presented a solar panel output model based on a clearness index; Karim, S. A. *et al.* [14] used a wavelet transform and curve fitting methods to model sun radiation.

This chapter presents a methodology for simulating the operation of multiple wind farms and PV stations. For wind power output, we first generate the correlated wind speeds of different wind farms by using a stochastic differential equation that accounts for marginal distribution and time sequencing. The wind power output is further obtained using the wind power output characteristic function. Wind speed seasonal rhythms and diurnal patterns, turbine output characteristics, wind turbine reliability, and wake effects are considered in the model. For PV output, we generate the time series of unshaded solar irradiation and the clearness index and then calculate the PV outputs considering the solar panel power output characteristics as well as their tilt and direction of deployment. The simulation is also able to consider the output correlation between multiple PV stations. The proposed methods provide a useful tool for renewable energy operation and planning

6.2 Multiple Wind Farm Output Simulation

A methodology to simulate correlated multiple wind farm operation is described in this section, which includes historical wind speed data processing, wind speed series generation, the setup of wind turbine output characteristics, the wind turbine reliability model and the wake effect. The aim of this section is to build continuous multiple wind farm output curves through a year which fit the given statistical indices, including the wind speed probability density function, wind speed autocorrelation, seasonal rhythms and diurnal patterns, and the wind speed correlation coefficient matrix among wind farms. The methodological framework is shown in Figure 6.1. The simulation model details are described as follows.

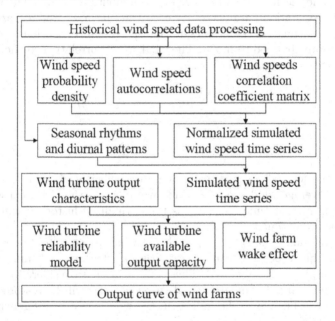

FIGURE 6.1 Methodological framework of simulation of multiple wind farm operations

6.2.1 Historical Wind Speed Data Processing

The main purpose of historical wind speed data processing is to obtain statistical indices that are proposed to fit simulated wind speeds. Historical wind speed data for a projected wind farm are primarily obtained from area anemometer towers.

For a simulation of a wind farm, n, p represents the serial number of the wind farm, the wind speed seasonal rhythms and diurnal patterns are extracted from historical data using Equation (6.1):

$$k_{nm} = \bar{v}_{nm}/\bar{v}_n, \; k_{nh} = \bar{v}_{nh}/\bar{v}_n \qquad (6.1)$$

in Equation (6.1), k_{nm} is the wind speed seasonal factor for month m, k_h is the wind speed diurnal factor for hour h, $\bar{v}_n$ is the average yearly wind speed, $\bar{v}_{nm}$ is the average wind speed for month m and $\bar{v}_{nh}$ is the average wind speed for hour h on every day. Using k_{nm} and k_{nh}, the normalized historical wind speed can be obtained from Equation (6.2):

$$\tilde{v}_{nt} = v_{nt}/(k_{nh}k_{nm}). \qquad (6.2)$$

The normalized historical wind speed $\tilde{v}_{nt}$ is more likely a stochastic process than $\tilde{v}_{nt}$ and has no scale difference between months or hours. The wind speed Weibull distribution parameters for each wind farm, c_n and d_n, are obtained from $\tilde{v}_{nt}$. Attenuation coefficients of wind speed autocorrelation a_n of different lag times are obtained from $\tilde{v}_{nt}$. The wind speed correlation coefficient matrix is obtained from Equation (6.3):

$$\rho_{np} = \frac{\sum\limits_{t=1}^{T} (\tilde{v}_{nt} - \bar{v}_n)(\tilde{v}_{pt} - \bar{v}_p)}{\sqrt{\sum\limits_{t=1}^{T} (\tilde{v}_{nt} - \bar{v}_n)}\sqrt{\sum\limits_{t=1}^{T} (\tilde{v}_{pt} - \bar{v}_p)}}. \qquad (6.3)$$

In Equation (6.3), ρ_{np} is the correlation coefficient of wind farms n and p. $\bar{v}_n$ is the average wind speed of wind farm n, v_{nt} is the historical wind speed of wind farm n at time t and $\tilde{v}_{nt}$ is the normalized historical wind speed of wind farm n at time t.

It is relevant that c_n, d_n, a_n and ρ of wind farms are obtained from $\tilde{v}_{nt}$, not v_{nt}. This is because wind speed seasonal rhythms and diurnal patterns have been primarily separated. These effects are not considered in the generation of normalized simulated wind speed $\tilde{v}_{nt}^*$ and will be restored after $\tilde{v}_{nt}^*$ is generated. Therefore, the indices in the wind speed series generation step should not reconsider these effects.

6.2.2 Generating Wind Speed Time Series

A wind speed series $\tilde{v}_{nt}^*$ is generated based on stochastic process theory. A stochastic differential equation is used to sample the hourly wind speed of proposed wind farms. The parameters c_n, d_n and a_n come from Subsection 6.2.1. When simulating multiple correlated wind speeds, a multivariate dW_t is used. The correlation coefficient matrix of dW_t is composed of ρ_{np} obtained by Equation (6.3).

The wind speed is defined within the interval $(0, +\infty)$. However, the stochastic process has the probability to obtain a sample value less than zero, which should be regulated to zero. The probability of this occurring is so low that regulation will not affect the probability density of $\tilde{v}_{nt}^*$.

Wind speed seasonal rhythms and diurnal patterns are not considered in $\tilde{v}_{nt}^*$. k_{nm} and k_{nh} are added for these effects; see Equation (6.4).

$$v_{nt}^* = k_{nh}k_{nm}\tilde{v}_{nt}^* \tag{6.4}$$

where v_{nt}^* is the simulated wind speed time series at time t.

6.2.3 Calculating Wind Turbine Output

Wind turbine output characteristics are employed to calculate the wind power output from simulated wind speeds v_{nt}^*. A typical characteristic function is given by Equation (6.5):

$$E_n(v) = \begin{cases} 0, \ 0 \leqslant v < v_{in}, \ v > v_{out} \\ \dfrac{v^3 - v_{in}^3}{v_{out}^3 - v_{in}^3}Cap_{rated}, \ v_{in} \leqslant v \leqslant v_{rated} \\ Cap_{rated}, \ v_{rated} \leqslant v \leqslant v_{out} \end{cases} \tag{6.5}$$

where $E_n(x)$ is the wind turbine output eigenfunction. v_{in}, v_{rated} and v_{out} are cut-in rated and cut-off wind speeds. Cap_{rated} is the rated capacity. The output capacity of each available wind turbine can be obtained from Equation (6.5).

6.2.4 Wind Turbine Reliability Model and Wake Effect

In actual wind farm operation, it is almost impossible that each turbine on the wind farm would be available at all times during the year. Unplanned outages and maintenance of wind turbines should be accounted for in an operation simulation. Unplanned outages and maintenance of wind turbines can be recognized as independent random events. The most direct way is to set up a two-state repairable model for each wind turbine. The continuous operation time and the continuous failure time of a wind turbine are recognized as exponential distributions and can be sampled from appropriate probability density functions. From the point of view of each moment, the number of available wind turbines in a wind farm is subject to a binomial distribution. A wind farm with 100 wind turbines and an availability ratio of 0.9 is supposed to correspond with $B(100, 0.9)$, which is shown in Figure 6.2. In this chapter, n_{nt} is the number of available wind turbines on a wind farm.

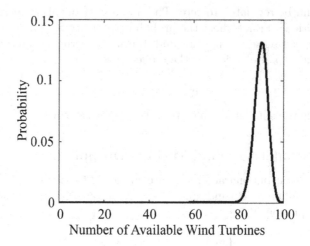

FIGURE 6.2 The number of available wind turbines and its probability for a wind farm with 100 wind turbines and an availability ratio of 0.9

Considering the time constant of the wind turbine reliability model is far larger than that of wind speed variation, the time sequencing of the wind turbine reliability model can be omitted for simplicity. The easiest way is to sample the number of available wind turbines in each wind farm from a binomial distribution at relatively long intervals (e.g. every week) during the simulation.

The wake effect should also be considered in an operation simulation. The wake effect is determined by the arrangement of wind turbines. A wind farm with a sparser arrangement has a smaller wake effect. The wake effect factor is defined as the total percentage of energy lost for a wind farm due to the wake effect. The typical value of the wake effect factor is 5% to 10%. In this chapter, η_n represents the wake effect factor of a wind farm.

To summarize, the final simulated output P_{nt}^w of wind farm n at time t is expressed as Equation (6.6):

$$P_{nt}^w = n_{nt}(1 - \eta_n)E_n\left(v_{nt}^* k_{nh} k_{nm}\right). \tag{6.6}$$

6.3 Multiple PV Power Station Output Simulation

6.3.1 PV Output Model

The basic principle of PV power generation is to use solar panels to convert solar energy into electricity through the photovoltaic effect. Studies show that P_t^v can be determined using Equation (6.7):

$$P_t^v = P_{stc}^v \frac{I(r_t, s_t, I_{0t})}{I_{stc}} \left[1 + \alpha_T \left(T_t - T_{STC}\right)\right] \tag{6.7}$$

where P_t^v is the solar panel output at time t, P_{stc}^v is the rated output of solar panels, whose definition is the solar panel output under standard conditions (solar irradiance $I_{stc} = 1000W/m^2$ and temperature $T_{stc} = 25°C$). When the actual solar irradiance is higher than that of the standard condition, the solar panel output will be larger than the rated output. I_{0t} represents the exo-atmospheric solar irradiance without considering the random factors of atmospheric attenuation such as the scattering of sunlight and cloud shading; this value is only related to the relative position change of the sun and Earth. s_t is the clearness index, which is defined as the ratio of total solar irradiance at ground level I_t to the exo-atmospheric solar irradiance I_{0t} : $s_t = I_t/I_{0t}$, where I_t is the solar irradiance at ground level at time t including both direct and scattered radiation. s_t is mainly affected by factors such as cloud cover, weather changes and elevation. r_t represents the ratio of the solar irradiance on the inclined surface to the total solar irradiance at ground level at time t, the value of which is roughly the ratio of the cosine of the solar incident angle onto the inclined surface to that onto the ground. Its variation pattern is related to the PV panel placement tilt angle and the PV panel tracking method, which can be fixed, horizontal tracking, tilted axis tracking or biaxial tracking. $I(r_t, s_t, I_{0t})$ represents the total solar irradiance on the PV panels after considering the solar irradiance, which is direct, scattered or reflected, the clearness index and the type of PV panel tracking system. T is atmospheric temperature, and α_T is the solar panel power temperature coefficient.

In the above equation, P_{stc}^v, I_{stc}, T_{STC} and α_T are constants. The factors that actually affect the solar panel output are I_{0t}, s_t, r_t and T. In the following sections, first, the output simulation framework for multi-PV power plants is given, and then, the simulation ideas for each variable are described separately.

6.3.2 PV Output Simulation Framework

The essence of PV output simulation lies in the simulation of I_{0t}, s_t, r_t and T. The horizontal unshaded solar irradiation I_{0t} is related to the geographical location of solar PV panels, the season and the time of the day and is calculated by the global solar irradiance model. The clearness index s_t is mainly affected by factors such as cloud coverage, weather changes, and elevation and features randomness and a strong intermittent nature. This index is the main source of PV output uncertainty and must be approximated by stochastic simulation. Additionally, the clearness indices for multiple PV stations with close geographic proximity have very strong correlations; thus, the stochastic differential equation model can be used for sampling considering this correlation. r_t is mainly related to the PV panel's own placement and tracking method. The incident angle of sunlight for different PV panel tracking methods must be derived. The temperature T has certain randomness; its simulation and information collection are quite difficult. Additionally, the value of α_T is usually small; thus, the impact of temperature on the PV output is also small. In the model put forth in this chapter, the impact of temperature on PV output will be ignored. Therefore, the simulation procedure of the proposed solar PV output is shown in Figure 6.3:

1. Acquire the basic PV station information, including the geographical information of the PV stations, the autocorrelation coefficient of solar irradiance and the basic parameters of the probabilistic model of the clearness index, etc.

2. Let there be N PV stations. We generate multiple non-independent, normally distributed time sequences $X_m(t) = x_{m1,t}, x_{m2,t}, \cdots, x_{mN,t}$ including correlation between the different PV stations and the autocorrelation of each PV station; m is the number of simulation times.

3. First, obtain the weather type probability distribution in each PV station and the maximum clearness index corresponding to the different types of day. Obtain the clearness index time sequence $S_{th,m}(t) = s_{m1,t}, s_{m2,t}, \cdots, s_{mN,t}$ for different PV stations by sampling according to the weather type probability distribution. Calculate the cumulative distribution function (CDF) of the solar irradiance for different clearness indices according to the clearness index probability model, expressed as $CDF_{S_{th,m}(t)}$.

4. Use $X_m(t)$ and $CDF_{S_{th,m}(t)}$ to sample and obtain the time sequence of the solar irradiance in different PV stations $I_m(t) = i_{m1,t}, i_{m2,t}, \cdots, i_{mN,t}$.

5. Use $I_m(t)$ to calculate the PV station output for each PV station according to the output formula for different PV array types.

6. Repeat the above procedure until the number of simulations m is equal to the given number M. Calculate the average value $P^v_{AV}(t)$ for the M calculations; select the set of calculation results that is closest to the average as the final simulation result $P^v(t)$, and output the result.

The following sections introduce the calculation methods for the unshaded solar irradiance I_{0t} at ground level, the solar irradiance ratio r_t and the clearness index s_t.

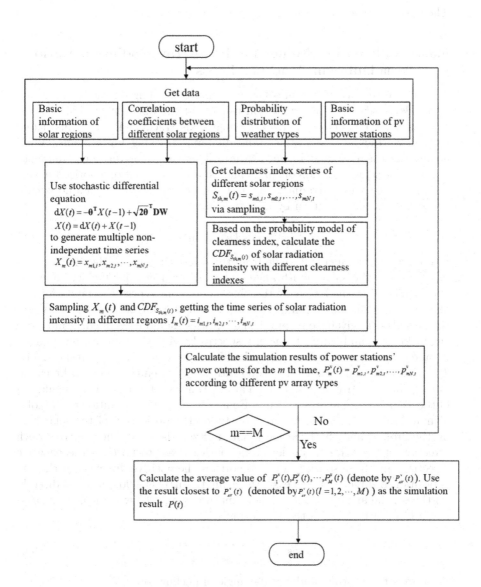

start

Get data

| Basic information of solar regions | Correlation coefficients between different solar regions | Probability distribution of weather types | Basic information of pv power stations |

Use stochastic differential equation
$$dX(t) = -\boldsymbol{\theta}^\mathrm{T} X(t-1) + \sqrt{2\boldsymbol{\theta}}^\mathrm{T} \mathbf{DW}$$
$$X(t) = dX(t) + X(t-1)$$
to generate multiple non-independent time series
$$X_m(t) = x_{m1,t}, x_{m2,t}, \cdots, x_{mN,t}$$

Get clearness index series of different solar regions
$$S_{th,m}(t) = s_{m1,t}, s_{m2,t}, \cdots, s_{mN,t}$$
via sampling

Based on the probability model of clearness index, calculate the $CDF_{S_{th,m}(t)}$ of solar radiation intensity with different clearness indexes

Sampling $X_m(t)$ and $CDF_{S_{th,m}(t)}$, getting the time series of solar radiation intensity in different regions $I_m(t) = i_{m1,t}, i_{m2,t}, \cdots, i_{mN,t}$

Calculate the simulation results of power stations' power outputs for the m th time, $P_m^v(t) = p_{m1,t}^v, p_{m2,t}^v, \cdots, p_{mN,t}^v$ according to different pv array types

m==M

No

Yes

Calculate the average value of $P_1^v(t), P_2^v(t), \cdots, P_M^v(t)$ (denote by $P_{av}^v(t)$). Use the result closest to $P_{av}^v(t)$ (denoted by $P_{av}^v(t) (l = 1, 2, \cdots, M)$) as the simulation result $P(t)$

end

FIGURE 6.3 Solar PV output simulation procedure

6.3.3 Calculation Model for Unshaded Irradiance at Ground Level I_{0t}

The unshaded solar irradiation model refers to Subsection 3.3.2.

6.3.4 Calculation Model for PV Array Radiation Ratio r_t for Different Tracking Types

To enhance the utilization of solar radiation, depending on the actual situation, a large-scale PV power station usually adopts different tracking modes, such as fixed PV arrays, horizontal single-axis tracking PV systems, tilted single-axis tracking PV systems, and dual-axis tracking PV systems. A PV array surface is usually an inclined surface. There is a relatively large difference between the solar radiation on an inclined surface and that of a horizontal surface. r_t is the ratio of irradiance on an inclined surface to that of a horizontal surface; its formula is given by Equation (6.8):

$$r_t = \cos Deg_c / \sin \alpha \tag{6.8}$$

where Deg_c is the incident angle from the direct sunlight to the inclined surface and α is the solar zenith angle. The variation in solar incident angle is closely related to the PV array tracking mode. Four types of formulas for calculating the solar incident angle for a PV array are introduced below. Before the formulas are given, some physical quantities should be defined. β and γ are the inclination and azimuth angle, respectively; A is the solar azimuth angle; β_r and γ_r are the tracking axis inclination and azimuth angle, respectively; Deg_r is the tracking angle. The solar altitude and zenith angle can be calculated from the astronomical formula at any time and location. The calculation of the solar zenith angle is shown in Equation (6.9). The definition of the solar azimuth angle A refers to the angle between the projection of the solar rays on the ground and the local meridian, and zero is set at the positive south direction of the azimuth angle. From south to east to north is the negative direction, and from south to west to north is the positive direction. If the sun is positioned directly to the east, the azimuth angle is $-90°$; if it lies directly northeast, the angle is $-135°$. The azimuth angle is $90°$ when it is positioned at west and $\pm180°$ at north. The formula is:

$$\cos A = \frac{\sin \alpha \sin \phi - \sin \delta}{\cos \alpha \cos \phi} \tag{6.9}$$

where ϕ is the latitude and δ is the angle of declination.

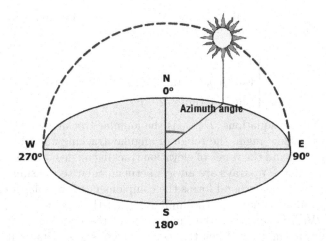

FIGURE 6.4 Solar azimuth angle diagram

The calculation formulas for the cosine of the incident angle $\cos Deg_c$, the tangent of the tracking angle $\tan Deg_r$, and the tangent of the corresponding inclination angle of PV panel relative to the horizontal plane $\tan \beta_t$ for 4 types of PV arrays are given as follows:

1) For fixed PV arrays,

$$\cos Deg_c = \sin \alpha \cos \beta + \cos \alpha \sin \beta \cos (\gamma - A). \tag{6.10}$$

Fixed PV arrays have no tracking angle, and the inclination angle relative to the horizontal plane is always β at any time; that is, $\tan \beta_t = \tan \beta$.

2) For horizontal single-axis tracking PV arrays,

$$\begin{cases} \cos Deg_c = \sqrt{1 - \cos^2\alpha\cos^2(\gamma_r - A)} \\ \tan Deg_r = \sin(A - \gamma_r)/tan\alpha \\ \tan \beta_t = \tan Deg_r \end{cases} \tag{6.11}$$

The inclination angle between the horizontal tracking axis of PV arrays and the ground forms the tracking angle. The range of tracking angle of the PV arrays is $\pm 90°$.

3) For tilted single-axis tracking PV arrays,

$$\begin{cases} \cos Deg_c = \sqrt{1 - \{\cos(\alpha - \beta_r) - \cos \beta_r \cos \alpha [1 - \cos(A - \gamma_r)]\}^2} \\ \tan Deg_r = \dfrac{\cos \alpha \sin(A - \gamma_r)}{\sin(\alpha - \beta_r) + \sin \beta_r \cos \alpha [1 - \cos(A - \gamma_r)]} \\ \tan \beta_t = \dfrac{\tan \beta \sqrt{\tan Deg_r^2 + \sin \beta^2}}{\sin \beta} \end{cases} \tag{6.12}$$

The range of tracking angle of PV arrays is usually $\pm 65°$.

4) For double-axis tracking PV arrays,

$$
\begin{cases}
Deg_c = 0 \\
\tan Deg_{r1} = \tan A, \tan Deg_{r2} = \tan \alpha \\
\tan \beta_t = \tan(\pi/2 - \alpha)
\end{cases}
\tag{6.13}
$$

In the above equations, Deg_{r1} is the angular tracking angle, Deg_{r2} is the elevation tracking angle, the range of angular tracking angle of PV arrays is usually $\pm 100°$, and the range of elevation tracking angle normally has no limit. Biaxial tracking PV arrays are always perpendicular to the sunlight; thus, its angle relative to the ground forms the complement of the solar incident angle.

In the above equations, the limit range of the rotational angle for each type of PV array refers to the limit range of the tracking angle Deg_r for a PV array tracking axis. When the value of Deg_r exceeds its limit at a certain time, the PV panel becomes the PV panel of fixed inclination and direction angle; at this point, let γ_r be the angle A of the solar azimuth angle when Deg_r reaches the limit and the calculation is carried out assuming the PV panel is fixed.

6.3.5   Solar Radiation Probability Density Model

In this subsection, the expression $I(r_t, s_t, I_{0t})$ is first derived, and then, the probability distribution of $I(r_t, s_t, I_{0t})$ is given based on the empirical probability distribution of s_t. Ultimately, the sampling method that considers the correlation among multiple PV power stations is given.

1) Solar radiation model that considers reflection and scattering

Reference [13] provides a model of PV panel solar irradiance after ground reflection and air scattering are considered. The total solar irradiance I_β of a PV panel with inclination β is expressed by the following equation:

$$
I(r_t, s_t, I_{0t}) = I_b r_t + I_d \left(\frac{1 + \cos \beta}{2} \right) + I_t \rho \left(\frac{1 - \cos \beta}{2} \right).
\tag{6.14}
$$

in which I_b represents direct solar irradiance of the horizontal surface; r_t represents the ratio of solar irradiance on the inclined surface to that on a horizontal surface; I_d represents the total solar irradiance on the horizontal surface; β is the PV panel tilt angle relative to the ground; ρ is the ground reflectivity. I_t is the total solar irradiance on a horizontal surface considering cloud cover, including direct solar irradiance I_b and scattered solar irradiance I_d, that is, $I_t = I_b + I_d$. I_β represents the total solar irradiance of a PV panel with inclination β. The equation above indicates that the total solar irradiance on the PV panel contains 3 parts: direct radiation, scattered radiation and reflected radiation, listed from the ground to the PV panel.

Rearranging the above equation, $I(r_t, s_t, I_{0t})$ can be expressed as follows:

$$I(r_t, s_t, I_{0t}) = (I_t - I_d)r_t + I_d \left(\frac{1 + \cos \beta}{2}\right) + I_t \rho \left(\frac{1 - \cos \beta}{2}\right)$$

$$= I_t r_t + I_d \left(\frac{1 + \cos \beta}{2} - r_t\right) + I_t \rho \left(\frac{1 - \cos \beta}{2}\right).$$

Scattered solar irradiance on the horizontal surface I_d and the total solar irradiance on the horizontal surface I_t have the following relationship:

$$I_d = I_t(p - q s_t)$$

where p and q are the parameters that relate to the air mass.

Since $I_t = s_t I_{t0}$ is known, $I(r_t, s_t, I_{0t})$ can be expressed as a function of r_t, s_t, I_{0t}:

$$I_\beta = \left[r_t + p - q s_t \left(\frac{1 + \cos \beta}{2} - r_t\right) + \rho \left(\frac{1 - \cos \beta}{2}\right)\right] s_t I_{0t}.$$

In the equation above, except s_t which is a random variable, all other variables can be determined by geographic or time information.

The above equation can be arranged into a quadratic function that relates to the random variable s_t:

$$I_\beta = A_1 s_t - A_2 s_t^2$$

in which:

$$A_1 = \left[r_t + p \left(\frac{1 + \cos \beta}{2} - r_t\right) + \rho \left(\frac{1 - \cos \beta}{2}\right)\right] I_{0t}$$

$$A_2 = q \left(\frac{1 + \cos \beta}{2} - r_t\right) I_{0t}.$$

Solving the above quadratic equation gives:

$$s_t = \frac{A_1 \pm \sqrt{A_1^2 - 4 A_2 I_\beta}}{2 A_2}$$

$$= \frac{\frac{A_1}{A_2} \pm \sqrt{\left(\frac{A_1}{A_2}\right)^2 - \frac{4 I_\beta}{A_2}}}{2}$$

$$= \frac{\alpha_1 \pm \alpha_2}{2}.$$

in which:

$$\alpha_1 = \frac{A_1}{A_2}$$

$$\alpha_2 = \sqrt{\alpha_1{}^2 - \frac{4 I_\beta}{A_2}}.$$

The physical meaning of s_t indicates that the value range should be $[0,1]$.

Under the premise that the equation has a real solution, there are several findings.

When $A_1 < 0$, then $s_t < 0$;

When $A_1 > 0$ and $A_2 > 0$, then $s_t \geqslant 0$, and in order to make $s_t \leqslant 1$, there should be $s_t = \frac{\alpha_1 - \alpha_2}{2}$;

When $A_1 > 0$ and $A_2 > 0$, to make $s_t \geqslant 0$, there should be $s_t = \frac{\alpha_1 + \alpha_2}{2}$.

According to [12], the probability density function of s_t can be expressed in the following form:

$$f(s_t) = \frac{J\left[e^{g s_t}\left(1 - \gamma_1 s_t\right) - 1\right]}{s_{th} g \gamma_1}$$

In which s_{th} is the maximum value of the clearness index s_t, which is determined by the weather type; J, g and γ_1 are constants determined by s_{th}.

The probability distribution of solar irradiance $I(r_t, s_t, I_{0t})$ can be expressed by the following equations on the premise that the probability density function of s_t is known:

When $A_1 > 0$ and $A_2 < 0$:

$$f(I(r_t, s_t, I_{0t})) = \frac{J\left(s_{th} - \frac{\alpha_1 + \alpha_2}{2}\right) e^{g \frac{\alpha_1 + \alpha_2}{2}}}{-s_{th} A_2 \alpha_2}. \tag{6.15}$$

When $A_1 > 0$ and $A_2 > 0$:

$$f(I(r_t, s_t, I_{0t})) = \frac{J\left(s_{th} - \frac{\alpha_1 - \alpha_2}{2}\right) e^{g \frac{\alpha_1 - \alpha_2}{2}}}{s_{th} A_2 \alpha_2}. \tag{6.16}$$

2) Solar radiation sampling method based on a stochastic differential equation

Radiation on PV power stations within close proximity shows a strong correlation. A non-independent time sequence is generated here by using a stochastic differential equation model. The function of the stochastic differential equation model in a PV output simulation is to create a time sequence that simultaneously obeys time variation, spatial correlation, and a normally distributed edge distribution. Its mathematical expression is:

$$\mathrm{d}X(t) = -\theta^{\mathbf{T}} X(t - 1) + \sqrt{2\theta}^{\mathbf{T}} \mathbf{DW}$$
$$X(t) = \mathrm{d}X(t) + X(t - 1). \tag{6.17}$$

where $\mathbf{DW}$ is the sampling vector of non-independent multidimensional normal distribution; its time correlational coefficient matrix between each dimension is the spatial correlation coefficient of solar radiation in each PV station. The parameter θ is the column vector composed of the exponential attenuation coefficient of the autocorrelational coefficient of the solar radiation in each PV station. $\theta^{\mathbf{T}}$ is the transposition of θ, and $\sqrt{2\theta}$ represents the sequence formed by the value obtained by taking the square root of two times each element in θ. $X(t)$ is the multidimensional time sequence obtained by simulation; $\mathrm{d}X(t)$

is the differential sequence of $X(t)$. Iterative computation of the two previous equations is applied during the entire simulation. According to the theoretical simulation of stochastic differential equations, the autocorrelational function of the time sequence for each sub-dimension of $X(t)$ obeys the exponential attenuation of the parameter θ:

$$corr(X_{s+t}, X_s) = e^{-\theta t}.$$

In the simulation, the equation above is used to generate a non-independent normal distribution time sequence, which is then applied to conduct solar PV output sampling:

$$\tilde{I}_t = H^{-1}\{\Phi[X(t)].\} \tag{6.18}$$

In the above equation, $H^{-1}(\cdot)$ is the inverse distribution function of $I(r_t, s_t, I_{0t})$, and $\Phi(\cdot)$ is the distribution function of a standard normal distribution.

3) Weather type and its corresponding clearness index

The various parameters of the random variable s_t are all directly related to the weather conditions. The method that calculates the probability distribution of hour-scale solar irradiance was given earlier. However, to determine the parameters for s_{th}, the weather factors need to be described.

As stated above, the probability characteristics of solar irradiance are closely related to weather factors. Weather factors not only affect the magnitude of the solar irradiance which is expectation value but also affects its fluctuation. Comparing a sunny day with a cloudy day, the former has a much greater expected value of solar irradiance than the latter, but in either case, the fluctuation of solar irradiance is small. For weather types with varying weather conditions, such as sunny to cloudy, sunny to rainy, and rainy to sunny, solar irradiance shows obvious fluctuation. Therefore, it is necessary to divide weather types rationally to determine the relevant parameters of s_t for different weather.

Theoretically, the parameter s_{th} should be statistically calculated by actual observational data. However, for the power-system dispatching plan, the value of future clearness index s_t cannot be accurately predicted. Nonetheless, it is easy to predict future weather type; therefore, the modeling concept in this section is to first rationally classify different weather types and then determine the value range of s_{th} for different weather.

There are 33 weather types in meteorology. The weather types are divided into 8 categories in the broad sense according to the typicality of different weathers and their probability of occurrence. The eight general weather types are divided into fixed weather and changing the weather, which depends on whether the weather changes during a day. For the fixed weather type, the variation pattern is similar during a day, and s_{th} can be represented by a fixed value. For changing weather types, the variation pattern may vary greatly at different times; thus, it is necessary to estimate the value variation of s_{th} with a focus on the weather conditions of different times. Typical values of s_{th} for the eight weather types are shown in Table 6.1.

TABLE 6.1 Weather type category

Broad-sense weather type	Metrological weather type	s_{th}
Fixed weather type	A:Sunny	0.864
	B:Cloudy, foggy	0.609
	C:Overcast, light rain, light snow, freezing rain	0.396
	D:Medium rain, heavy rain, torrential rain, downpour, extraordinary storm, medium snow, heavy snow, blizzard, sandstorm	0.178
Changing weather type	E:Mostly sunny, partly sunny	0.737
	F:Overcast cloudy, cloudy overcast	0.503
	G:Showers, sleet, thundershowers, thundershowers with hail, snow showers, light to moderate rain, light to medium snow	0.287
	H:Medium to heavy rain, heavy rain to torrential rain, torrential rain to downpour, downpour to extraordinary storm, moderate to heavy snow, heavy snow to blizzard	0.089

6.4 Case Study

6.4.1 Wind Power Output Simulation

This section verifies the validity and applicability of the proposed model. Two cases are studied. In the first case, the operation of three wind farms with wind speed correlation is simulated. The statistical indices of each wind farm are evaluated. The same three wind farms with a relatively weak correlation are investigated in the second case and compared with the statistical indices for the wind farms' total output in the first case.

1) Operation simulation of three correlated wind farms

The operation simulation of three correlated wind farms is conducted for one year, with a resolution of 1 hour. The parameters c, d, and a for the three wind farms are shown in Table 6.2. The correlation coefficient matrix of $\tilde{v}_{nt}$ is shown in Table 6.3, representing a strong correlation. The wake effect factor of each wind farm is set to 5%. Each wind farm has 50 identical 2.0 MW wind turbines, which have availability ratios of 0.9. The v_{in}, v_{rated} and v_{out} for each wind turbine are 3 m/s, 12 m/s and 25 m/s, respectively.

TABLE 6.2 Wind speed parameters for the three wind farms

	c	d	a
Wind Farm #1	8.9	2.4	0.04
Wind Farm #2	8.5	2.0	0.05
Wind Farm #3	8.7	1.7	0.06

TABLE 6.3 Proposed wind speed correlation coefficient matrix in the first case

	Wind Farm #1	Wind Farm #2	Wind Farm #3
Wind Farm #1	1	0.8	0.8
Wind Farm #2	0.8	1	0.8
Wind Farm #3	0.8	0.8	1

The operation simulation results are shown in Figure 6.5, Figure 6.6 and Table 6.4. The probability densities and autocorrelation functions of the normalized simulated wind speed $\tilde{v}_{nt}^*$ for each wind farm are shown in Figure 6.5. To contrast, these with the theoretical probability densities and autocorrelation functions, the proposed probability densities and autocorrelation functions are also presented. The stochastic behavior coincides well with the anticipated behavior. The correlation coefficient matrix of $\tilde{v}_{nt}^*$ is shown in Table 6.4. There is a little error between the simulated and proposed values because of stochastic sampling. This error is expected to decrease as the operation simulation time increases.

TABLE 6.4 Simulated wind speed correlation coefficient matrix in the first case

	Wind Farm #1	Wind Farm #2	Wind Farm #3
Wind Farm #1	1	0.8114	0.7755
Wind Farm #2	0.8114	1	0.7951
Wind Farm #3	0.7755	0.7951	1

The power output probability density of each wind farm is shown in Figure 6.6. It is remarkable that each wind farm has a greater probability of output less than 10% and more than 90% than moderate output. However, the conclusion may not be correct for the total output of three wind farms, a point which will be discussed in the following paragraph.

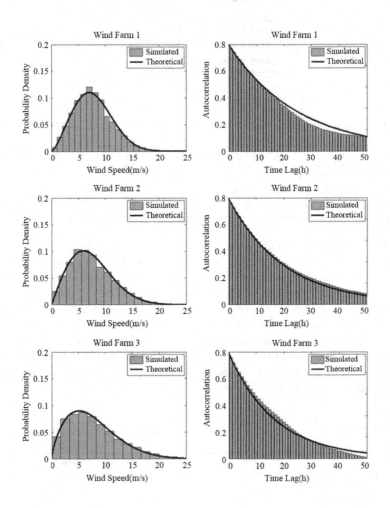

FIGURE 6.5 Probability densities (left) and autocorrelation functions (right) of the normalized simulated wind speed for each wind farm

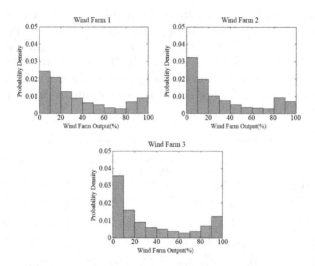

FIGURE 6.6 Probability densities of output power for each wind farm

2) Comparison of wind farms with and without correlation

Three wind farms are studied in this case. The parameters of wind farms are extracted from Subsection 6.4.1 item 2. The wind speed correlation coefficient matrix is shown in Table 6.5.

TABLE 6.5 Proposed wind speeds correlation coefficient matrix in second case

	Wind Farm #1	Wind Farm #2	Wind Farm #3
Wind Farm #1	1	0.1	0.1
Wind Farm #2	0.1	1	0.1
Wind Farm #3	0.1	0.1	1

Segments of three simulated wind farm output curves are presented in Figure 6.7. The length of the segment is 168 hours (1 week). The strongly correlated case in Subsection 6.4.1 item 1. is also presented for comparison. It can be noticed that in the strongly correlated case, wind farms outputs are also correlated. When the power of wind farm #1 increases, the power of wind farms #2 and #3 also increases with great probability. This regular pattern does not appear in the weakly correlated case.

To better understand the two cases, the probability densities of the total output of the three wind farms are analyzed, as shown in Figure 6.8. The distribution of the weakly correlated case is more centralized than the other case, indicating that the total output of wind farms with relatively weak correlation have a larger probability of generating medium output than either extremely high or low because output variations of wind farms are more likely to counterbalance each other. The opposite effect is more likely in the strongly correlated case, so the total probability density stays the same as for a single wind farm. The stochastic behavior of wind farm total output is important in reliability studies and operation planning. The correlation among wind farms has such a significant influence that it must be evaluated in a related study.

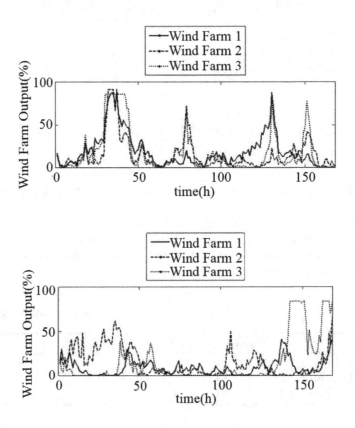

FIGURE 6.7 Segments of three simulated wind farm power output curves: the strongly correlated case (upper side) and the weakly correlated case (bottom)

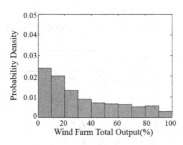 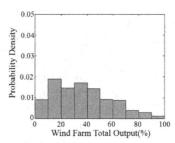

FIGURE 6.8 Probability densities of total output of three wind farms: the strongly correlated case (left) and the weakly correlated case (right)

3) Operation simulation of 13 GW Gansu wind farms based in China

The Gansu 13 GW wind farm, which is one of the largest wind farm bases in China, is located in a 200×200 km area in Jiuquan, Gansu province. Its planned wind generation capacity reached 13 GW by the end of 2015. In this case study, an operation simulation of this wind farm base in 2015 is conducted using the proposed model.

According to [15], the correlation of wind speeds is closely related to the geographical location of wind farms. The Gansu wind farm base consists of more than 100 wind farms which are mainly located in eight wind zones. The installed capacity of each wind zone is outlined in Table 6.6. The wind speed correlation coefficient matrix of the eight wind zones is calculated using the formula proposed in [15], which is presented in Table 6.7.

The simulation results are as follows: The total wind power production chronological time series and the duration curve are shown in Figure 6.9. The probability densities of the simulated wind farm total output are shown in Figure 6.10. The two graphs show that the wind speeds of wind zones are so closely related that the probability of total output less than 10% of the installed capacity is more than 30%.

TABLE 6.6 Wind zones and relevant wind turbine installed capacity in Gansu province, China

	Wind Zones	Installed capacity(MW)
1	South of Beidaqiao	3550
2	North of Beidaqiao	3200
3	Ganhekou	1800
4	Liuyuan	100
5	Qiaowan	600
6	Changma	1110
7	Qixingtan and Mahuangtan	1200
8	Mazongshan	1600

TABLE 6.7 Wind speeds correlation coefficient matrix among different wind zones in Gansu province, China

	1	2	3	4	5	6	7	8
1	1.00	0.92	0.90	0.90	0.87	0.83	0.83	0.86
2	0.92	1.00	0.85	0.87	0.89	0.82	0.86	0.92
3	0.90	0.85	1.00	0.91	0.79	0.76	0.74	0.77
4	0.90	0.87	0.91	1.00	0.79	0.74	0.76	0.82
5	0.87	0.89	0.79	0.79	1.00	0.89	0.95	0.92
6	0.83	0.82	0.76	0.74	0.89	1.00	0.87	0.82
7	0.83	0.86	0.74	0.76	0.95	0.87	1.00	0.91
8	0.86	0.92	0.77	0.82	0.92	0.82	0.91	1.00

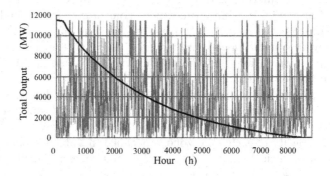

FIGURE 6.9 Simulation results for total wind power production chronological time series and duration curve for the 13-GW Gansu wind farm base

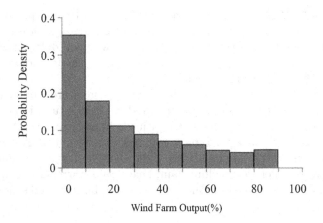

FIGURE 6.10 Simulated total output probability density for wind farms in the 13-GW Gansu wind farm base

6.4.2   PV Power Output Simulation

1) Boundary conditions for the PV operation simulation

Choosing Yunnan province as an example for PV simulation, the planned installation scale of Yunnan province in 2020 is approximately 1957 MW. The PV is divided into 6 PV stations: Kunming, Chuxiong, Yuxi, Dali, Baoshan and Honghe according to the Yunnan Power Grid PV power station planned program. The reported internal weather conditions, for example, PV shading factor, for each PV station are assumed to be the same at any moment. The theoretical PV power station output, for instance, in sunny conditions, can be calculated according to the geographical location. When the stochastic difference equation is used to simulate the PV shading factor, for each time period, six PV shading factors considering correlation are simultaneously generated. According to the geographic locations and historical output data of Yunnan PV power stations, the PV panel inclination angles are set by the location latitude when facing south, and the correlation coefficient settings for different PV stations range from 0.2 to 0.4 when correlation coefficients between adjacent regions are larger than that between widely separated regions.

Since the clearness index is mainly affected by cloud block, weather changes, and altitude, it features intense intermittency and randomness. It is also the main source of PV output uncertainty which needs to be modeled by stochastic simulation. The probability distribution of the weather type for each PV station during simulation is shown in Table 6.8. The clearness index time sequence $S_{th,m}(t) = s_{m1,t}, s_{m2,t}, \cdots, s_{mN,t}$ for different PV stations through sampling is obtained according to the weather type probability distribution. The solar irradiance CDF for different clearness indices is calculated according to the clearness index probability model, expressed as $CDF_{S_{th,m}(t)}$. The normally distributed time sequence of the PV station $X_m(t) = x_{m1,t}, x_{m2,t}, \cdots, x_{mN,t}$ and $CDF_{S_{th,m}(t)}$ are used for sampling, and the time sequence of solar irradiance for different PV stations $I_m(t) = i_{m1,t}, i_{m2,t}, \cdots, i_{mN,t}$ is obtained. Finally, the PV power station output for each PV station is calculated.

In Table 6.8, G, λ and Ω are the clearness index probability distribution parameters; east longitude is positive and west longitude is negative when north latitude is positive and south latitude is negative; weather type 1 includes sunny days; weather type 2 includes cloudy and foggy days; weather type 3 includes overcast, light rain, light snow and freezing rain days; and weather type 4 includes medium rain, heavy rain, torrential rain, downpour, extraordinary storm, medium snow, heavy snow, blizzard and sandstorm days.

TABLE 6.8 Geographical distribution and weather parameters for Yunnan PV power stations

PV station name	Probability J	Probability g	Probability γ_1	Geographic longitude(°)	Geographic latitude(°)
Kunming	0.2994	5.062	0.0343	102.42	25.04
Yuxi	0.2994	5.062	0.0343	102.32	24.22
Chuxiong	0.2994	5.062	0.0343	101.32	25.01
Dali	0.2994	5.062	0.0343	100.13	25.34
Baoshan	0.2994	5.062	0.0343	99.1	25.08
Honghe	0.2994	5.062	0.0343	103.09	23.21

PV station name	Location altitude (m)	Weather type 1 probability	Weather type 2 probability	Weather type 3 probability	Weather type 4 probability
Kunming	1895	0.752	0.123	0.065	0.06
Yuxi	1630	0.752	0.123	0.065	0.06
Chuxiong	1773	0.752	0.123	0.065	0.06
Dali	2090	0.752	0.123	0.065	0.06
Baoshan	1800	0.752	0.123	0.065	0.06
Honghe	1080	0.752	0.123	0.065	0.06

2) Analysis of simulation results and output characteristics of PV operation

The time sequence curves and continuous curves for total PV output are shown in Figure 6.11. Figure 6.12 illustrates the probability density distribution of the total PV output.

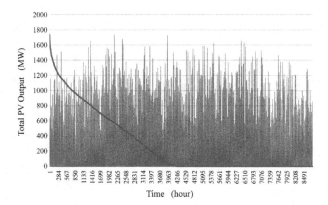

FIGURE 6.11 Simulated total PV output time sequence curve and continuous curve for the Yunnan Power Grid in 2020

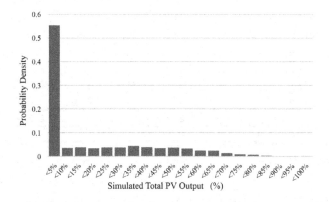

FIGURE 6.12 Simulated total PV output probability density curve for the Yunnan Power Grid in 2020

The simulation results clearly indicate that the average PV simulation output of Yunnan is 294 MW, and the number of utilization hours is 1318. The maximum output at 95% confidence is 1427 MW, and the minimum output is 0 MW.

Figure 6.13 shows the monthly statistics of the PV simulation results in Yunnan, and the daily statistical characteristics are shown in Figure 6.14.

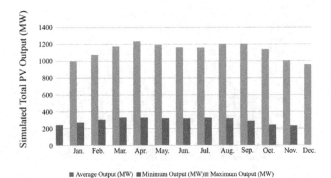

FIGURE 6.13 Simulated total PV output monthly characteristic curves for the Yunnan Power Grid

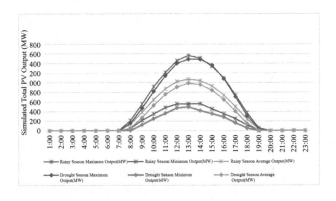

FIGURE 6.14 Simulated total PV output daily characteristic curve for the Yunnan Power Grid

The simulation results indicate that Yunnan PV output is low during the rainy season and is higher in April, August, and September. The daily PV characteristics are evident. The output begins to gradually increase at 7:00 in the morning, reaching a maximum at approximately 13:00, then begins to decline and drops to zero at approximately 19:00 in the evening.

6.5   Summary

This chapter presents methodologies for simulating the operation of multiple wind farms with given statistical indices. For wind power, correlated wind speed time series for wind farms are generated using a stochastic differential equation to achieve the proposed distribution and autocorrelation functions. Wind speed seasonal rhythms and diurnal patterns, turbine output characteristics, wind turbine reliability, and wake effects are considered in the model.

For photovoltaic power, we separately simulate its deterministic and stochastic components. The global solar radiation intensity model is used to generate the deterministic part. The concept of photovoltaic shielding factor is introduced to model and simulate its stochastic part.

The simulation of the wind farm and PV power station output in this chapter is able to consider the spatio-temporal correlation characteristics of the wind farm and the output of the PV power station, a suitable approach for the joint simulation of the wind farms and the output of the PV power station. It is not only applicable to large-scale, centralized wind power / PV power generation bases, but can also be applied to distributed wind power / PV. The method of the wind farm and PV power output simulation proposed in this chapter will provide input data for power system planning and operation simulation in the following chapters.

Bibliography

[1] A. M. Silva, W. S. Sales, L. A. Manso, and R. Billinton. Long-term probabilistic evaluation of operating reserve requirements with renewable sources. *IEEE Transactions on Power Systems*, 25(1):106–116, 2010.

[2] Y. Sun, J. Wu, J. Li, and J. He. Dynamic economic dispatch considering wind power penetration based on wind speed forecasting and stochastic programming. *Proceedings of the CSEE*, 57(5):346–7, 2009.

[3] H. Wang, Z. Lu, and S. Zhou. Research on the capacity credit of wind energy resources. *Proceedings of the CSEE*, 25(10):103–106, 2005.

[4] K. R. Voorspools and W. D. D'Haeseleer. Critical evaluation of methods for wind-power appraisal. *Renewable & Sustainable Energy Reviews*, 11(1):78–97, 2007.

[5] B. Martin and J. Carlin. Wind-load correlation and estimates of the capacity credit of wind power - an empirical investigation. *Wind Engineering*, 7(2):79–84, 1983.

[6] R. Billinton, H. Chen, and R. Ghajar. A sequential simulation technique for adequacy evaluation of generating systems including wind energy. *Energy Conversion IEEE Transactions on*, 11(4):728–734, 1996.

[7] R. Karki, P. Hu, and R. Billinton. Reliability evaluation considering wind

and hydro power coordination. *IEEE Transactions on Power Systems*, 25(2):685–693, 2010.

[8] Y. Wu, M. Ding, and S. Li. Reliability assessment of wind farms in generation and transmission systems. *Transactions of China Electrotechnical Society*, 68(3):1075–1081, 2004.

[9] A. S. Dobakhshari and M. Fotuhi-Firuzabad. A reliability model of large wind farms for power system adequacy studies. *IEEE Transactions on Energy Conversion*, 24(3):792–801, 2009.

[10] M. Doquet. Use of a stochastic process to sample wind power curves in planning studies. *Power Tech, 2007 IEEE Lausanne*, pages 663–670, 2008.

[11] M. B. Bo, I. M. Skovgaard, and M. Sørensen. Diffusion-type models with given marginal distribution and autocorrelation function. *Bernoulli*, 11(2):191–220, 2005.

[12] K. G. Hollands and R. G. Huget. A probability density function for the clearness index, with applications. *Solar Energy*, 30(3):195–209, 1983.

[13] G. M. Tina, C. Brunetto, and C. Nemes. Adequacy indices to evaluate the impact of photovoltaic generation on balancing and reserve ancillary service markets. *International Conference and Exposition on Electrical and Power Engineering*, pages 945–950, 2013.

[14] S. A. Karim, B. S. Singh, R. Razali, and N. Yahya. Data compression technique for modeling of global solar radiation. *IEEE International Conference on Control System, Computing and Engineering*, pages 348–352, 2012.

[15] H. Holttinen. Impact of hourly wind power variations on the system operation in the Nordic countries. *Wind Energy*, 8(2):197–218, 2005.

7

Finding Representative Renewable Energy Scenarios

The uncertainty and intermittency of renewable energy generation make it necessary to consider multiple renewable energy output scenarios in power system operation and planning problems. Due to the limit of computation capability, only a limited number of scenarios can be considered. This raises concerns about finding the representative scenarios and whether the selected scenarios can fully represent the uncertain nature of renewable energy generation. This chapter reviews several typical wind power scenario reduction techniques and categorizes them by both the scenario clustering approach and scenario reduction criterion. The performance of reduced scenarios is quantified by the statistical quality, and its economic value is quantified based on the performance in stochastic unit commitment (SUC) problems. Finally, the proposed method is verified through a case study on the IEEE RTS-79 system using the real wind power data from the NREL database.

7.1 Overview

Large-scale integration of renewable energy brings significant challenges to power system operation and planning due to the inherent intermittency and uncertainty of renewable energy generation. A stochastic programming approach is thus applied in power system decision-making problems to cope with the uncertainty of renewable energy generation [1]. Because of the limitation of the calculation capability, only a limited number of scenarios of renewable energy output can be considered. A scenario reduction technique is thus employed to select the finite representative outputs of renewable energy generation, which leads to an imperfect approximation of the original uncertainty and intermittency. Therefore the quality of the selected scenarios directly drives the effectiveness of the stochastic programming method. It thus raises concerns about how to find the representative scenarios and whether the selected scenarios can fully represent the uncertain nature of renewable energy output.

Currently, methods of generating scenarios to represent the uncertainty of wind power have been widely studied. Botterud *et al.* [2] present a framework of applying scenario techniques to solve unit commitment and reserve the scheduling problem of hedging wind power uncertainty. Pinson *et al.* [3] use the data of wind power observation and point forecast to train the wind uncertainty model in order to generate multiple scenarios. Instead of using a probabilistic forecast. Sun *et al.* [4] apply the empirical cumulative distribution function to describe the wind power uncertainty and approximate the variability using the exponential covariance structure during wind power scenario generation. Intelligence algorithms like the Shuffled Frog Leaping Algorithm (SFLA) are developed to generate and reduce wind power scenarios in [5]. Metrics for evaluating the quality of generated scenarios are also proposed [6].

Compared with the wind power scenario generation technique, the scenario reduction technique receives relatively less attention. Scenarios reduction contains two steps: scenario clustering and selecting representative scenarios. Various clustering techniques have been proposed such as k-means, hierarchical clustering, and fuzzy clustering. The principle of these methods is to select the scenarios with a higher probability. However, some extreme scenarios such as the minimum possible output are usually associated with low density and would probably be excluded, although they may result in load shedding and much higher system operation costs. Consequently, it is recommended to include such extreme scenarios in the reduced scenario set [7]. For representative scenario selecting in the cluster, the discussion is continuing as to whether selecting the highest probability scenario or averaging all the scenarios into one scenario is more suitable [8]. Metrics of measuring the similarity between different scenarios are usually used in scenario reduction. Dupacova *et al.* [9] present the procedure for generating scenario reduction trees based on Kantorovich distance. Sumaili *et al.* [8] propose a reduction methodology that selects the scenario with the highest number of neighbors according to Euclidean distance.

This chapter evaluates the impacts of scenario reduction techniques on stochastic unit commitment (SUC) performance from two aspects: the statistical quality of the reduced scenarios and their economic value to the SUC. The statistical quality of reduced scenarios is defined as the quantity of missed characteristics during scenario reduction process, and two metrics are proposed to quantify the stochastic characteristics of wind power from the perspective of output uncertainty and ramp diversity. The economic value of reduced scenarios is regarded as the optimality of the overall operating cost associated with SUC decisions [10]. Then, we compared different wind power scenario reduction approaches and evaluated their impacts on SUC. In order to capture the primary difference of commonly applied scenario reduction methods, combinations with three typical cluster approaches and two common scenario selection criteria are adopted as scenario reduction methods for comparison. The statistical quality of each method is tested on the real wind power data from the NREL database, and the economic value of each method is demonstrated on the performance of a two-stage SUC model using a modified IEEE RTS-79 system.

7.2 Framework of Modeling Wind Power Uncertainty

The uncertainty of wind power can be represented using a set of scenarios. A scenario is regarded as one of the possible realizations of the stochastic process of wind power output. The process of generating wind power scenario for SUC consists of three procedures: wind power probabilistic forecasting, scenario generation, and scenario reduction.

Probabilistic forecasting: Instead of providing single-value point forecasts, probabilistic forecasting provides estimations of multiple possible outputs of wind power with a probability to each outcome. The forecast output may be expressed by many forms such as probability density functions, a set of quantiles and interval forecasts. A large number of methods have been proposed to forecast the distribution of wind power [11, 12].

Scenario generation: Due to the difficulty of using probabilistic forecasts directly, the scenario technique is introduced and embedded in SUC model to represent the uncertainty of wind power. The process of scenario generation is to reconstruct a series of realizable wind power scenarios based on probabilistic forecasting results. Techniques of scenario generation include Monte Carlo methods, the ARMA model [11], and stochastic differential equation model [13]. The scenarios can also be obtained using historical measurement or resemble forecast model [3].

Scenario reduction: The scenario reduction technique is used to limit the number of scenarios considered in the SUC because of calculation capability. The main principle of scenario reduction is to divide the origin generated scenario database into several sub-sets by clustering similar wind power scenarios and to choose the most representative scenario in each sub-set. Every selected representative scenario is associated with a probability determined by the relative weight of the sub-set it represents. Obviously, the scenario reduction would further bring approximation in modeling the stochastic characteristics of wind power. The accuracy and representativeness of the reduction approximation will, therefore, affect the performance of the SUC model.

7.3 Wind Power Scenario Reduction Techniques

The scenario reduction technique can be divided by how the initial scenarios are clustered and how the representative scenario is selected in each cluster. In this study, we relatively review and compare three scenario clustering methods and two selection criteria as shown in Figure. 7.1. Six scenario reduction methods can be established through cross combination.

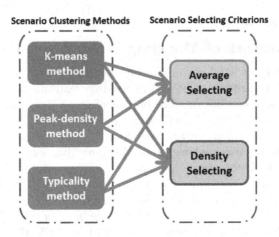

FIGURE 7.1 Different scenario clustering methods and representative scenario selecting criteria.

7.3.1   Scenario Clustering Methods

Given the number of clusters, the purpose of the clustering process is to merge similar scenarios and divide original wind power scenarios into sub-sets. The similarity is usually measured using Euclidean distance.

K-means method picks center scenarios and associates each scenario with the nearest center. The procedure is as follows: 1) Select initial scenarios as cluster centers randomly; 2) calculate the distance between each origin scenario and each center scenario, and allocate each origin scenario to the nearest center; 3) update the center scenario by calculating the mean value of all the corresponding associated scenarios, namely updating the centroid center for each cluster; 4) repeat steps 2) and 3) until cluster allocation is stable.

Peak-density method utilizes a newly proposed cluster technique in [14]. It defines the concept of density as the number of nearby scenarios in its neighborhood. The scenario with high density and far from other scenarios with higher density will be identified as scenario center. The procedure of is as follows: 1) Calculate the density for each scenario and calculate the distance from itself to the nearest scenario with higher density; 2) regard the scenarios with top values of index as cluster center scenarios, namely peak-density scenarios; 3) every origin scenario is allocated to the nearest peak-density scenario with higher density.

Typicality method regards the pre-defined typical scenarios as cluster center vectors and allocates each origin scenario to the nearest center. In this chapter, the pre-defined typical scenarios include the one with maximum or minimum wind power output, the one with maximum or minimum aggregate wind power output, the one with greatest or least peak-to-valley values, the one with maximum up-ramp or down-ramp, the one resulting in least variance from the total sampled mean value, and the one with the greatest sum of the absolute ramp.

Comparison: The above three scenario clustering methods are differentiated in terms of how they value the representativeness of scenarios. K-means method identifies centroid centers for clustering. It is commonly applied but sensitive with initial random center scenarios. Besides, it is an iterative and time-consuming algorithm when the initial scenario set is huge. The peak-density method identifies density centers for clustering. It is fast and effective but pays relatively less attention to extreme scenarios with low probability. Typicality method overcomes the shortcoming of neglecting extreme crucial scenarios, yet it requires a subjectively pre-defined typical scenario and may not guarantee reasonable discrimination between clusters.

7.3.2 Scenario Selection Criteria

Criteria for selecting the most representative scenario from a scenario cluster can be categorized into two types, namely average selecting and density selecting. The average selecting criterion calculates the centroid of each cluster as the most representative scenario, while the density selecting criterion selects the one with the highest density as the most representative scenario for each cluster. The average selecting process will create a new wind power scenario and bring a smoothing effect, therefore the representative scenario may not be realizable and will result in a conservative estimate of the variation of wind power. The density selecting process maintains the primary fluctuation of wind power but sometimes loses the diversity of the scenario, since it tends to select the most representative scenario.

For clarity of description, the scenarios before reduction are denoted as full scenarios, while the scenarios after reduction are named reduced scenarios.

7.4 The Statistical Quality of Reduced Scenarios

Reduced scenarios will inevitably lose the uncertainty information of wind power compared with the full scenarios. This section quantifies such losses from two perspectives: output uncertainty and ramp diversity.

7.4.1   Measurement of Losses on Output Uncertainty

Output uncertainty carried by a set of scenarios can be quantified by:

$$Q_{OU} = \int_{t=1}^{T} \left(\begin{array}{l} \displaystyle\int_{p_w^t=0}^{\tilde{p}_w^t} F(p_w^t)\,(\tilde{p}_w^t - p_w^t)\,dp_w^t + \\[2em] \displaystyle\int_{p_w^t=\tilde{p}_w^t}^{+\infty} (1 - F(p_w^t))\,(p_w^t - \tilde{p}_w^t)\,dp_w^t \end{array} \right) dt \qquad (7.1)$$

$$\approx \frac{1}{100 * T} \sum_{t=1}^{T} \sum_{a/100=1}^{100} \left\{ \begin{array}{ll} a(\tilde{p}_w^t - p_a^t) & if(p_a^t \le \tilde{p}_w^t) \\ (1-a)(p_a^t - \tilde{p}_w^t) & if(p_a^t \ge \tilde{p}_w^t) \end{array} \right.$$

$$F(p_w^t = p) = Pro(p_w^t \ge R) \qquad (7.2)$$

where p_w^t is a random variable representing the value of wind power output for time period t, and $F(p_w^t)$ is the cumulative probabilistic density function for p_w^t. p_a^t is the quantile for $F(p_w^t)$ with a as the target quantile. The change value of Q_{OU} (denoted as ΔQ_{OU}) is thus used to quantify the loss of output uncertainty during scenario reduction. In general, the full scenarios will perform better on sharpness calibration than the reduced scenarios. Hence, the value of Q_{OU} will generally increase after scenario reduction.

7.4.2   Measurement of Losses on Ramp Diversity

Ramp diversity carried by a set of scenarios can be quantified by:

$$Q_{RU} = \int_{R=0}^{+\infty} F(R)dR \approx \frac{1}{100} \sum_{a/100=1}^{100} R_a \qquad (7.3)$$

$$F(R = Ramp) = Pro(Ramp \ge R) \qquad (7.4)$$

where R is a random variable representing the absolute value of the wind power ramp, and $F(R)$ is the cumulative probabilistic density function for R. R_a is the quantile for $F(R)$ with a as the target quantile. Q_{RU} is defined as the area surrounded by $F(R)$. The purpose of Q_{RU} is to evaluate how many ramps are abandoned during scenario reductions. The change value of Q_{RU} (denoted as ΔQ_{RU}) is thus used to quantify the loss of ramp diversity during scenario reduction. Because some extreme scenarios (e.g. scenario with maximum ramp) always are discarded during scenario reduction, the value of Q_{RU} will generally decrease after scenario reduction.

7.4.3 Quality of Reduced Scenarios

The overall quality of reduced scenarios, denoted by Q_{RS}, is defined using both Q_{OU} and Q_{RU} :

$$Q_{RS} = \alpha \left(1 - \frac{|\Delta Q_{OU}|}{Q_{OU}^0}\right) + (1 - \alpha)\left(1 - \frac{|\Delta Q_{RU}|}{Q_{RU}^0}\right) \qquad (7.5)$$

where Q_{OU}^0 and Q_{RU}^0 are the quantities of uncertainty of the full scenarios. α is the weight coefficient of output uncertainty information. The value of Q_{RS} lies in the range $[0,1]$, representing the percentage of remaining uncertainty information.

7.5 The Economic Value of Reduced Scenarios

In this section, the mathematical formulation of the SUC model is given. The economic value of reduced scenarios is defined as the optimality of SUC results.

7.5.1 Notations

Indices

i	Index of thermal units from 1 to I
j	Index of wind farms from 1 to J
m	Index of buses from 1 to M
l	Index of transmission lines from 1 to L
t	Index of time periods from 1 to T
s	Index of scenarios from 1 to S
k	Index of segments of piecewise liner cost function from 1 to K

Parameters

$a_{i,k}^c, b_{i,k}^c$	Coefficient of the kth segment of piecewise linear generating cost for generator i
c_i^{up}/c_i^{down}	Cost of generator i offering the up/down spinning reserve
$c_i^{nonspin}$	Cost of generator i offering the non-spinning reserve
θ^{eens}	Cost coefficient of load curtailment
θ^{wind}	Cost coefficient of wind curtailment

o_i^{on}/o_i^{down}	Start up / shut down cost of generator i
$\underline{R}_t^{up}/\underline{R}_t^{down}$	Fixed up/down spinning reserve requirement during period t
$\underline{P}_i/\bar{P}_i$	Min/max generation limits for generator i
$\Delta P_i^{up}/\Delta P_i^{down}$	Upward / downward ramping rate of generator i
$\underline{t}_i^{on}/\underline{t}_i^{off}$	Minimum on / off time intervals of generator i
$D_{m,t}$	Forecast load of bus m during period t
$P_{j,t}^{fore}$	Forecast output of wind farm j during period t
$\tilde{P}_{j,t,s}^{fore}$	Potential output of wind farm j during period t in scenario s
$\bar{F}_l$	Power flow limitation of branch l
$g_{l,m}$	Power flow distribution factor of the branch l with respect to the net injection of bus m
$g_{l,i}/g_{l,j}$	Power flow distribution factor of the branch l with respect to the output of generator i/ wind farm j

Variable

$P_{i,t}^c$	Scheduled generation of thermal generator i during period t
$P_{j,t}^w$	Scheduled generation curtailment of wind farm j during period t
$D_{m,t}^d$	Scheduled load shedding of node m during period t
$C_{i,t}$	Generating cost of unit i during period t
$U_{i,t}$	State variable of unit i during period t
$U_{i,t}^{on}/U_{i,t}^{off}$	State variable indicating the on and off of unit i during period t
$R_{i,t}^{up}/R_{i,t}^{down}$	Up/down spinning reserve scheduled on unit i during period t
$R_{i,t}^{nonspin}$	Non-spinning reserve scheduled on unit i during period t
$S_{i,t}^{on}$	Start-up cost of generator i during period t
$\tilde{P}_{i,t,s}^c$	Scheduled generation of unit i during period t in scenario s
$\tilde{P}_{j,t,s}^w$	Scheduled generation wind farm j during period t in scenario s
$\tilde{D}_{m,t,s}^d$	Scheduled load shedding of node m during period t in scenario s
$\tilde{R}_{i,t,s}^{up}/\tilde{R}_{i,t,s}^{down}$	Deployed up/down spinning reserve of unit i during period t in scenarios
$\tilde{R}_{i,t,s}^{nonspin}$	Deployed non-spinning reserve of unit i during period t in scenario s

7.5.2 Formulation of SUC

The general formulation is given as follows. The unit commitment and reserve scheduling decisions are made before the wind power uncertainty unfolds, in the first stage. Those generation scheduling decisions are always called here-and-now decisions. Revealing the uncertainty of wind power as a series of scenarios, optimal generation dispatch, and reserve deploy decisions for each realizable scenario are made in the second stage. Those generation deploy scheduling decisions are always called wait-and-see decisions. The objective of the SUC model is to minimize both generation scheduling costs and generation deploying expectation costs. The mathematical formulation is given as follows.

1) Objective function

$$\min \sum_{t=1}^{T}\sum_{i=1}^{I}\left(C_{i,t}+c_i^{up}r_{i,t}^{up}+c_i^{down}r_{i,t}^{down}+c_i^{nonspin}r_{i,t}^{nonspin}+s_{i,t}^{on}\right)$$

$$+\sum_{j=1}^{J}\theta^{wind}p_{j,t}^{w}+\sum_{m=1}^{M}\theta^{eens}d_{m,t}^{d} \tag{7.6}$$

$$+\sum_{s=1}^{S}\pi_s\left[\sum_{t=1}^{T}\sum_{i=1}^{I}\begin{pmatrix}\tilde{c}_i^{up}\tilde{r}_{i,t,s}^{up}+\tilde{c}_i^{down}\tilde{r}_{i,t,s}^{down}+\tilde{c}_i^{nonspin}\tilde{r}_{i,t,s}^{nonspin}\end{pmatrix}\\ +\sum_{j=1}^{J}\theta^{wind}\tilde{p}_{j,t,s}^{w}+\sum_{m=1}^{M}\theta^{eens}\tilde{d}_{m,t,s}^{d}\right]$$

2) First-stage constraints (not depending on scenario s)

$$\sum_{i=1}^{I}p_{i,t}^{c}+\sum_{j=1}^{J}w_{i,t}-\sum_{j=1}^{J}p_{j,t}^{w}+\sum_{m=1}^{M}d_{m,t}^{d}=\sum_{m=1}^{M}d_{m,t} ,\forall\, t \tag{7.7}$$

$$p_{i,t}^{c}-p_{i,t-1}^{c}+u_{i,t-1}\left(\underline{p}_{i}-\Delta p_i^{up}\right)+u_{i,t}\left(\bar{p}_i-\underline{p}_{i}\right)\le\bar{p}_i ,\forall\, t,i \tag{7.8}$$

$$p_{i,t-1}^{c}-p_{i,t}^{c}+u_{i,t}\left(\underline{p}_{i}-\Delta p_i^{down}\right)+u_{i,t-1}\left(\bar{p}_i-\underline{p}_{i}\right)\le\bar{p}_i ,\forall\, t,i \tag{7.9}$$

$$\left(u_{i,t}-u_{i,t-1}\right)\underline{t}_{i}^{on}+\sum_{j=t-\underline{t}_{i}^{on}-1}^{t-1}u_{i,j}\ge 0,\ \forall t,i \tag{7.10}$$

$$\left(u_{i,t-1}-u_{i,t}\right)\underline{t}_{i}^{off}+\sum_{j=t-\underline{t}_{i}^{off}-1}^{t-1}\left(1-u_{i,j}\right)\ge 0,\ \forall t,i \tag{7.11}$$

$$-\bar{f}_l \le \sum_{i=1}^{I} g_{l,i} p_{i,t}^c + \sum_{j=1}^{J} g_{l,j} \left(w_{j,t} - p_{j,t}^w\right)$$

$$-\sum_{l=1}^{L} g_{l,m}\left(d_{m,t} - d_{m,t}^d\right) \le \bar{f}_l, \ \forall t, \ l \tag{7.12}$$

$$r_{i,t}^{up} + p_{i,t}^c \le u_{i,t}\bar{p}_i, \ \forall t, i \tag{7.13}$$

$$p_{i,t}^c - r_{i,t}^{down} \ge u_{i,t}\underset{-i}{p}, \ \forall t, i \tag{7.14}$$

$$r_{i,t}^{up} \ge 0, \ r_{i,t}^{down} \ge 0, \ \forall t, i \tag{7.15}$$

$$0 \le r_{i,t}^{nonspin} \le \bar{p}_i(1 - u_{i,t}), \ \forall t, i \tag{7.16}$$

$$\sum_{i=1}^{I} r_{i,t}^{up} \ge \underset{-t}{r^{up}}, \sum_{i=1}^{I} r_{i,t}^{down} \ge \underset{-t}{r^{down}}, \sum_{i=1}^{I} r_{i,t}^{nonspin} \ge \underset{-t}{r^{nonspin}}, \ \forall \tag{7.17}$$

$$0 \le p_{j,t}^w \le w_{j,t}, \ \forall t, j \tag{7.18}$$

$$C_{i,t} \ge a_{i,t,k}^c p_{i,t}^c + b_{i,t,k}^c u_{i,t}, \ \forall t, i, k \tag{7.19}$$

$$s_{i,t}^{on} \ge o_i^{on}\left(u_{i,t} - u_{i,t-1}\right), s_{i,t}^{on} \ge 0, \ \forall t, i \tag{7.20}$$

For first-stage constraints depending on the point forecast of wind power, Equation (7.7) is the load-generation balance constraints on each period for all the buses. Equations (7.8) and (7.9) form the ramping rate constraints for each generator. Equations (7.10) and (7.11) ensure the minimum on and minimum off time periods for each generator. Equation (7.12) sets the power flow limit for each branch. Equations (7.13)-(7.16) limit the output and the scheduled reserve for each generator. Equation (7.17) ensures satisfaction of up / down / non-spinning reserve requirements. Equation (7.18) limits the schedule of wind farms no more than its forecast. Equation (7.19) sets the price-wise linear generating cost of each generator using multiple linear function constraints. Equation (7.20) identifies the start-up cost of generators.

3) Second-stage constraints (depending on scenario s)

$$u_{i,t,s}\underset{-i}{p} \le \tilde{p}_{i,t,s}^c \le u_{i,t,s}\bar{p}_i, \ \forall t, i, s \tag{7.21}$$

$$0 \le \tilde{p}_{j,t,s}^w \le \tilde{w}_{j,t,s}, \ \forall t, j, s \tag{7.22}$$

$$\tilde{p}^c_{i,t,s} - \tilde{p}^c_{i,t-1,s} + u_{i,t-1,s}\left(\underset{-i}{p} - \Delta p^{up}_i\right)$$

$$+ u_{i,t,s}\left(\bar{p}_i - \underset{-i}{p}\right) \le \bar{p}_i, \ \forall t, i, s \tag{7.23}$$

$$\tilde{p}^c_{i,t-1,s} - \tilde{p}^c_{i,t,s} + u_{i,t,s}\left(\underset{-i}{p} - \Delta p^{down}_i\right)$$

$$+ u_{i,t-1,s}\left(\bar{p}_i - \underset{-i}{p}\right) \le \bar{p}_i, \ \forall t, i, s \tag{7.24}$$

$$-\bar{f}_l \le \sum_{i=1}^{I} g_{l,i}\tilde{p}^c_{i,t,s} + \sum_{j=1}^{J} g_{l,j}\left(\tilde{w}_{j,t,s} - \tilde{p}^w_{j,t,s}\right)$$

$$-\sum_{l=1}^{L} g_{l,m}\left(d_{m,t} - \tilde{d}^d_{m,t,s}\right) \le \bar{f}_l, \ \forall t, l, s \tag{7.25}$$

$$\sum_{i=1}^{I} \tilde{p}^c_{i,t,s} + \sum_{j=1}^{J} \tilde{p}^w_{j,t,s} - \sum_{j=1}^{J} \tilde{p}^w_{j,t,s} + \sum_{m=1}^{M} \tilde{d}^d_{m,t,s} = \sum_{m=1}^{M} d_{m,t}, \ \forall t, s \tag{7.26}$$

For second-stage constraints depending on realizable scenarios, Equation (7.21) limits the output for each generator. Equation (7.22) limits the dispatch of wind farms to no more than its maximum output in each scenario. Equations (7.23) and (7.24) set the ramping limit for each generator. Equation (7.25) forms the power flow limit for each branch. Equation (7.26) gives the load-generation balance constraints on each period.

4) Constraints Linking the First- and Second-Stage Variables

$$\tilde{p}^c_{i,t,s} = p^c_{i,t} + \tilde{r}^{up}_{i,t,s} - \tilde{r}^{down}_{i,t,s} + \tilde{r}^{nonspin}_{i,t,s}, \ \forall t, i, s. \tag{7.27}$$

$$0 \le \tilde{r}^{up}_{i,t,s} \le r^{up}_{i,t}, \ 0 \le \tilde{r}^{down}_{i,t,s} \le r^{down}_{i,t},$$

$$0 \le \tilde{r}^{nonspin}_{i,t,s} \le r^{nonspin}_{i,t}, \ \forall t, i, s \tag{7.28}$$

For linking constraints, Equation (7.27) indicates that the dispatched power output of the generator in each scenario should be equal to the scheduled output in the first stage plus deployed up / down / non-spinning reserves. Equation (7.28) limits the deployed up / down / non-spinning reserves no more than their scheduled values.

7.5.3   The Economic Value of Reduced Scenarios

It should be noted that a reduced scenario set with lower SUC cost cannot justify a better scenario reduction method. Over-optimistic scheduling decisions may result in a significant cost in real-time dispatch. Therefore, the optimality of SUC results should be evaluated by combining the cost associated with SUC and the expected cost of real-time dispatch among all the full scenarios:

$$C_{O\_S}^{Total} = C^{UC} + C_{O\_S}^{ED} = C^{UC} + \sum_{\substack{s \in \, full \\ scenario\ set}} \pi_s C_S^{ED}(\tilde{w}_s). \tag{7.29}$$

where the cost of UC decision C^{UC} can be formulated as:

$$C^{UC} = \sum_{t=1}^{T} \sum_{i=1}^{I} \left(C_{i,t} + c_i^{up} r_{i,t}^{up} + c_i^{down} r_{i,t}^{down} + c_i^{nonspin} r_{i,t}^{nonspin} + s_{i,t}^{on} \right)$$

$$+ \sum_{j=1}^{J} \theta^{wind} p_{j,t}^w + \sum_{m=1}^{M} \theta^{eens} d_{m,t}^d. \tag{7.30}$$

The cost associated with real-time dispatch $C_S^{ED}(\tilde{w}_s)$ (when the wind power output scenario $\tilde{w}_s$ is realized) can be evaluated using the following economic dispatch model:

$$\min\ C_S^{ED}(\tilde{w}_s) = \sum_{t=1}^{T} \sum_{i=1}^{I} \left(\tilde{c}_i^{up} \tilde{r}_{i,t,s}^{up} + \tilde{c}_i^{down} \tilde{r}_{i,t,s}^{down} + \tilde{c}_i^{nonspin} \tilde{r}_{i,t,s}^{nonspin} \right)$$

$$+ \sum_{j=1}^{J} \theta^{wind} \tilde{p}_{j,t,s}^w + \sum_{m=1}^{M} \theta^{eens} \tilde{d}_{m,t,s}^d \tag{7.31}$$

$s.t. constraints\ (15)^{\sim}(21).$

$C_{O\_S}^{Total}$ reflects the expected actual cost when wind power output is realized. Therefore, it is used to quantify the value of the reduced scenarios. Lower $C_{O\_S}^{Total}$ denotes the higher value of the reduced scenarios.

Furthermore, the expected cost combining the C^{UC} and $C_S^{ED}(\tilde{w}_s)$ with only reduced scenarios can also be evaluated:

$$C_{R\_S}^{Total} = C^{UC} + C_{R\_S}^{ED} = C^{UC} + \sum_{\substack{s \in \, reduced \\ scenario\ set}} \pi_s C_S^{ED}(\tilde{w}_s). \tag{7.32}$$

The difference between $C_{O\_S}^{Total}$ and $C_{R\_S}^{Total}$ shows how much underlying risks are underestimated/overestimated by the reduced scenarios.

7.6 Data and Results

In this section, different reduced scenario sets are generated from the above-mentioned six scenario reduction methods, with the same origin wind power scenario set. The qualities of these six reduced scenario sets are calculated and compared, and their values are demonstrated with the proposed metric on a modified IEEE RTS79 system [15].

7.6.1 Probabilistic Forecast and Scenario Generation

The wind power history data are from the Wind Integration Datasets of NREL [16]. We choose the aggregate output of ten sites with a total capacity of 1700 MW along the east coast of the United States in 2004 with 1-h resolution. The probabilistic forecast of wind power is modeled using the copula-based technique proposed in [14]. One thousand possible scenarios are generated according to the probabilistic forecast using the stochastic differential equation-based technique described in [13]. A daily probabilistic forecast and generated scenarios are shown in Figure 7.2. The green curves represent the generated scenarios, while the gradient red lines are the quantiles of wind power probabilistic distribution from 5% to 95% with a step of 5%. The quantities of output uncertainty information carried by probability forecasts and 1000 generated scenarios are, respectively, 37.13 and 49.56. This reflects that some information has been lost in the scenario generation process.

7.6.2 Reduced Scenario Sets from Different Methods

The above-mentioned six scenario reduction methods are adopted to reduce initial the 1000 generated scenarios to 10 representative scenarios. The results, namely six reduced scenario sets, are compared in Figure 7.3. The green curves represent the generated 1000 scenarios and the lines with different colors show the reduced 10 representative scenarios. Figure 7.4 provides two-dimensional (2D) views illustrating how the scenarios are clustered using each technique. The left three sub-figures show where the cluster center locates in the $\rho_i - d_i$ plot for different cluster methods with the x-axis denoting the density of each scenario (ρ_i) and the Y-axis denoting distance value (δ_i) proposed in the peak-density method. The right three figures map the Euclidean distance between all scenarios on the 2D view and categorize the clusters using different colors.

   The comparisons in Figure 7.3 and Figure 7.4 show that the peak-density and k-means cluster methods select a cluster center based mainly on density and lead to less diversity between clusters, while the typicality method shows stronger diversity. An average selection criterion introduces a strong smoothing effect which causes less fluctuation but brings better approximation on uncertainty distribution. A density selection criterion results in little diversity between representative scenarios and ignores extreme scenarios. Figure 7.5 compares the performances of the approximation of the variation of wind power. The average selection criterion shows a weaker variation approximation compared with the density selection criterion, because of the smoothing effect.

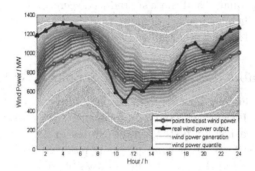

FIGURE 7.2 Wind power probability forecast and scenario generation.

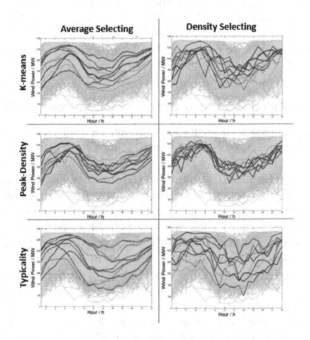

FIGURE 7.3 Wind power scenario reduction results with different methods.

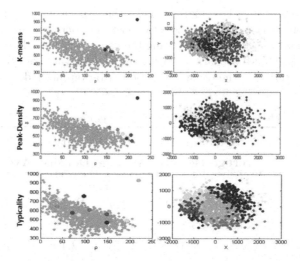

FIGURE 7.4 Wind power scenario clustering results with different methods.

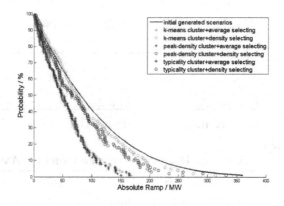

FIGURE 7.5 The cumulative probability density distribution of an absolute ramp.

7.6.3   Comparison of the Quality of Reduced Scenarios

Table 7.1 calculates the measurement of output uncertainty and ramp diversity for each reduced scenario set and compares the index of Q_{RS} among them. The results of output uncertainty measurement show that the k-means method has the best performance, while the peak-density method performs worst from the perspective of scenario clustering. The average criterion always performs better than the density criterion from the perspective of representative scenario selection. For ramp diversity, the impact of the scenario selection criterion dominate: the density selection criterion performs much better than the average selection criterion. Setting the value of the weight coefficient α as 0.5, the best-reduced scenario set with the highest value of Q_{RS} comes from k-means clustering + the density selection criterion method.

TABLE 7.1 The statistical quality of each reduced scenario set

Metrics	K-means		Peak-Density		Typicality	
	Average	Density	Average	Density	Average	Density
Q^0_{OU}			49.56			
Q_{OU}	54.906	58.979	60.117	60.180	58.978	59.007
ΔQ_{OU}	5.346	9.419	10.557	10.620	9.418	9.447
$Q_{RS}(\alpha = 1)$	90.3%	84.0%	82.4%	82.4%	84.0%	84.0%
Q^0_{RU}			98.10			
Q_{RU}	55.27	90.28	63.46	89.49	55.88	90.04
ΔQ_{RU}	42.83	7.82	34.64	8.61	42.22	8.06
$Q_{RS}(\alpha = 0)$	56.3%	92.0%	64.7%	91.2%	56.9%	92.2%
$Q_{RS}(\alpha = 0.5)$	73.3%	88.0%	73.6%	86.8%	70.5%	87.9%

TABLE 7.2 The economic value of each reduced scenario set

Credible Results(M$)	K-means		Peak-Density		Typicality	
	Average	Density	Average	Density	Average	Density
C^{UC}	5.0329	4.9848	4.8747	4.7874	5.1901	5.1261
$C^{ED}_{R\_S}$	0.0675	0.1097	0.1123	0.1727	0.0309	0.1127
$C^{Total}_{R\_S}$	5.1004	5.0945	4.9870	4.9601	5.2210	5.2388
$C^{ED}_{O\_S}$	0.2775	0.4687	0.5882	0.9510	0.1181	0.1860
$C^{Total}_{O\_S}$	5.3764	5.5470	5.5768	5.8799	5.3237	5.3812
$\frac{C^{Total}_{R\_S}}{C^{Total}_{O\_S}}$	94.87%	91.84%	89.42%	84.36%	98.07%	97.35%

7.6.4   Comparison on the Value of Reduced Scenarios

The results in Table 7.2 summarize the impact of reduced scenarios on the optimality SUC. From the perspective of scenario clustering, the typicality method pays more attention to extreme scenarios, and leads to the highest generation scheduling cost and total operating cost on reduced scenarios, but results in the lowest "actual" overall cost on full scenarios. The peak-density method performs the opposite and the k-means method has intermediate performance. From the perspective of representative scenario selection, the average selection method always has higher generation scheduling costs, but results in better performance on the overall cost of full scenarios, compared with the density selection method. The results suggest that the "actual" operation cost is more sensitive in the extreme scenarios. The Peak-density clustering and the density selecting method usually eliminate extreme scenarios and thus lose relatively more information on the output uncertainty of wind power. This, in turn, results in higher total operation cost on full scenarios.

7.6.5   Discussions

Comparing the above evaluation results shows that the evaluation of the quality and the value for reduced scenarios are not totally consistent: 1) The scenario reduction method cannot maintain the character of uncertainty and ramp well at the same time, e.g. the density selection criterion always performs oppositely on the proposed two quality metrics. 2) The high quality of the output uncertainty representation will often lead to more cost-efficient decisions, suggesting that the economic value of reduced scenarios is more sensitive to the performance of the output uncertainty representation. 3) Only evaluating the performance from the statistical calibration of the scenario reduction technique is not enough for scenario-based SUC.

7.7   Summary

This chapter compares the performance of different scenario reduction techniques and analyzes their impacts on the result of scenario-based SUC. The metrics of statistical quality and economic value are proposed to quantify the ability of the reduced scenarios in representing the original stochastic nature of wind power. The statistical quality of reduced scenarios is defined as the information loss of output uncertainty and ramp diversity during the scenario reduction process. In addition, a two-stage scenario-based SUC model is proposed to quantify the economic value of reduced scenarios. The expected overall operating cost, namely the sum of the generation scheduling cost (from SUC) and the expected generation deploying cost under all the realizable scenarios, is used to compare the economic value of different scenario reduction techniques.

For capturing the primary difference of commonly applied scenario reduction methods, combinations with three typical cluster approaches (k-means, peak-density clustering, and typicality clustering) and two common scenario selection criteria (density selection and average selection) are adopted as scenario reduction methods for comparison. The statistical quality of each method is tested on the real wind power data from the NREL database, and the economic value of each method is demonstrated on the performance of a two-stage SUC model using a modified IEEE RTS-79 system.

Some interesting observations are found from the results: The scenario reduction methods that rely on identifying high-probability scenarios (i.e. typicality clustering + the density selection criterion method) would be more likely to discard extreme scenarios and failed to represent the output uncertainty of wind power. However, they may achieve good performance in presenting the ramp diversity. Reduced scenarios with good statistical quality may not obtain a favorable SUC decision. The optimality of the SUC decision is more sensitive to the quality of output uncertainty representation rather than ramp diversity. The results suggest that extreme scenarios should be given more attention to making a good balance between typicality and extremity during the scenario reduction process for SUC.

Bibliography

[1] Q. Zheng, J. Wang, and A. Liu. Stochastic optimization for unit commitment—a review. *IEEE Transactions on Power Systems*, 30(4):1913–1924, 2015.

[2] A. Botterud, Z. Zhou, J. Wang, J. Valenzuela, J. Sumaili, R. J. Bessa, H. Keko, and V. Miranda. Unit commitment and operating reserves with probabilistic wind power forecasts. *PowerTech, 2011 IEEE Trondheim*, pages 1–7, 2011.

[3] P. Pinson, G. Papaefthymiou, B. Klockl, and H. A. Nielsen. Generation of statistical scenarios of short-term wind power production. *Power Tech IEEE Lausanne*, 12(1):491–496, 2007.

[4] X. Ma, Y. Sun, and H. Fang. Scenario generation of wind power based on statistical uncertainty and variability. *IEEE Transactions on Sustainable Energy*, 4(4):894–904, 2013.

[5] K. Shaloudegi, A. Alimardani, and S.H. Hosseinian. Stochastic unit commitment with significant wind penetration using novel scenario generation and reduction. *Thermal Power Plants (CTPP), 2011 Proceedings of the 3rd Conference on*, pages 1–6, 2011.

[6] P. Pinson and R. Girard. Evaluating the quality of scenarios of short-term wind power generation. *Applied Energy*, 96:12–20, 2012.

[7] A. Papavasiliou, S. S. Oren, and R. P. O'Neill. Reserve requirements for wind power integration: A scenario-based stochastic programming framework. *IEEE Transactions on Power Systems*, 26(4):2197–2206, 2011.

[8] J. Sumaili, H. Keko, V. Miranda, Z. Zhou, A. Botterud, and J. Wang. Finding representative wind power scenarios and their probabilities for stochastic models. *Intelligent System Application to Power Systems (ISAP), 2011 16th International Conference on*, pages 1–6, 2011.

[9] J. Dupačová, N. Gröwe-Kuska, and W. Römisch. Scenario reduction in stochastic programming. *Mathematical Programming*, 95(3):493–511, 2003.

[10] J. M. Morales, A. J. Conejo, and J. Pérez-Ruiz. Economic valuation of reserves in power systems with high penetration of wind power. *IEEE Transactions on Power Systems*, 24(2):900–910, 2009.

[11] C. Monteiro, R. Bessa, V. Miranda, A. Botterud, J. Wang, and G. Conzelmann. A quick guide to wind power forecasting: State-ofthe-art. *Argonne Nat. Lab., Tech. Rep. ANL/DIS-10-2*, 2009.

[12] N. Zhang, C. Kang, Q. Xia, and J. Liang. Modeling conditional forecast error for wind power in generation scheduling. *IEEE Transactions on Power Systems*, 29(3):1316–1324, 2014.

[13] N. Zhang, C. Kang, C. Duan, X. Tang, J. Huang, Z. Lu, W. Wang, and J. Qi. Simulation methodology of multiple wind farms operation considering wind speed correlation. *International journal of power & energy systems*, 30(4):264–273, 2010.

[14] A. Rodriguez and A. Laio. Clustering by fast search and find of density peaks. *Science*, 344(6191):1492–1496, 2014.

[15] Probability Methods Subcommittee. IEEE reliability test system.

[16] M. Milligan, E. Ela, D. Lew, D. Corbus, Y. Wan, and B. Hodge. Assessment of simulated wind data requirements for wind integration studies. *IEEE Transactions on Sustainable Energy*, 3(4):620–626, 2012.

Part III

Short-Term Operation Optimization

8

Probabilistic Load Flow under Uncertainty

The stochastic nature of renewable energy resources injects the complex uncertainties of power flow into both the transmission system and the distribution system. The large-scale integration of distributed PV or wind power has a significant impact on the distribution of power flow and voltage in the active distribution system (ADS). This chapter proposes a discrete convolution methodology for probabilistic load flow (PLF) of the ADS considering correlated uncertainties. First, the uncertainties of load and renewable energy are modeled using the distribution of the corresponding forecast error, and the correlation is formulated using a copula function. A novel reactive power-embedded linearized power flow model with high accuracy in both branch flow and node voltage is introduced into the ADS. Finally, the distribution of power flow is calculated using dependent probabilistic sequence operation (DPSO), which is capable of handling non-analytical probability distribution functions. In addition, a reduced dimension approximation method is proposed to further reduce the computational burden. The proposed PLF algorithm is tested on the IEEE 33-node system and 123-node system, and the results show that the proposed methodology requires less computation and produces higher accuracy compared with current methods.

8.1 Introduction

With the increasing penetration of distributed generation (DG), the active distribution system (ADS) has revolutionized electrical generation and consumption through a bi-directorial power flow [1]. The traditional deterministic analysis approach is not well suited to addressing the uncertainty and variability of renewable energy resources such as wind and photovoltaics (PV). Additionally, the mismatch between load and renewable energy might influence the voltage quality of feeders, including over-voltage and voltage flicker [2]. A better understanding of the impact of the uncertainty of load and renewable energy is of great significance for renewable energy accommodation and assurance of power quality [3]. Therefore, probabilistic load flow (PLF) is proposed to evaluate how the uncertainties of renewable energy and load influence the power flow and the underlying operational risks of the system.

The calculation of PLF is essentially a process of estimating the probability distribution of the function of multiple random variables. For AC PLF, the relationship between the system inputs and the status to be solved is nonlinear. Commonly used methods include numerical methods and point estimate methods (PEMs) [4]. Monte Carlo is the most frequently used stochastic simulation method [5], but such a method could be time-consuming because the power flow must be calculated for each sampling. To improve the sampling efficiency, selected stratified sampling methods such as sampling reduction methods [6] and Latin hypercube sampling [7] have been proposed. PEMs estimate the statistical moments of a random quantity that is a function of one or several random variables by calculating the value of this random quantity on several sampled points of those random variables. According to the number of points to be calculated, PEMs can be divided into 2m PEM, 2m+1 PEM, and km+1 PEM [8]. For DC PLF, the relationship between the system inputs and status to be solved is linear. However, only the branch power flow and node voltage phase can be calculated. By linearizing the AC power flow equations around the expected operation status, the probability distribution of a linear combination of multiple random variables can be calculated using convolution [9] and the cumulant method [10]. The Fast Fourier transform (FFT) has been adopted to accelerate the process of convolution [9]. Applications of cumulant methods such as Gram-Charlier, Edgeworth, and Laguerre have been used to estimate the probability distributions of the system status [11]. The calculation error depends on the orders of the cumulants considered in the PLF. Inputs that are far different from Gaussian distributions might cause large errors.

Earlier methods can only be used to handle the independent uncertainties, but research shows that the uncertainties of loads and renewable energy are correlated, which makes it necessary to take the correlations into consideration. The simplest method treats the inputs as linearly (or completely) dependent. Thus, multiple random variables can be transformed into a single random variable in the PLF. The correlation coefficient is equal to 1.0 in this situation, which is not consistent with reality. Another approach of considering correlations is to calculate their covariance matrix. PEM with consideration of the covariance matrix is proposed in [12], in which the weights of the estimated points are determined by the decomposed covariance matrix. The covariance matrix is not affected by the marginal distribution of each variable, although it can only quantify the linear correlation between variables. Based on the Gaussian distribution hypothesis, multiple correlated variables can be transformed into several independent Gaussian distributions using Cholesky decomposition according to the properties of the Gaussian distribution. PLF with correlation consideration can be computed by combining the transformation with the cumulant method and the Monte Carlo method [13]. For the situation in which variables have non-Gaussian marginal distributions, the inverse transformation is conducted after Cholesky decomposition and Monte Carlo [14] or Latin hypercube sampling [15] to ensure that the variables follow their respective marginal distributions. However, the correlations of transformed variables show minimal differences from their original correlations. Copula functions have also been used to characterize the joint distribution and marginal distribution of multiple variables [16].

The PLF algorithms mentioned above are mainly implemented for transmission systems. However, two challenges remain for PLF calculations in distribution systems.

Challenge #1: The r/x ratio of a branch in the ADS is greater than that in a transmission system, which enhances the coupling of active and reactive power and causes a larger error for DC PLF. Additionally, the voltage regulation and power supply reliability are of greater concern in the ADS, and accuracy of the node voltage and branch power flow is thus necessary for PLF calculation in the ADS.

Challenge #2: The uncertainty of renewable energy and loads in the ADS are fairly large, and the probability distributions are likely unable to be characterized using standard probability distributions [17]. In addition, the correlations of renewable energy and load uncertainties are more significant than those of the transmission system because the coverage area of the ADS is much smaller.

To address these challenges, this chapter implements a PLF method for the ADS based on dependent probability sequence operation (DPSO) [18] and the improved DC power flow model. To address Challenge #1, the improved DC power flow model with consideration of reactive power and voltage is proposed to express the feeder voltage and branch power flow as a linear expression of the nodal injections without loss of accuracy. To address Challenge #2, the DPSO method, an extension of traditional discrete convolution, is applied to calculate the distribution of the sum of the correlated variables with arbitrary probability distributions. The overall model thus facilitates an efficient PLF method for ADS.

8.2 PLF Formulation for ADS

This section introduces two formulation techniques for the ADS to facilitate the use of DPSO in the PLF calculation. First, the uncertainties in the ADS are modeled using discrete probability distributions (DPDs) and copula functions. Second, an improved linearized power flow that considers reactive power and nodal voltage is proposed to linearize the relationship of the nodal voltage and injections.

Note that because the proposed uncertainty analysis method is able to handle arbitrary probability distributions, both the output wind power and PV can be modeled. This chapter only focuses on wind power to simplify the description.

8.2.1 Uncertainty of Loads and Wind Power

Short-term load and wind power forecasting are usually conducted on a day-ahead basis and deliver estimates of the load and wind farm output for the next day. For the ADS, the load forecast error of each feeder cannot be ignored because its relative error in percentage is much larger than that of the system load. Naturally, the forecast error of the renewable energy output must also be considered because of the strong uncertainty. Both the renewable energy output P_w and load P_l can be treated as a superposition of the point-forecasted value (P_{wf}, P_{lf}) and the corresponding error (P_{we}, P_{le}):

$$P_w = P_{wf} + P_{we}$$
$$P_l = P_{lf} + P_{le}. \tag{8.1}$$

This chapter focuses on analysis and modeling of the uncertain component, i.e., the forecast error of the load and wind power output. The probability density functions (PDFs) of the forecast errors of all feeder injections (loads and wind power) can be defined as follows:

$$f_{P_{we}}(x) \quad x \in [\underline{P}_{we}, \overline{P}_{we}]$$
$$f_{P_{le}}(x) \quad x \in [\underline{P}_{le}, \overline{P}_{le}] \tag{8.2}$$

where $\underline{P}_{we}$, $\overline{P}_{we}$, $\underline{P}_{le}$, and $\overline{P}_{le}$ are the lower and upper bounds of the forecast errors of renewable energy and load.

If there is no load or wind power injection to the feeder, the corresponding PDF of the load or wind power forecast error is a unit pulse function. To facilitate computerized calculation, the PDFs are discretized with a step of Δp, and the discrete approximations of PDFs, i.e., the DPDs, are obtained:

$$P_{p_{we}}(i) = \int_{\underline{P}_{we}+i\Delta p-\Delta p/2}^{\underline{P}_{we}+i\Delta p+\Delta p/2} f_{P_{we}}(x)dx$$
$$P_{P_{le}}(i) = \int_{\underline{P}_{le}+i\Delta p-\Delta p/2}^{\underline{P}_{le}+i\Delta p+\Delta p/2} f_{P_{le}}(x)dx. \tag{8.3}$$

Note that the step Δp should be the same in both the load and wind power discretization (for example, 1 kW). If the uncertainty is characterized by a non-standard probability distribution, the DPDs can be obtained by simple statistics:

$$P_{p_{we}}(i) = I_{Wi}/N_W$$
$$P_{p_{le}}(i) = I_{Li}/N_L. \tag{8.4}$$

where N_W and N_L denote the total number of samples of renewable energy and load, respectively, and I_{Wi} and I_{Li} denote the number of samples of renewable energy and load, respectively, for the ith interval.

Thus, the uncertainty of each injection can be neatly formulated by a DPD. It is worth noting that the distribution of the forecast error might change at different times of the day or at different ranges of the forecasted levels. Thus, the statistical range of the forecast error can be flexibly modeled at different times or forecasted ranges separately. For example, the distribution of load forecast error varies with different periods and load levels. By dividing the time axis and load level axis into several segments, the uncertainty can be characterized conditionally for different partitions. The correlation between the forecast error and the forecasted value of wind power is also investigated [19].

8.2.2 Copula-Based Uncertainty Correlation Modeling

As stated in Chapter 2, the modeling of dependence structures among multiple variables can be transformed into the fitting of copula functions and determination of parameters. Kendall's rank correlation coefficient matrix $\mathbf{C}_D$ can be defined for the uncertainty correlations of loads and wind power:

$$
\mathbf{C}_D = \begin{bmatrix} \mathbf{C}_{WP} & \mathbf{C}_{WP-WQ} & \mathbf{C}_{WP-LP} & \mathbf{C}_{WP-LQ} \\ \mathbf{C}_{WP-WQ}^T & \mathbf{C}_{WQ} & \mathbf{C}_{WQ-LP} & \mathbf{C}_{WQ-LQ} \\ \mathbf{C}_{WP-LP}^T & \mathbf{C}_{WQ-LP}^T & \mathbf{C}_{LP} & \mathbf{C}_{LP-LQ} \\ \mathbf{C}_{WP-LQ} & \mathbf{C}_{WQ-LQ}^T & \mathbf{C}_{LP-LQ}^T & \mathbf{C}_{LQ} \end{bmatrix}. \tag{8.5}
$$

where $\mathbf{C}_{WP}$, $\mathbf{C}_{WQ}$, $\mathbf{C}_{LP}$, and $\mathbf{C}_{LQ}$ denote Kendall's rank correlation coefficients among the active and reactive power of different wind farms and loads, respectively, and the denotations of other block matrices are consistent with the corresponding subscripts.

Theoretically, all the block matrices can be obtained from the statistics of practical data. However, it is difficult to collect the reactive power of load and wind power in practice such that the coefficients related to reactive power are estimated by the correlation of the accordant active power.

$\mathbf{C}_{WQ}$ and $\mathbf{C}_{LQ}$ approximately take the same value as $\mathbf{C}_{WP}$, $\mathbf{C}_{LP}$.

The diagonal entries of $\mathbf{C}_{WP-WQ}$ can be assumed to be 1.0 if the wind farms operate under a constant power factor. Each non-diagonal entry of $\mathbf{C}_{WP-WQ}$, c_{ij}^{WP-WQ} denotes the correlations between the active power forecast error of one wind farm and the reactive power forecast error of another wind farm, which can be estimated indirectly:

$$
c_{ij}^{WP-WQ} = c_{ij}^{WP} c_{jj}^{WP-WQ} \tag{8.6}
$$

where c_{ij}^{WP} and c_{jj}^{WP-WQ} are the entries of $\mathbf{C}_{WP}$ and $\mathbf{C}_{WP-WQ}$, respectively.

$\mathbf{C}_{LP-LQ}$ can be estimated in the same manner as $\mathbf{C}_{WP-WQ}$.

$\mathbf{C}_{WQ-LQ}$ can take the same value as $\mathbf{C}_{WP-LP}$.

$\mathbf{C}_{WQ-LP}$ can be estimated in the same manner as $\mathbf{C}_{WP-LQ}$:

$$
c_{ij}^{WQ-LP} = c_{ii}^{WP-WQ} c_{ij}^{WP-LP}. \tag{8.7}
$$

Each entry of $\mathbf{C}_{WP-LQ}$, c_{ij}^{WP-LQ} can be approximately estimated by the production of the corresponding entries of $\mathbf{C}_{WP-LP}$ and $\mathbf{C}_{LP-LQ}$:

$$c_{ij}^{WP-LQ} = c_{ij}^{WP-LP} c_{jj}^{LP-LQ}. \tag{8.8}$$

8.2.3 Linearized Power Flow Considering Nodal Voltage

The power flow model is required in the PLF calculation to estimate the distribution of branch power flow and nodal voltages using power injections. The AC power flow equations in polar coordinates are proposed as shown in Equation (8.9).

$$P_{wk} - P_{lk} = V_k \sum_{j \in k} V_j \left(G_{kj} \cos \theta_{kj} + B_{kj} \sin \theta_{kj} \right)$$
$$Q_{wk} - Q_{lk} = V_k \sum_{j \in k} V_j \left(G_{kj} \sin \theta_{kj} - B_{kj} \cos \theta_{kj} \right). \tag{8.9}$$

where G_{kj} and B_{kj} denote the real and imaginary components of the admittance of line $k - j$, respectively, θ_{kj} denotes the phase angle difference of the kth and j^{th} node voltages, and V_j denotes the magnitudes of the jth bus voltage.

To linearize the AC power flow equations, two approximations are applied.

Approximation 1: For small θ_{kj}, $\cos \theta_{kj} \approx 1$ and $\sin \theta_{kj} \approx \theta_{kj}$. Two constant coefficients λ_1 and λ_2 are added to minimize the approximation errors. Both λ_1 and λ_2 can be set to 0.95 [20].

$$G_{kj} \cos \theta_{kj} + B_{kj} \sin \theta_{kj} \approx \lambda_1 \left(G_{kj} + B_{kj} \theta_{kj} \right)$$
$$G_{kj} \sin \theta_{kj} - B_{kj} \cos \theta_{kj} \approx \lambda_2 \left(G_{kj} \theta_{kj} - B_{kj} \right). \tag{8.10}$$

Approximation 2: The second-order terms of the production of two voltages are neglected.

$$V_k(V_k - V_j)$$
$$= (V_k - V_j) + (1 - V_k)^2 - (1 - V_k)(1 - V_j) \tag{8.11}$$
$$\approx (V_k - V_j).$$

Taking the active power flow equations as an example, the AC power flow equations can be linearized into DC power flow equations:

$$W_{Pk} - L_{Pk} = V_k \sum_{j \in k} V_j \left(G_{kj} \cos \theta_{kj} + B_{kj} \sin \theta_{kj} \right)$$
$$\approx \lambda_1 V_k \sum_{j \in k} V_j \left(G_{kj} + B_{kj} \theta_{kj} \right)$$
$$\approx \lambda_1 g_{kk} V_k + \lambda_1 \sum_{j \in k} \left(g_{kj}(V_k - V_j) + b_{kj}(\theta_k - \theta_j) \right) \tag{8.12}$$
$$= \lambda_1 \sum_{j \in k} G_{kj} V_j - \lambda_1 \sum_{j \in k} B_{kj} \theta_j.$$

The second and third lines of Equation (8.12) are approximated according to Equations (8.10) and (8.11), respectively. Equation (8.12) and the reactive power flow equations can be rewritten into matrix form:

$$\begin{bmatrix} P_w - P_l \\ Q_w - Q_l \end{bmatrix} = - \begin{bmatrix} B' & -G' \\ G'' & B'' \end{bmatrix} \begin{bmatrix} \theta \\ V \end{bmatrix} \qquad (8.13)$$

where $G'+jB'=\lambda_1(G+jB)$ and $G''+jB''=\lambda_2(G+jB)$.

Note that the net injections are transformed into a linear combination of the voltage magnitudes and phase angles. PV nodes will be more common in future ADN, such as nodes with reactive compensator, distributed generation (DG) under constant voltage mode. No matter whether the node is a PV node or PQ node, the manner in which the injection influences the voltages of the feeders can always be obtained by Gaussian elimination. Particularly, if the root node of the ADS is chosen as a slack bus, and all other nodes are assumed to be PQ types, the voltage magnitudes of the PQ nodes can be represented as follows:

$$V = MP_{net} + NQ_{net} + [1]. \qquad (8.14)$$

where M and N are the coefficient matrices of the active power and reactive power injections, respectively, which are determined by the line parameters and topology structure of the ADS:

$$\begin{aligned} M &= -B_{nn}^{-1}G_{nn}H_{nn}^{-1} \\ N &= -(B_{nn}^{-1} + B_{nn}^{-1}G_{nn}H_{nn}^{-1}G_{nn}B_{nn}^{-1}). \end{aligned} \qquad (8.15)$$

This observation suggests that the voltage magnitude of each node is essentially the linear combination of the active and reactive power of the load and wind power of all PQ nodes. Furthermore, by substituting Equation (9.1) into Equation (9.14) and after selected manipulations, we obtain

$$\begin{aligned} V &= MP_{wf} - P_{lf} + NQ_{wf} - Q_{lf} + [1] \\ &+ Mp_{we} - p_{le} + Nq_{we} - q_{le} \end{aligned} \qquad (8.16)$$

The first three terms of Equation (8.16) are the deterministic components of voltage (recorded as V_f) and can be calculated directly. The last two terms are the uncertain components of the voltage magnitude (recorded as V_e) and are analyzed by DPSO in the next section. P_{we}, P_{le}, Q_{we}, and Q_{le} are correlated stochastic variables that can be modeled using DPDs and the copula function.

8.3   DPSO-Based Algorithm for PLF

This section introduces the algorithm for calculation of ADS power flow using the model proposed in the previous section. The proposed algorithm improves

traditional discrete convolution by introducing a copula function to enable it
to handle the dependencies in both loads and wind power.

8.3.1 DPSO-Based PLF Calculation

The DPD of each voltage can be obtained by calculating the traditional discrete convolution of DPSO among the active and reactive injections of all
nodes. Thus, the proposed DPSO can be used to calculate the DPDs of all
voltage magnitudes if their own marginal PDFs and dependence structures
are well modeled.

$$V_{ke} = \sum_{j=1}^{n} \overset{c}{\otimes} \left[M_{kj}p_{jwe} \overset{c}{\otimes} (-M_{kj}p_{jle}) \overset{c}{\otimes} N_{kj}q_{jwe} \overset{c}{\otimes} (-N_{kj}q_{jle}) \right] \qquad (8.17)$$

When we implement PLF for the ADS using DPSO, two main issues should
be addressed. The first issue is the change of the discretization interval for
the distribution because of the coefficients M_{kj} and N_{kj} in the power flow
equation. **Rescaling** of DPD should be conducted. As illustrated in Figure
8.1, the DPD of active power p_{jwe}, originally with a discrete step of Δp_{p0}, is
rescaled with a discrete step of $\Delta p_{p1} = M_{kj}\Delta p_{p0}$ after multiplication by M_{kj}.

The second issue is the inconsistency of the discrete steps of the injections
and voltages. For example, the discrete step of the injections is 0.001 p.u.,
whereas the discrete step of the voltages might be 0.0015 p.u. Thus, **reshaping** of DPD should be applied. As shown in the shadow area in Figure 8.1,
the horizontal coordinate is discretized with a step of Δp_{pV} such that the
corresponding probability in each interval is calculated by the shadow area
that it covers.

8.3.2 Dimension Reduction

If feeders are connected with wind power, the number of items to be summed
in Equation (8.17) is $2(a + b)$. It is difficult to model the $2(a + b)$ dimensional dependence structure and the corresponding copula function due to
the "dimension curse". The dimension can be reduced according to several
characteristics of the loads and wind power.

First, the correlation between the loads and wind power forecast error is
usually sufficiently small to be ignored because the loads are primarily affected
by human activity and temperature, whereas the output of wind power is
mainly determined by wind speed. Thus, the joint DPDs of the loads and
wind power uncertainties can be calculated separately by DPSO, and the
DPDs of the voltage magnitude can be calculated by convolving the joint
DPDs of the loads and wind power, as expressed in Equation (8.18). This
process transforms a $2(a + b)$-dimensional problem into two lower dimensional

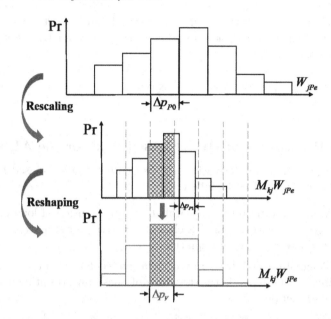

FIGURE 8.1 Rescaling and reshaping of DPSO implementation for PLF.

problems (2a and 2b).

$$V_{ke} = \left[-\sum_{j=1}^{n} \overset{c}{\otimes} \left(M_{kj} p_{jle} \overset{c}{\otimes} N_{kj} q_{jle} \right) \right] \otimes \left[\sum_{j=1}^{n} \overset{c}{\otimes} \left(M_{kj} p_{jwe} \overset{c}{\otimes} N_{kj} q_{jwe} \right) . \right]$$

(8.18)

Second, the distributions of the load forecast error and dependence can be well fit by a Gaussian distribution and Gaussian copula, respectively. Thus, the term $\sum_{j=1}^{n} \overset{c}{\otimes} \left(M_{kj} p_{jle} \overset{c}{\otimes} N_{kj} q_{jle} \right)$ is also Gaussian distributed. In this way, we need only to calculate the expectation and variance. The correlations do not influence the expectation, which is the sum of each of the load expectations. For the variance, the variance of the sum of two variables can be calculated by:

$$D \left(\alpha + \beta \right) = \sigma_{\alpha}^2 + \sigma_{\beta}^2 + 2\rho_{\alpha\beta}\sigma_{\alpha}\sigma_{\beta}.$$

(8.19)

where $\rho_{\alpha\beta}$ denotes their linear correlation coefficient. For multiple variables, Equation (8.19) can be extended.

$$D \left(\sum_{k=1}^{K} a_k \beta_k \right) = \sum_{k1=1}^{K} \sum_{k2=1}^{K} \rho_{\beta_{k1}\beta_{k2}} a_{k1} \sigma_{\beta_{k1}} a_{k2} \sigma_{\beta_{k2}}.$$

(8.20)

Third, the active and reactive power of the wind farm can be treated as completely correlated if the wind turbines work in constant power factor (φ_j) mode. Thus, the 2b-dimensional problem for wind power DPSO is reduced to a b-dimensional problem:

$$\sum_{j=1}^{n} \overset{c}{\otimes} \left(M_{kj} W_{jPe} \overset{c}{\otimes} N_{kj} W_{jQe} \right) = \sum_{j=1}^{n} \overset{c}{\otimes} (M_{kj} + N_{kj} \tan \varphi_j) W_{jPe}. \qquad (8.21)$$

8.3.3   Procedures of DPSO-Based PLF for the ADS

According to the above derivation, the procedure of DPSO-based PLF for the ADS can be summarized as follows.

1. Model and discretize the probability distribution of load and wind power forecast errors of each node by fitting or statistics according to Equation (8.2) or Equation (8.4).

2. Calculate Kendall's rank correlation coefficient matrix $\mathbf{C}_D$ and model the dependence structure of these uncertainties using copula functions.

3. Obtain the coefficient matrices of active power and reactive power injections M and N through the proposed DC power flow equations according to Equation (8.15).

4. Rescale and reshape the marginal DPD of each injection according to the discrete step shown in Figure 8.1.

5. Conduct traditional discrete convolution based on the reduced dimension approximation to implement PLF for the ADS according to Equation (8.21).

8.4   Numerical Examples

The numerical experiments on a DPSO-based PLF for the ADS are implemented on a modified IEEE 33-node distribution system and a modified IEEE 123-node distribution system with MATLAB R2015a on a standard PC with an Intel Core^{TM} i7-4770MQ CPU running at 2.40 GHz and 8.0 GB RAM.

8.4.1   Description of Basic Data

The IEEE 33-node distribution system consists of 33 nodes, 37 branches, and 4 paths, as shown in Figure 8.2 [21]. Node #1 is the slack node, the voltage, and angle of which are set to 1 and 0, respectively. The green nodes are nodes with wind power. The forecasted loads for each node are given in [22]. The wind power forecast data used in this chapter are from the Wind Integration

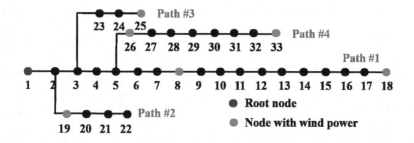

FIGURE 8.2 Modified IEEE 33-node distribution system.

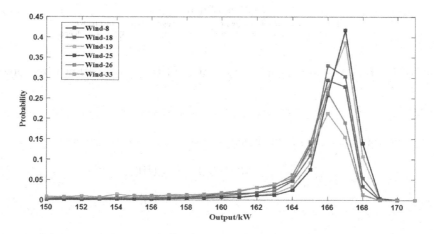

FIGURE 8.3 PDFs of the outputs of the six wind farms.

Datasets of NREL [23], where the forecasted outputs of wind power in Nodes 8, 18, 19, 25, 26, and 33 are set to 0.603, 0.597, 0.604, 0.598, 0.569, and 0.5 MW, respectively.

All the load forecast errors are assumed to be Gaussian distributed with standard deviations of 10% of the forecasted values, and their Kendall's rank correlation coefficients are set to 0.3. The distributions of the output of the six wind farms are obtained using the conditional forecast error modeling method proposed in [24], as shown in Figure 8.3, and their Kendall's rank correlation coefficients are set from 0.3 to 0.7. All the outputs of wind power have constant power factors of 0.9.

To evaluate the performance of the proposed algorithm, the results from the Monte Carlo simulation method with large numbers of samples (10^6) are treated as "exact solutions", and comparisons with traditional linearized power flow are also conducted. The errors of the obtained PDFs are quantified

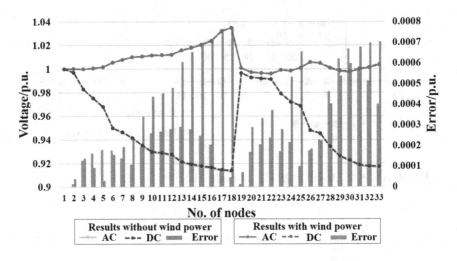

FIGURE 8.4 Comparison between AC and the proposed linearized power flow.

by the absolute error of the expected value and standard deviation:

$$\varepsilon_\mu = |E\left(x'(i)\right) - E\left(x(i)\right)| \tag{8.22}$$

$$\varepsilon_\sigma = |\sigma\left(x'(i)\right) - \sigma\left(x(i)\right)|. \tag{8.23}$$

8.4.2   Accuracy of the Proposed Linearized Power Flow

Power flows with and without wind power integration are calculated to validate the proposed linearized power flow model. Figure 8.4 shows the voltage magnitude of all nodes under the expected wind power and load for both cases. The absolute error of the proposed linearized model compared with that of the AC model is also calculated. The results show that the voltage magnitudes without wind power integration obtained from the improved linearized power flow are highly accurate with a maximum error of 0.00078 p.u. The maximum error with wind power integration is only 0.00069 p.u. Comparison between the two cases shows that the error is much smaller if the voltage magnitudes are closer to 1.0 p.u.

8.4.3   Comparative Studies

Figure 8.5 displays the quantiles for all node voltages calculated by the proposed algorithm. It can be observed that the further away the node is located from the root node, the larger the variances of the nodes will be. Among these four paths, the voltages on Path #2 show smaller uncertainties, especially

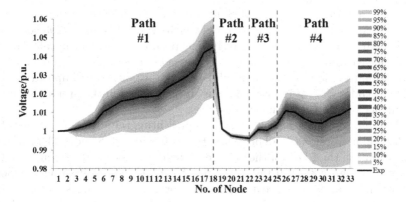

FIGURE 8.5 Quantiles for the voltage magnitude uncertainties of all nodes.

TABLE 8.1 Performance of DC-DPSO and DC-MC for IEEE 33-Nodes

Method	$\varepsilon_{\mu,ave}$	$\varepsilon_{\mu,\max}$	$\varepsilon_{\sigma,ave}$	$\varepsilon_{\sigma,\max}$	T(s)
DC-DPSO	0.00044	0.00078	0.00052	0.00093	5.8915
DC-MC	0.00027	0.00053	0.00012	0.00034	23.603
AC-MC	/	/	/	/	1730.32

compared with those of Path #3. The difference between Path #2 and Path #3 is the connection node of the wind farm, which shows that the large uncertainties of wind power influence the voltage of the preceding nodes but have a smaller impact on the subsequent nodes.

The normal operating range of voltage magnitude for the ADS is 0.92-1.04 p.u. It can be observed that wind power integration can ease the burden of low voltage, but it might cause over-voltage as well. For example, Nodes #17 and #18 run the risk of over-voltage. Measures such as reactive voltage control should be applied to ensure voltage quality.

The accuracy of the proposed method was tested by comparing the results obtained from two Monte Carlo methods: the AC power flow-based Monte Carlo method (AC-MC) and the linearized power flow-based Monte Carlo method (DC-MC). We treat the proposed method as DC-DPSO. The absolute errors of the expected value and standard deviation of DC-MC and DC-DPSO are obtained according to Equations (8.22) and (8.23), where AC-MC is used as a benchmark. The mean and maximum absolute error of the expected value and standard deviation of DC-MC and DC-DPSO ($\varepsilon_{\mu,ave}$, $\varepsilon_{\mu,\max}$, $\varepsilon_{\sigma,ave}$, $\varepsilon_{\sigma,\max}$) and the computation time (T) are summarized in Table 8.1.

The maximum errors of the proposed method are 0.00078 p.u. and 0.00093 p.u. for the expected value and standard deviation, respectively. The performance of DC-DPSO is comparable to that of DC-MC and fully satisfies the practical requirement for the ADS. In addition, DC-DPSO has the advantages of lower computing-time consumption, which is only 1/4 and 1/293 consumed by DC-MC and AC-MC, respectively. The proposed method can calculate the PDF of each node voltage individually instead of calculating all the node voltages, as in the Monte Carlo method, i.e., only the voltages of important nodes such as Nodes #17, #18 and #33 must be calculated in the proposed method to further reduce the computation times.

Limitations of space prevent us from presenting the PDFs of voltages of all nodes and power flows of all branches. Because Nodes #18 and #33 have larger uncertainties, the PDFs of the voltage of these two nodes are presented in this chapter in Figure 8.6 and Figure 8.7 to offer more detailed information. Some key branches are also of great importance for system operators and monitors. Thus, the power flow distribution of the branches should also be calculated. Figure 8.8 shows the active power flow of Branch #2-#3 (the active power flow from Node #2 to Node #3). In these three figures, the PDFs obtained by both DC-MC and DC-DPSO only have a minor error from the "accurate" PDF obtained by AC-MC. Note that the error of the proposed DC-DPSO method can be decomposed into two aspects. One aspect is the discretization error in the DPSO calculation, which can be measured by the difference between DC-DPSO and DC-MC. The other aspect is the linearization error in the power flow calculation, which can be measured by the difference between DC-MC and AC-MC. The discretization error can be decreased by reducing the discretization step Δp, whereas the linearization error can be further reduced by adjusting the value of two constant coefficients λ_1 and λ_2 to obtain the optimal approximation.

8.4.4 Scalability Tests

To verify the effectiveness and efficiency of the proposed method, case studies on the IEEE 123-node distribution system [25] where 10 nodes are integrated with renewable energy are conducted in this part.

The maximum error of the proposed method on the IEEE 123-node system are 0.00137 p.u. and 0.00037 p.u. for the expected value and standard deviation, respectively. It shows that the proposed method can also ensure high accuracy in the larger distribution system. The time consumed by the proposed method is only 8.6652 seconds, which is only 1/3 and 1/423 consumed by DC-MC and AC-MC, respectively. The results show that the proposed method is extendable to larger systems with high accuracy and efficiency, which is of significance for short-term and real-time PLF.

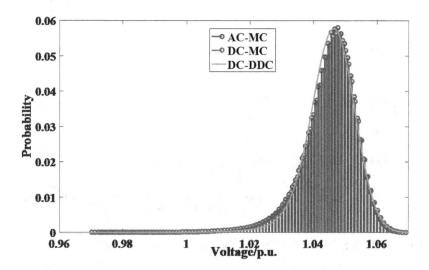

FIGURE 8.6 PDF of the voltage of Node #18.

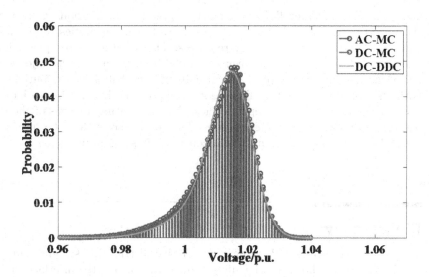

FIGURE 8.7 PDF of the voltage of Node #33.

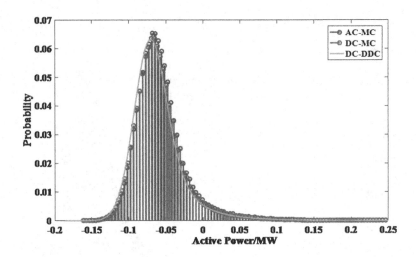

FIGURE 8.8 PDF of the active power of Branch #2-#3.

8.5   Summary

This chapter proposes a novel PLF algorithm for the ADS by combining DPSO and a novel DC power flow with reactive power considerations to analyze the uncertainties of node voltages caused by the uncertainties of the load and distributed renewable energy resources. The novel linearized power flow linearizes the relationship between the nodal voltages and nodal injections. DPSO is subsequently applied to calculate the probability distributions of both the voltages and power flows. The proposed method is able to handle arbitrary distribution with dependencies, thus facilitating analysis of the impact of renewable energy uncertainties. The result shows that the proposed method achieves higher accuracy and consumes less computing time.

Bibliography

[1] A. Saint-Pierre and P. Mancarella. Active distribution system management: A dual-horizon scheduling framework for dso/tso interface under uncertainty. *IEEE Transactions on Smart Grid*, 8(5):2186–2197, 2017.

[2] M. N. Kabir, Y. Mishra, and R. C. Bansal. Probabilistic load flow for dis-

tribution systems with uncertain pv generation. *Applied Energy*, 163:343–351, 2016.

[3] R. Tonkoski, Luiz A. C. Lopes, and T. El-Fouly. Coordinated active power curtailment of grid connected pv inverters for overvoltage prevention. *IEEE Transactions on Sustainable Energy*, 2(2):139–147, 2011.

[4] P. Chen, Z. Chen, and B. Bak-Jensen. Probabilistic load flow: A review. *International Conference on Electric Utility Deregulation and Restructuring and Power Technologies*, pages 1586–1591, 2008.

[5] S. Conti and S. Raiti. Probabilistic load flow using Monte Carlo techniques for distribution networks with photovoltaic generators. *Solar Energy*, 81(12):1473–1481, 2007.

[6] J. Tang, F. Ni, F. Ponci, and A. Monti. Dimension-adaptive sparse grid interpolation for uncertainty quantification in modern power systems: Probabilistic power flow. *IEEE Transactions on Power Systems*, 31(2):907–919, 2016.

[7] Y. Chen, J. Wen, and S. Cheng. Probabilistic load flow method based on Nataf transformation and Latin hypercube sampling. *IEEE Transactions on Sustainable Energy*, 4(2):294–301, 2013.

[8] C. L. Su. Probabilistic load-flow computation using point estimate method. *IEEE Transactions on Power Systems*, 20(4):1843–1851, 2005.

[9] R. N. Allan, A. M. Leite Dasilva, and R. C. Burchett. Evaluation methods and accurary in probabilistic load flow solutions. *Power Apparatus & Systems IEEE Transactions on*, PER-1(5):43–44, 2010.

[10] J. Usaola. Probabilistic load flow with wind production uncertainty using cumulants and Cornish–Fisher expansion. *International Journal of Electrical Power & Energy Systems*, 31(9):474–481, 2009.

[11] T. Williams and C. Crawford. Probabilistic load flow modeling comparing maximum entropy and gram-charlier probability density function reconstructions. *IEEE Transactions on Power Systems*, 28(1):272–280, 2013.

[12] A. M. Leite Da Silva and V. L. Arienti. Probabilistic load flow by a multilinear simulation algorithm. *IEE Proceedings C - Generation, Transmission and Distribution*, 137(4):276–282, 1990.

[13] W. Wu, K. Wang, G. Li, X. Jiang, and Z. Wang. Probabilistic load flow calculation using cumulants and multiple integrals. *IET Generation Transmission & Distribution*, 10(7):1703–1709, 2016.

[14] J Usaola. Probabilistic load flow in systems with wind generation. *IET Generation Transmission & Distribution*, 3(12):1031–1041, 2009.

[15] H. Yu, C. Y. Chung, K. P. Wong, H. W. Lee, and J. H. Zhang. Probabilistic load flow evaluation with hybrid Latin hypercube sampling and Cholesky decomposition. *IEEE Transactions on Power Systems*, 24(2):661–667, 2009.

[16] Q. Fu, D. Yu, and J. Ghorai. Probabilistic load flow analysis for power systems with multi-correlated wind sources. *Power and Energy Society General Meeting*, pages 1–6, 2011.

[17] J. An and J. Yang. Reliability evaluation for distribution system considering the correlation of wind turbines. *International Conference on Advanced Power System Automation and Protection*, pages 2110–2113, 2012.

[18] N. Zhang, C. Kang, C. Singh, and Q. Xia. Copula based dependent discrete convolution for power system uncertainty analysis. *IEEE Transactions on Power Systems*, 31(6):5204–5205, 2016.

[19] J. Tastu, P. Pinson, E. Kotwa, H. Madsen, and H. A. Nielsen. Spatio-temporal analysis and modeling of short-term wind power forecast errors. *Wind Energy*, 14(1):43–60, 2011.

[20] S. M. Fatemi, S. Abedi, G. B. Gharehpetian, S. H. Hosseinian, and M. Abedi. Introducing a novel DC power flow method with reactive power considerations. *IEEE Transactions on Power Systems*, 30(6):3012–3023, 2014.

[21] S. Ganguly and D. Samajpati. Distributed generation allocation on radial distribution networks under uncertainties of load and generation using genetic algorithm. *IEEE Transactions on Sustainable Energy*, 6(3):688–697, 2015.

[22] M. E. Baran and F. F. Wu. Network reconfiguration in distribution systems for loss reduction and load balancing. *IEEE Transactions on Power Delivery*, 4(2):1401–1407, 1989.

[23] M. Milligan, E. Ela, D. Lew, D. Corbus, Y. H. Wan, and B. Hodge. Assessment of simulated wind data requirements for wind integration studies. *IEEE Transactions on Sustainable Energy*, 3(4):620–626, 2012.

[24] N. Zhang, C. Kang, Q. Xia, and J. Liang. Modeling conditional forecast error for wind power in generation scheduling. *IEEE Transactions on Power Systems*, 29(3):1316–1324, 2014.

[25] S. Bolognani and S. Zampieri. On the existence and linear approximation of the power flow solution in power distribution networks. *IEEE Transactions on Power Systems*, 31(1):163–172, 2016.

9

Risk-Based Stochastic Unit Commitment

The integration of renewable energy requires the power system to be sufficiently flexible to accommodate its forecast errors. In the market-clearing process, the scheduling of flexibility relies on the manner in which the renewable energy uncertainty is addressed in the unit commitment (UC) model. Comparing with the existing research. In this chapter, a RUC model is proposed that can identify the underlying risks of the scheduling imposed by renewable energy uncertainty, including loss of load, wind curtailment and branch overflow. In the RUC model the risks are formulated as functions of the scheduling decisions. The formulation facilitates a strategic scheduling that is able to be aware of the associated cost and the limits of the risks. The RUC model is reformulated as a MILP problem without adding extra variables. Such compact modeling contributes to improving the calculation efficiency of the RUC model.

9.1 Overview

The market clearing of major generators is performed one day ahead before the renewable energy output can be accurately forecasted. Thus, the uncertainty of renewable energy must be considered together with the generation-demand balance and transmission congestion management [1–3]. Traditionally, spinning reserves are scheduled using deterministic and static criteria, e.g., by a fixed amount or a fixed ratio of the load. Such criteria are not suitable for coping with the uncertainty of renewable energy because the predictability of renewable energy is much poorer and more variable than the load demand and the use of deterministic and static criteria may not be economical or reliable for limiting the risk of uncertainty [4]. Transmission congestion management should also be reconsidered when incorporating large-scale renewable energy [5].

To more strategically accommodate the uncertainty of renewable energy in day-ahead scheduling, significant work has been performed in unit commitment models through stochastic optimization techniques. Various unit commitment (UC) models have been proposed and can be generally divided into four categories based on the manner in which they address the uncertainty, namely, scenario-based stochastic programming [6–11], robust optimization [12–17], chance-constrained optimization [18–21] and risk-based optimization [22–28].

The scenario-based UC (SUC) model incorporates the dispatching of multiple renewable energy realization scenarios with day-ahead scheduling. This approach is also known as the two-stage UC model. The day-ahead point forecast of the renewable energy output is used to form the variables and constraints corresponding to the day-ahead stage (here-and-now decisions), and the renewable energy scenarios form the variables and constraints corresponding to the real operation stage (wait-and-see decisions) [8]. The objective function of SUC is usually the combination of the cost of here-and-now decisions and the expected cost associated with the redispatch of generators, the load shedding and wind curtailment in the wait-and-see decisions [7, 8]. The model optimizes the decisions on both stages and thus guarantees that the scheduling of conventional generation is sufficiently flexible for the uncertainties. Proposed SUC models vary in the scenarios considered, e.g., equipment failure, renewable energy, and load uncertainty. The scenario tree technique can also be used in such models [6, 7, 9]. SUC problems are always large-scale problems and may face computational efficiency issues as the number of scenarios increases. Some strategic simplification of the models may be required to improve the calculating efficiency. The choice of representative scenarios is also challenging when aiming to avoid arbitrariness since the conservativeness and robustness of the solution depends on how far the selected scenarios are from the worst case [11].

The robust UC model considers the randomness of renewable energy by identifying and minimizing the scheduling cost of the worst-case scenario. The model is usually a complex programming model and is generally solved using dualization techniques and Bender's decomposition method [12, 13]. The interval optimization approach has also been proposed and can be regarded as a special type of robust optimization [15]. The robust UC model does not require detailed information regarding the uncertainty; however, the worst-case scenario should be carefully chosen to obtain a reasonable tradeoff between the coverage of uncertainty and the cost-effectiveness of the scheduling. The concept of uncertainty budget has been proposed to avoid an over-conservative solution [12, 13].

The chance-constrained UC model sets the constraints with stochastic variables based on a certain probability limit at which the constraints should hold. The scheduling is optimized while guaranteeing that the violations of the real operation constraints, e.g., the generation deficiency and branch overload due to uncertainty are limited within a small probability. In general, such chance constraints lead to non-convex problems, and the model is solved using a sample-average approximation approach [18–20]. Only certain forms of the chance-constrained UC model can be transformed to an equivalent deterministic UC problem [21].

The risk-based UC (RUC) model introduces the concept of risk and considers the operational risks of the power system in the objective function and the constraints of the UC model. The risk is formulated by multiplying the cost of the uncertain event by its occurrence probability. The RUC model allows for a tradeoff between generating costs and underlying costs of uncertainties [26]. The risks considered are usually expected energy not served (EENS) and the expected renewable energy curtailment [22]. In [28] the risks of load shedding are considered by introducing the demand curve of operating reserve. It is quantified by the cost of unserved energy and the expected loss of load. The operating reserve demand curve is modeled by a stepwise function. Such modeling of risks maintains the UC model to be a MILP problem.

Each of the models mentioned above has its own merits in modeling the uncertainty; however, a rigorous representation of the uncertainty usually complicates the model, e.g., by multiplying the scale of the problem or by transforming the model into a nonlinear problem. The solution efficiency of the model is still of concern when applied to practical market clearing. This chapter proposes an RUC model accounting for the risks caused by wind power and load uncertainty.

9.2 Modeling Risks of Renewable Energy Integration

9.2.1 Model Assumptions and Notations

1. The probability density functions (PDFs) of the renewable energy forecast error are assumed to be known. The PDFs can either be obtained from a probabilistic forecast [29] or from historical statistics [30]. The PDF of the load forecast is also considered. Equipment failures are not considered.

2. Wind farms are not considered competitive and offer their forecast generation at zero price. Other generators offer their energy cost curve, on-off cost and reserve cost to the market. The cost curve of a generator is assumed to follow a convex form, e.g., a quadratic function.

3. Load demands are considered to be inelastic and do not provide the spinning reserve to the market.

4. The transmission grid is modeled through a linear DC power flow model, and net losses are neglected.

5. The time horizon considered in the model is one day ahead, although the model can be extended to longer time periods.

9.2.2   Notations

Indices

i	Index of generators from 1 to I.
j	Index of wind farms from 1 to J.
m	Index of buses from 1 to M.
l	Index of transmission lines from 1 to L.
t	Index of time periods from 1 to T.

Functions

$C_i(\cdot)$	Generation cost function of generator i.
WSF_t'	Original wind speed forecast value in time period t.
$f_t^{net}(\cdot)$	Probability density function of net load during period t (considering the uncertainty of load and wind).
$f_{l,t}^{flow}(\cdot)$	Probability density function of the power flow of branch l during period t (considering the uncertainty of load and wind).

Parameters

$\underline{R}_t^{up}/\underline{R}_t^{down}$	Fixed up/down spinning reserve requirement during period t.
$\underline{P}_i/\overline{P}_i$	Min/max generation limits for generator i.
$\underline{t}_i^{on}/\underline{t}_i^{off}$	Minimum on/off time intervals of generator i.
$c_i^{up}/\underline{c}_i^{down}$	Cost of offering the up/down spinning reserve by generator i.
o_i^{on}	Start-up cost of generator i.
$D_{m,t}$	Load forecast for bus m during period t.
$\hat{P}_{j,t}^w$	Forecast output of wind farm j during period t.
$\Delta P_i^{up}/\Delta P_i^{down}$	Upward/downward ramping rate of generator i.
$\overline{L}_l$	Power flow limitation of branch l.
$g_{l,m}$	Power flow distribution factor of branch l with respect to the net injection of bus m.
$g_{l,i}/g_{l,j}$	Power flow distribution factor of branch l with respect to the output of generator i/ wind farm j.
θ^{eens}	Cost coefficient of load shedding.
θ^{wind}	Cost coefficient of wind curtailment.
θ^{flow}	Cost coefficient of branch overflow.
$\overline{R}_t^{load}$	Limit of load shedding during period t.
$\overline{R}_t^{wind}$	Limit of wind curtailment during period t.
$\overline{R}_{l,t}^{wind}$	Overflow risk limit of branch l during period t.

Variable

$P_{i,t}^c$	Scheduled generation of generator i during period t.
$P_{j,t}^w$	Scheduled wind farm curtailment of wind farm j during period t.
$u_{i,t}$	State variable of generator i during period t (equals 1 if the generator is on during period t; equals 0 otherwise).
$R_{i,t}^{up}/R_{i,t}^{down}$	Up / down spinning reserve provided by generator i during period t.
$s_{i,t}^{on}$	Start-up cost of generator unit i during period t.
$L_{l,t}$	Scheduled power flow of branch l during period t.
R_t^{eens}	Expected energy not served during period t.
R_t^{flow}	Expected overflow of branch l during period t.

9.2.3 Modeling Risks

The purpose of the RUC model is to schedule the generators while being aware of the underlying risks of misestimating renewable energy output and load. The risks of loss of load, wind curtailment, and branch overflow are considered in the proposed RUC model. The EENS and expected wind curtailment are modeled using the integration of the probability of net load ($f_t^{net}(x)$) multiplied by the deficiency of up and down reserves as follows

$$E_t^{eens} = \int_{D_t + \sum\limits_{i=1}^{I} R_{i,t}^{up}}^{\overline{D}_t} \left(x - D_t - \sum_{i=1}^{I} R_{i,t}^{up} \right) f_t^{net}(x)\mathrm{d}x,$$

$$E_t^{wind} = \int_{\underline{D}_t}^{D_t + \sum\limits_{i=1}^{I} R_{i,t}^{down}} \left(D_t - \sum_{i=1}^{I} R_{i,t}^{down} - x \right) f_t^{net}(x)\mathrm{d}x, \qquad (9.1)$$

$$D_t = \sum_{m=1}^{M} D_{m,t} - \sum_{q=1}^{Q} P_{q,t}^w, \; t = 1, 2, ..., T,$$

where $\overline{D}_t$ and $\underline{D}_t$ are the possible maximum and minimum net load, respectively. D_t is the scheduled net load during time period. $f_t^{net}(\cdot)$ is probability density function of net load during period t (considering the uncertainty of load and wind). Branch overflow risk is measured by a similar form using the exceeded power flow and its corresponding probability.

$$\cdot E_{l,t}^{flow} = \int_{\overline{L}_l}^{\overline{L}_{l,t}} \left(x - \overline{L}_l \right) f_{l,t}^{flow}(x)\mathrm{d}x + \int_{\underline{L}_{l,t}}^{-\overline{L}_l} \left(-\overline{L}_l - x \right) f_{l,t}^{flow}(x)\mathrm{d}x,$$

$$t = 1, 2, ..., T, \; l = 1, 2, ..., B. \qquad (9.2)$$

$f_{l,t}^{flow}(x)$ can be calculated using the probabilistic load flow (PLF) technique [31]. In the RUC model, the cost of the risks are considered as penalties and are minimized together with generating costs. At the same time, the hard constraints of the risks can also be included to ensure the risk of the scheduling is limited within a certain amount.

9.2.4   Risk-Based Stochastic Unit Commitment

The rest of the RUC model is formulated as

$$\text{minimize} \sum_{t=1}^{T}\sum_{i=1}^{I}\left(c_i(P_{i,t}^c)+c_i^{up}R_{i,t}^{up}+c_i^{down}R_{i,t}^{down}+s_{i,t}^{on}\right)+\sum_{t=1}^{T}\sum_{q=1}^{Q}\theta^{wind}P_{q,t}^w$$

$$+\sum_{t=1}^{T}\left(\theta^{eens}E_t^{eens}+\theta^{wind}E_t^{wind}\right)+\sum_{t=1}^{T}\sum_{l=1}^{L}\theta^{flow}E_{l,t}^{flow} \tag{9.3}$$

$$\text{s.t.}\ \sum_{i=1}^{I}P_{i,t}^c+\sum_{q=1}^{Q}(\hat{P}_{q,t}^w-P_{q,t}^w)=\sum_{m=1}^{M}D_{m,t},\ \ t=1,2,...,T \tag{9.4}$$

$$\sum_{i=1}^{I}R_{i,t}^{up}\geq\underline{R}_t^{up},\ \sum_{i=1}^{I}R_{i,t}^{down}\geq\underline{R}_t^{down},\ \ t=1,2,...,T \tag{9.5}$$

$$R_{i,t}^{up}+P_{i,t}^c\leq u_{i,t}\overline{P}_i,\ P_{i,t}^c-R_{i,t}^{down}\geq u_{i,t}\underline{P}_i,$$
$$R_{i,t}^{up}\geq 0,\ R_{i,t}^{down}\geq 0\ \ t=1,2,...,T,\ i=1,2,...,I \tag{9.6}$$

$$0\leq P_{q,t}^w\leq\hat{P}_{q,t}^w,\ t=1,2,...,T,\ q=1,2,...,Q \tag{9.7}$$

$$P_{i,t}^c-P_{i,t-1}^c+u_{i,t}\left(\overline{P}_i-\underline{P}_i\right)+u_{i,t-1}\left(\underline{P}_i-\Delta P_i^{up}\right)\leq\overline{P}_i,$$
$$P_{i,t-1}^c-P_{i,t}^c+u_{i,t-1}\left(\overline{P}_i-\underline{P}_i\right)+u_{i,t}\left(\underline{P}_i-\Delta P_i^{down}\right)\leq\overline{P}_i,$$
$$t=1,2,...,T,\ i=1,2,...,I \tag{9.8}$$

$$(u_{i,t}-u_{i,t-1})\,\underline{t}_i^{on}+\sum_{q=t-\underline{t}_i^{on}}^{t-1}u_{i,q}\geq 0,$$
$$(u_{i,t-1}-u_{i,t})\,\underline{t}_i^{off}+\sum_{q=t-\underline{t}_i^{off}}^{t-1}(1-u_{i,q})\geq 0,$$
$$t=1,2,...,T,\ i=1,2,...,I \tag{9.9}$$

$$s_{i,t}^{on}\geq o_i^{on}\left(u_{i,t}-u_{i,t-1}\right),s_{i,t}^{on}\geq 0,\ t=1,2,...,T,\ i=1,2,...,I \tag{9.10}$$

$$L_{l,t}=\sum_{i=1}^{I}g_{l,i}P_{i,t}^c+\sum_{q=1}^{Q}g_{l,q}(\hat{P}_{q,t}^w-P_{q,t}^w)-\sum_{m=1}^{M}g_{l,m}D_{m,t}$$
$$-\overline{L}_l\leq L_{l,t}\leq\overline{L}_l,\ t=1,2,...,T,\ l=1,2,...,B \tag{9.11}$$

$$E_t^{eens} \le \overline{E}_t^{eens}, E_t^{wind} \le \overline{E}_t^{wind}, \ t = 1, 2, ..., T \qquad (9.12)$$

$$E_{l,t}^{flow} \le \overline{E}_{l,t}^{flow}, \ t = 1, 2, ..., T, \ l = 1, 2, ..., B. \qquad (9.13)$$

The constraints considered are as follows: Equation (9.4) is the market balance equation. Equation (9.5) presents the up / down fixed spinning reserve constraint. Equation (9.6) limits the output and the reserve offered by the generators except for renewable energy. Equation (9.7) limits the market schedule of the wind farms to no more than its forecast. Equation (9.8) forms the ramping rate constraint; it also guarantees that the generator is scheduled at the minimum output for the first time period after starting up and for the last time period before shutting down. Equation (9.9) ensures the minimum on and off time periods for the generators. Equation (9.10) identifies the start-up cost of generators. Equation (9.11) calculates the power flow and enforces the capacity limit for each branch. Equations (9.12) and (9.13) ensure the risks are limited within a certain amount.

9.3 Solving Method of Risk-Based Unit Commitment

9.3.1 Problem Reformulation

In the proposed RUC model, the nonlinear terms include the expression of E_t^{eens}, E_t^{wind} and $E_{l,t}^{flow}$, and the energy cost function, $C_i\left(P_{i,t}^c\right)$. We now describe how these nonlinear terms can be linearized.

1) Linearizing E_t^{eens} and E_t^{wind}

These two variables are given in the form of integration terms, which contain the decision variables $R_{i,t}^{up}$ and $R_{i,t}^{down}$. Their values are not known a priori before solving the RUC. We first relax Equation (9.1) as an inequality and then approximate the integration terms by a piecewise linear function of decision variables $R_{i,t}^{up}$ and $R_{i,t}^{down}$. Taking E_t^{eens} as an example, the relaxed form is

$$E_t^{eens} \ge \int_{D_t + \sum\limits_{i=1}^{I} R_{i,t}^{up}}^{\overline{D}_t} \left(x - D_t - \sum_{i=1}^{I} R_{i,t}^{up} \right) f_t^{net}(x) \mathrm{d}x. \qquad (9.14)$$

When using Equation (9.14) instead of Equation(9.1), the optimization will end in a solution in which the original equation holds; otherwise, the objective function can be further minimized. In other words, this relaxation does not change the optimal solution to the problem.

Futhermore, the integration terms of Equation (9.14) correspond to a convex function with respect to each $R_{i,t}^{up}$, $i = 1, 2, ..., I$, and the proof is as follows:

In Equation (9.1), E_t^{eens} and E_t^{wind} vary with the scheduled spinning reserves $R_{i,t}^{up}$ and $R_{i,t}^{down}$, $i = 1, 2, ..., I$ and can be seen as a multivariate function with respect to $R_{i,t}^{up}$ and $R_{i,t}^{down}$, respectively. To demonstrate their convexity, we must prove that their second-order partial derivative matrixes (Hessian matrix) with respect to the corresponding decision variables are positive semidefinite matrixes. Taking E_t^{eens} as an example, the second-order partial derivatives with respect to $R_{i,t}^{up}$ are

$$
\frac{\partial^2 E_t^{eens}}{\left(\partial R_{i,t}^{up}\right)^2} = \frac{\partial^2}{\left(\partial R_{i,t}^{up}\right)^2} \int_{D_t+\sum\limits_{i=1}^{I} R_{i,t}^{up}}^{\overline{D_t}} \left(x - D_t - \sum_{i=1}^{I} R_{i,t}^{up}\right) f_t^{net}(x) \mathrm{d}x
$$

$$
= \frac{\partial}{\partial R_{i,t}^{up}} \left[\begin{array}{c} -\int_{D_t+\sum\limits_{i=1}^{I} R_{i,t}^{up}}^{\overline{D_t}} f_t^{net}(x)\mathrm{d}x \\[2ex] -\left(D_t + \sum\limits_{i=1}^{I} R_{i,t}^{up}\right) f_t^{net}(D_t + \sum\limits_{i=1}^{I} R_{i,t}^{up}) \\[2ex] +\left(D_t + \sum\limits_{i=1}^{I} R_{i,t}^{up}\right) f_t^{net}(D_t + \sum\limits_{i=1}^{I} R_{i,t}^{up}) \end{array} \right] \tag{9.15}
$$

$$
= f_t^{net}\left(D_t + \sum_{i=1}^{I} R_{i,t}^{up}\right) \geq 0, \ t=1,2,...,T, i=1,2,...,I.
$$

According to Equation (9.15), $\partial^2 E_t^{eens}/\left(\partial R_{i,t}^{up}\right)^2$ for all $i = 1, 2, ..., I$ is non-negative because $f_t^{net}(\cdot)$ is a PDF. All of the $\partial^2 E_t^{eens}/\left(\partial R_{i,t}^{up}\right)^2, i = 1, 2, ..., I$ terms form the diagonal elements of the Hessian matrix, while the off-diagonal elements can be calculated through a similar process:

$$
\frac{\partial^2 E_t^{eens}}{\partial R_{i_1,t}^{up}\partial R_{i_2,t}^{up}} = f_t^{net}(D_t + \sum_{i=1}^{I} R_{i,t}^{up}) \equiv \frac{\partial^2 R_t^{up}}{\left(\partial R_{i,t}^{up}\right)^2}, \tag{9.16}
$$

$$
t=1,2,...,T, \ i, i_1, i_2 = 1,2,...,I, i_1 \neq i_2.
$$

Equations (9.15) and (9.16) show that all of the elements of the Hessian matrix are non-negative and identical. Therefore, this Hessian matrix is positive semidefinite, and $E_{l,t}^{eens}$ with respect to all $R_{i,t}^{up}, i=1,2,...,I$ is therefore convex. The convexity of E_t^{wind} can also be shown in a similar manner.

Therefore, Equation (9.14) can be further linearized using a K-block piecewise linear function by using the constrained cost variable (CCV) technique originally utilized in MATPOWER [32].

$$
E_t^{eens} \geq a_{t,k}^{eens} \sum_{i=1}^{I} R_{i,t}^{up} + b_{t,k}^{eens}, \ t=1,2,...,T, k=1,2,...,K_r \tag{9.17}
$$

where

$$a_{t,k}^{eens} = \frac{y_{t,k} - y_{t,k+1}}{x_{t,k} - x_{t,k+1}}, \quad b_{t,k}^{eens} = -x_{t,k} \frac{y_{t,k} - y_{t,k+1}}{x_{t,k} - x_{t,k+1}} + y_{t,k}$$

$$x_{t,k} = (k-1)\Delta r, \quad y_{t,k} = \int_{D_t + x_{t,k}}^{\overline{D}_t} (x - D_t + x_{t,k}) f_t^{net}(x) dx, \tag{9.18}$$

$$t = 1, 2, ..., T, k = 1, 2, ..., K_r.$$

In Equation (9.18), Δr is the width of each linear block and is determined by the number of piecewise blocks K_r and the range of $f_t^{net}(x)([D_t, \overline{D}_t])$. Figure 9.1 provides a visualization of the piecewise linearization using CCV. The constraint of E_t^{wind} is linearized by a similar approach:

$$E_t^{wind} \geq a_{t,k}^{wind} \sum_{i=1}^{I} R_{i,t}^{down} + b_{t,k}^{wind}, \quad t = 1, 2, ..., T, k = 1, 2, ..., K_r. \tag{9.19}$$

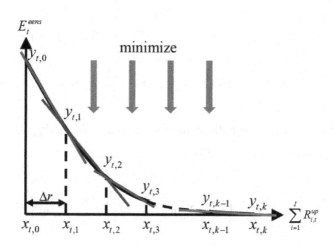

FIGURE 9.1 Piecewise linearization of using the CCV technique.

2) Linearizing $R_{l,t}^{flow}$

We will prove in the following that the $E_{l,t}^{flow}$ is convex with respect to all the $P_{q,t}^W, q = 1, 2, ..., Q$, so that the constraint of $R_{l,t}^{flow}$ can be linearized using CCV technique.

Before demonstrating the convexity of $E_{l,t}^{flow}$, its formula must be reformulated because $f_{l,t}^{flow}(\cdot)$ cannot be determined before the UC problem is solved. Assuming that the probabilistic load flow only involves the uncertainty of load and renewable energy, we introduce a variable $L_{l,t}'$ for the partial load of branch l during period t:

$$L'_{l,t} = \sum_{q=1}^{Q} g_{l,q} \hat{P}^w_{q,t} - \sum_{m=1}^{M} g_{l,m} D_{m,t}. \tag{9.20}$$

We assume that $f'_{l,t}(\cdot)$ is the PDF of $L'_{l,t}$ when considering the uncertainty of the bus load and renewable energy. We also assume that there is no wind curtailment in the day-ahead scheduling $\left(P^w_{q,t} = 0\right)$. A comparison of Equation (9.20) and Equation (9.11) shows that the difference between $L'_{l,t}$ and $L_{l,t}$ is $\sum_{i=1}^{I} g_{l,i} p^c_{i,t}$, which is a deterministic term in the probabilistic load flow calculation. Therefore, $f^{flow}_{l,t}(x)$ and $f'_{l,t}{}^{flow}(\cdot)$ have a deterministic relationship:

$$f^{flow}_{l,t}(x) = f'_{l,t}{}^{flow}\left(x - \sum_{i=1}^{I} g_{l,i} P^c_{i,t}\right). \tag{9.21}$$

By introducing an auxiliary variable $m = x - \sum_{i=1}^{I} g_{l,i} p^c_{i,t}$, $E^{flow}_{l,t}$ can be reformulated as

$$
\begin{aligned}
E^{flow}_{l,t} &= \int_{\overline{L}_l}^{\overline{L}_{l,t}} \left(x - \overline{L}_l\right) f'_{l,t}{}^{flow}\left(x - \sum_{i=1}^{I} g_{l,i} P^c_{i,t}\right) \mathrm{d}x \\
&\quad + \int_{\underline{L}_{l,t}}^{-\overline{L}_l} \left(-\overline{L}_l - x\right) f'_{l,t}{}^{flow}\left(x - \sum_{i=1}^{I} g_{l,i} P^c_{i,t}\right) \mathrm{d}x.
\end{aligned}
\tag{9.22}
$$

By introducing an auxiliary variable $m = x - \sum_{i=1}^{I} g_{l,i} P^c_{i,t}$, $E^{flow}_{l,t}$ can be reformulated as

$$
\begin{aligned}
E^{flow}_{l,t} &= \int_{\overline{L}_l - \sum_{i=1}^{I} g_{l,i} P^c_{i,t}}^{\overline{L}_{l,t} - \sum_{i=1}^{I} g_{l,i} P^c_{i,t}} \left(m + \sum_{i=1}^{I} g_{l,i} P^c_{i,t} - \overline{f}_l\right) f'_{l,t}{}^{flow}(m) \mathrm{d}\left(m + \sum_{i=1}^{I} g_{l,i} P^c_{i,t}\right) \\
&\quad + \int_{\underline{L}_{l,t} - \sum_{i=1}^{I} g_{l,i} P^c_{i,t}}^{-\overline{L}_l - \sum_{i=1}^{I} g_{l,i} P^c_{i,t}} \left(-\overline{L}_l - m - \sum_{i=1}^{I} g_{l,i} P^c_{i,t}\right) L'^{flow}_{l,t}(m) \mathrm{d}\left(m + \sum_{i=1}^{I} g_{l,i} P^c_{i,t}\right) \\
&= \int_{\overline{L}_l - \sum_{i=1}^{I} g_{l,i} P^c_{i,t}}^{\overline{L}'_{l,t}} \left(m + \sum_{i=1}^{I} g_{l,i} P^c_{i,t} - \overline{L}_l\right) L'^{flow}_{l,t}(m) \mathrm{d}m \\
&\quad + \int_{\underline{L}'_{l,t}}^{-\overline{L}_l - \sum_{i=1}^{I} g_{l,i} P^c_{i,t}} \left(-\overline{L}_l - m - \sum_{i=1}^{I} g_{l,i} P^c_{i,t}\right) L'^{flow}_{l,t}(m) \mathrm{d}m
\end{aligned}
\tag{9.23}
$$

where $\left[\overline{F}'_{l,t}, \underline{F}'_{l,t}\right]$ is the domain of $f'_{l,t}{}^{flow}(\cdot)$. The second-order partial derivatives of $E^{flow}_{l,t}$ with respect to $f'_{l,t}{}^{flow}(\cdot)$ are

$$\frac{\partial^2 E_{l,t}^{flow}}{\partial P_{i_1,t}^c \partial P_{i_2,t}^c} = g_{l,i_1} g_{l,i_2} \begin{bmatrix} L_{l,t}^{'flow}(\overline{L}_l - \sum_{i=1}^{I} g_{l,i} P_{i,t}^c) \\ + L_{l,t}^{'flow}(-\overline{L}_l - \sum_{i=1}^{I} g_{l,i} P_{i,t}^c) \end{bmatrix},$$

$$t = 1, 2, ..., T, \ i_1, i_2 = 1, 2, ..., I \tag{9.24}$$

Equation (9.24) shows that the Hessian matrix of $E_{l,t}^{flow}$ is a Gramian matrix, which is the inner product of two vectors:

$$\mathbf{H}_{E_{l,t}^{flow}} = \begin{bmatrix} s^2 g_{l,1} g_{l,1} & \cdots & s^2 g_{l,1} g_{l,I} \\ \vdots & \ddots & \vdots \\ s^2 g_{l,I} g_{l,1} & \cdots & s^2 g_{l,I} g_{l,I} \end{bmatrix} = G_l G_l^T \tag{9.25}$$

where

$$G_l = [sg_{l,1}, sg_{l,2}, \cdots, sg_{l,I}]^T$$

$$s = \sqrt{L_{l,t}^{'flow}(\overline{L}_l - \sum_{i=1}^{I} g_{l,i} P_{i,t}^c) + L_{l,t}^{'flow}(-\overline{L}_l - \sum_{i=1}^{I} g_{l,i} P_{i,t}^c)} \ . \tag{9.26}$$

According to the characteristics of Gramian matrixes, $H_{E_{l,t}^{flow}}$ is a positive semidefinite matrix, and therefore, $E_{l,t}^{flow}$ is convex with respect to all $P_{i,t}^c$, $i = 1, 2, ..., I$.

Similarly, we can proof that $E_{l,t}^{flow}$ is convex with respect to all the $P_{q,t}^w$, $q = 1, 2, ..., Q$.

Based on the above proof, the constraints of $R_{l,t}^{flow}$ can be linearized as piecewise linear functions of $P_{i,t}^c$, $i = 1, 2, ..., I$:

$$E_{l,t}^{flow} \geq a_{l,t,k}^{flow} \left(\sum_{i=1}^{I} g_{l,i} P_{i,t}^c + \sum_{q=1}^{Q} g_{l,q}(\hat{P}_{q,t}^w - P_{q,t}^w) - \sum_{m=1}^{M} g_{l,m} D_{m,t} \right) + b_{l,t,k}^{flow},$$

$$a_{l,t,k}^{flow} = \frac{y_{l,t,k} - y_{l,t,k+1}}{x_{l,t,k} - x_{l,t,k+1}}, \ b_{l,t,k}^{flow} = -x_{l,t,k} \frac{y_{l,t,k} - y_{l,t,k+1}}{x_{l,t,k} - x_{l,t,k+1}} + y_{l,t,k}$$

$$x_{l,t,k} = \underline{L}_{l,t} + (k-1)\Delta L, \ x_{l,t,k} \in [\underline{L}_{l,t} + \overline{L}_{l,t}]$$

$$y_{l,t,k} = \int_{\overline{L}_l}^{x_{l,t,k}} (x - \overline{L}_l) f_{l,t}^{flow}(x) dx + \int_{x_{l,t,k}}^{-\overline{L}_l} (-\overline{L}_l - x) f_{l,t}^{flow}(x) dx$$

$$t = 1, 2, ..., T, \ l = 1, 2, ..., I, k = 1, 2, ..., K_l. \tag{9.27}$$

3) Linearizing $C_i(p_{i,t}^c)$

The cost function is also piecewise linearized using CCV technique. Note that the generator state variable $u_{i,t}$ is employed to account for the no-load cost:

$$c_{i,t} \geq a^c_{i,t,k} P^c_{i,t} + b^c_{i,t,k} u_{i,t},$$
$$t = 1, 2, ..., T, \ i = 1, 2, ..., I, k = 1, 2, ..., K_c. \tag{9.28}$$

Equations (9.3)-(9.11) and (9.12)-(9.28) form a MILP problem for the RUC model.

9.3.2 Discussion

Note that the proposed model does not have strong assumptions on the shape of the PDF for the uncertainties ($f^{net}_t(x)$ and $f^{flow}_{l,t}(x)$). The detailed derivation shows that the only precondition for the convexity of the problem is that $f^{net}_t(x)$ and $f^{flow}_{l,t}(x)$ are non-negative, which applies for all of the PDFs, irrespective of their continuity and symmetry.

Linearization of the model does not require additional auxiliary variables and maintains the compactness of the model. Compared to the linearization model proposed in [27] which adds a series of binary variables, the proposed approach only adds several linear constraints and thus would be more computationally efficient. Indeed, compared with the currently operating unit commitment model, the proposed model only adds a minor complexity (a series of continuous variables and linear constraints) and would, therefore, be more applicable for practical large-scale power systems.

As the output of the generator is unknown before solving the UC problem, instead, when formulating the $f^{flow}_{l,t}(x)$ we calculate the partial probabilistic load flow caused only by the load and renewable energy. The spatial correlation of renewable energy uncertainties can be taken into account in the SLF calculation to more accurately evaluate the overflow risks of branches.

9.4 Case Study

9.4.1 Illustrative Example

1) Settings

The three-bus system shown in Figure 9.2 is used as an illustrative example. The system data were extracted from [8]. The generation parameters are modified as shown in Table 9.1, where P is the output of the generator. A four-hour market clearing is considered. The load is located on bus #3 with an hourly load of 30, 80, 110 and 50 MW. Two wind farms are connected to buses #1 and #2. The probabilistic forecasts for each hour are shown in Figure 9.3. The rank correlation of their forecast error is 0.71. θ^{eens}, θ^{wind} and θ^{flow} are all set as \$50/MW, which is ten times the value of the highest reserve offering cost. Only the uncertainty of renewable energy is considered in this example. A 10-block piecewise linear function is considered for the constraints of R_t^{eens}, R_t^{wind} and $R_{l,t}^{flow}$ (Equations (9.17), (9.19) and (9.27)).

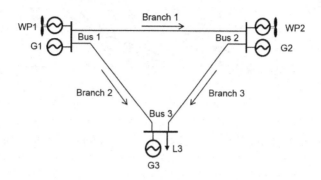

FIGURE 9.2 Three-bus system.

TABLE 9.1 Generator parameters of the three-bus system

Generator	1	2	3
$\overline{P}_i$/MW	90	90	50
$\underline{P}_i$/MW	10	10	10
o_i^{on}/\$	100	100	100
$C_i(P)$/\$/MWh	$40P + 450$	$40P + 360$	$30P + 250$
c_i^{up}/\$/MW	5	4	4
c_i^{down}/\$/MW	5	4	4

2) Unit commitment results

The unit commitment results are listed in Table 9.2. The results show that the up and down spinning reserves are dynamically scheduled according to the distribution of the renewable energy forecast error. In the first hour, only 0.2 MW of down reserve is contracted because the probabilistic wind forecast is biased and the probability of underestimating the renewable energy is very low. The scheduled up and down reserves increase from the 1st hour to the 4th hour due to the increasing uncertainty of the renewable energy.

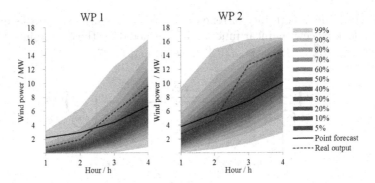

FIGURE 9.3 Probabilistic forecast of wind power and the actual output.

TABLE 9.2 Unit commitment results of the three-bus system

		$1^{st}h$	$2^{nd}h$	$3^{rd}h$	$4^{th}h$
	G1	0	0	0	0
Generation/ MW	G2	23.98	61.48	69.93	23.22
	G3	0	10	28.25	10
	G1	0	0	0	0
Up reserve/ MW	G2	5.3	6.7	7.7	8.3
	G3	0	0	0	0
	G1	0	0	0	0
Down reserve/ MW	G2	0.2	4	8.7	12.1
	G3	0	0	0	0
	Branch #1	−8.51	−21.36	−24.34	−8.86
Branch power flow/ MW	Branch #2	10.74	24.32	28.71	15.57
	Branch #3	19.26	45.68	53.04	24.43

3) Sensitivity analysis of the cost coefficients

Figure 9.4 shows the piecewise linear function constraints of R_t^{eens} and R_t^{wind} used in the model. They clearly demonstrate how the risks are recognized according to the probabilistic forecast and how they vary with the reserve scheduling. We change the values of the cost coefficient for both load shedding and wind curtailment (θ^{eens} and θ^{wind}) from \$50/MWh to $+\infty$ and list the results of the up and down reserve schedules in Figure 9.5. Both the up and down reserves increase as the cost coefficient increases, suggesting that the amount of the committed reserve is determined by how we value the risks.

The value of θ^{wind} can be chosen as the Value of Loss Load (VOLL), e.g. \$3500/MWh for MISO [33]. The value of θ^{wind} can be chosen as the expected social welfare losses when curtailing renewable energy.

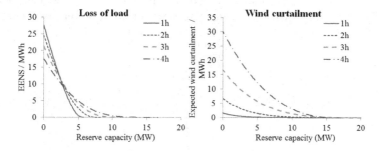

FIGURE 9.4 Risks of loss of load and wind curtailment for different scheduled reserve capacities.

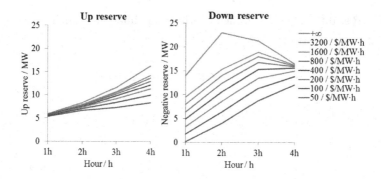

FIGURE 9.5 Up and down reserve schedules under different cost coefficients of load shedding and wind curtailment.

We now explore the effect of introducing the overflow risk $E_{l,t}^{flow}$. Due to the hard constraints of the power flow of Branch #3, G2 is contracted by only 69.93 MW for the 3rd hour instead of 90 MW. In the conventional deterministic UC model, the power flow of Branch #3, in this case, would be scheduled at its capacity limit. However, in this result, the scheduled power flow of Branch #3 is only 53.04 MW, which has a margin of 1.96 MW compared to its capacity limit. This margin is held for Branch #3 so that the renewable energy at Buses #1 and #2 will be more likely be higher than forecasted and Branch #3 may, therefore, be congested. This outcome is also verified by the pricewise linear risk constraints of the power flow shown in Figure 9.6 where the capacity of all the branches is the same and equal to 55 MW in Figure 9.6. Branch #3 is at risk when its power flow is higher than 50 MW. Figure 9.7 compares the scheduling of G2 under different overflow cost coefficients θ^{flow}. The contract for G2 decreases from 72.86 MW (when ignoring the risk) to 66.02 MW as the cost of overflow increases. The total generating cost is therefore increased as the cost coefficients increases. The results imply that the cost coefficient θ^{flow} can be tuned to control the conservativeness of power flow constraints.

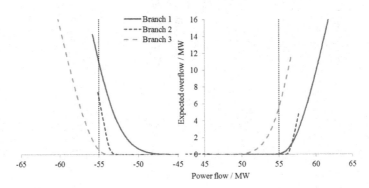

FIGURE 9.6 Expected branch overflow for different branch power flows in the 3rd hour.

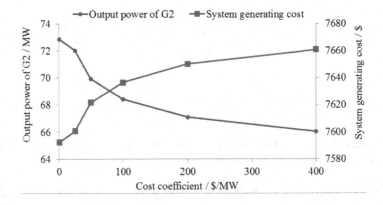

FIGURE 9.7 Output of G2 and the system generating cost under different overflow cost coefficients.

It should be noted that we can also use the hard constraints of $\overline{E}_t^{eens}$, $\overline{E}_t^{wind}$, and $\overline{E}_{l,t}^{flow}$ to limit the expected risks. The value of $\overline{E}_t^{eens}$ can be chosen according to the reliability requirement of the power system. The value of $\overline{E}_t wind$ can be set according to the regulations that may require a minimum purchase ratio (or maximum curtailment ratio) for wind energy. Sometimes it makes more sense to set one risk constraint over all the hours, i.e. $\displaystyle\sum_{t=1}^{T} E_t^{eens} \leq$ $\overline{E}_t^{eens}$. The value of $\overline{E}_{l,t}^{flow}$ can be set according to the maximum allowed short-term overflow limits of the transmission lines.

4) Sensitivity analysis of the segment number of piecewise linear approximation

Different numbers of the blocks in linearizing E_t^{eens} , E_t^{wind} and $E_{l,t}^{flow}$ were chosen to test the sensitivity of the solution on the precision of the piecewise linear approximation. The results of up and down reserve schedules were chosen to demonstrate how the solution changes with the numbers of the blocks, as shown in Table 9.3. The results with 1000 blocks can be seen as the benchmarks of the precise solution. The results show that the solution would deviate from the benchmark when we choose less piecewise blocks. However, the results are quite stable when the number of blocks is above 10. The results suggest that choosing the number between 10 and 50 would be acceptable for the practical application. It should also be noted that we can use variable intervals in the linearization to improve the approximation precision. A higher resolution of approximation can be applied where the risk function has greater nonlinearity.

TABLE 9.3 Comparison of reserve schedules with different number of block for the piecewise linear function

Up reserve					
Number of segment	5	10	50	100	1000
$1^{st} h$	6.0500	5.1830	5.2204	5.2395	5.2000
$2^{nd} h$	6.8647	6.6000	6.8935	6.7069	6.7060
$3^{rd} h$	7.7000	7.2500	7.8115	7.6500	7.7033
$4^{th} h$	8.0500	8.2500	8.4000	8.2500	8.3059
Down reserve					
Number of segment	5	10	50	100	1000
$1^{st} h$	0.0000	0.2000	0.3071	0.1500	0.2045
$2^{nd} h$	3.3500	3.3000	3.2958	3.1500	3.2144
$3^{rd} h$	9.0500	8.7000	8.7137	8.7028	8.7085
$4^{th} h$	12.220	11.950	12.010	12.021	12.050

9.4.2   Case Study

The model was also tested and compared with deterministic UC (DUC) and the SUC model using IEEE RTS-79 system. Grid and generator data were extracted from [34]. Two wind farms are connected to Buses #16 and #17 with a capacity of 180 MW for each bus. The probabilistic forecast of wind power was extracted from the case study of [30]. Only the uncertainty of wind power is considered. The power flow limitation of each branch is set to 50% of the value given by MATPOWER [32].

In the RUC model, θ^{eens}, θ^{wind} and θ^{flow} are all set as \$50/MW. A 10-block piecewise linear function is considered for the function of E_t^{eens}, E_t^{wind} and $E_{l,t}^{flow}$. Besides the proposed RUC model, a DUC model and two SUC models were studied using the same test system for comparison. In the DUC model, the up and down reserve requirements for wind power are set according to the 1% and 99% quantiles of the probabilistic wind forecast, similar to the method used in [35]. For the SUC models, we consider both models from [7] (namely SUC1) and from [8] (namely SUC2). The difference between SUC1 and SUC2 is that SUC1 only includes the first-stage unit state variable, while SUC2 considers both first stage and second stage unit state variables. In other words, SUC1 assumes that the states of all the generators stay the same in all the scenarios, while SUC2 allows the changes of generation states in different scenarios. The number of binary variables of SUC2 is therefore much larger than that of SUC1. In both models, 3000 scenarios of wind power output are generated and are reduced to 10 scenarios. The penalties of the load shedding, wind curtailment and branch overflow for the second stage decision are considered case-wise in the objective function of the SUC models. The penalty coefficients (θ^{eens}, θ^{wind} and θ^{flow}) are identical with those in the RUC. A 24-hour market clearing was conducted using each model. All four models were run using CPLEX 12.5 under MATLAB® on a Windows-based PC with four threads clocking at 2.5 GHz and 4 GB RAM.

Table 9.4 compares the performance of the four UC models in terms of problem scales, computing time and unit commitment results. The results show that the RUC has 960 additional continuous variables and 9600 additional constraints compared to the DUC. The computing time with both MILP gap targets of 1% and 0.1% shows that the RUC doubles the computing expense of DUC. The computing time of RUC is significantly lower than that of SUC1 and SUC2. The reasons are twofold: 1) The computation efficiency of the MILP problem is sensitive to the number of variables, especially binary variables. Though the number of binary variable in RUC and SUC1 is the same, SUC1 still has 24000 more continuous variables than RUC. 2) There are also more constraints in SUC than in RUC, especially the constraints that link the first stage variables and second stage variables. These variables jeopardize the structure of the coefficient matrix from being a block diagonal matrix, which would raise the difficulty of solving the problem.

TABLE 9.4 Comparison of results and statistics among different models

	DUC	RUC	SUCI	SUC2
Problem scales				
Number of binary variables	624	624	624	6864
Number of continuous variables	3120	4080	28080	28080
Total number of variables	3744	4704	28704	34944
Number of constraints	5924	15524	70924	77164
Computing time /s				
MILP gap = 1%	5.52	10.40	286.39	518.75
MILP gap = 0.1%	56.76	98.96	6562.55	> 10000*
Unit commitment results				
$Total cost/10^3$\$	1183.5764	1182.5962	1184.3576	1182.6458
Probability of load shedding / p.u.	0.0253	0.0274	0.0299	0.0264
Probability of load shedding / p.u.	0.0253	0.0274	0.0299	0.0264
Average hourly EENS/MWh	0.3975	0.2316	0.2805	0.2241
Probability of wind curtailment / p.u.	0.0231	0.0171	0.0161	0.0162
Expected hourly wind power curtailment / MWh	0.5045	0.3913	0.3713	0.3582
Probability of branch overflow / p.u.	0.0745	0.0584	0.0689	0.0576
Probability of branch overflow / p.u.	0.7514	0.6101	0.6821	0.6052

* The SUC2 converges with the gap of 0.39% when calculation has been running for 10000 seconds.

The unit commitment results are compared in terms of the total cost and risk levels. The total cost includes the generating cost, on-off cost, and reserve offering cost. The cost associated with the risks in the RUC model and the cost associated with the second-stage variable in the SUC are neglected for a fairer comparison. The metrics of evaluating the risks of the unit commitment solution include the probability and expectation of the load shedding, wind curtailment and branch overflow are calculated on an hourly average scale. The results show that the market clearing obtained from these four models varies in cost and risk levels. The scheduling of RUC, SUC1, and SUC2 are less risky than that of the DUC because of the detailed modeling of the uncertainty of wind power. The risk level of RUC, SUC1, and SUC2 are similar. However, SUC1 has a higher cost than the other two models. The reason why RUC has a lower cost than SUC1 is that RUC has a more precise scheduling of reserves and power flow margins by considering the detailed PDF of the wind power, compared with the scenarios considered in SUC1. Therefore RUC would obtain a more cost-effective scheduling decision with a similar risk level. The reason why SUC2 also has lower cost than SUC1 is that it allows the change of unit status between different scenarios. To sum up, the result confirms that the RUC facilitates a strategic market clearing in terms of bring down the operation risk of the power system.

9.5 Summary

This chapter proposes an RUC model for market clearing considering renewable energy uncertainty. The risks of loss of load, wind curtailment, and branch overflow are considered in both the objective function (as penalties) and the constraints (with a limit). The terms associated with the risks in the UC model are relaxed and linearized relying on their convex function features. The entire RUC model is then transformed into a MILP problem. An illustrative example and case study show how the model improves the generated schedule by recognizing the risks with different reserve and branch power flow margins. The linearization of the model does not make strong assumptions on the distribution of the uncertainty and is thus widely applicable. The proposed UC model addresses the uncertainty of renewable energy in a compact way, such that the model can be solved with high efficiency and can be adapted to larger power systems.

It is worth noting that the proposed model does not consider the wait-and-see variables and constraints associated with actual system operation and may overlook the detailed modeling of corrective controls in actual system operation. Nevertheless, the proposed risk modeling approach can also be used when incorporating wait-and-see variables in UC models. This topic deserves further research.

Bibliography

[1] C. Lowery and M. O'Malley. Impact of wind forecast error statistics upon unit commitment. *IEEE Transactions on Sustainable Energy*, 3(4):760–768, 2012.

[2] M. Milligan, E. Ela, D. Lew, D. Corbus, Y. Wan, B. M. Hodge, and B. Kirby. Operational analysis and methods for wind integration studies. *IEEE Transactions on Sustainable Energy*, 3(4):612–619, 2012.

[3] J. C. Smith, S. Beuning, H. Durrwachter, and E. Ela. The wind at our backs. *IEEE Power and Energy Magazine*, 8(5):63–71, 2010.

[4] M. Ahlstrom, D. Bartlett, C. Collier, J. Duchesne, D. Edelson, A. Gesino, M. Keyser, D. Maggio, M. Milligan, and C. Möhrlen. Knowledge is power: Efficiently integrating wind energy and wind forecasts. *IEEE Power and Energy Magazine*, 11(6):45–52, 2013.

[5] K. R. W. Bell, D. P. Nedic, and L. A. Salinas San Martin. The need for interconnection reserve in a system with wind generation. *IEEE Transactions on Sustainable Energy*, 3(4):703–712, 2012.

[6] L. Wu, M. Shahidehpour, and T. Li. Stochastic security-constrained unit commitment. *IEEE Transactions on Power Systems*, 22(2):800–811, 2007.

[7] F. Bouffard and F. D. Galiana. Stochastic security for operations planning with significant wind power generation. *IEEE Transactions on Power Systems*, 23(2):306–316, 2008.

[8] J. M. Morales, A. J. Conejo, and J. Perez-Ruiz. Economic valuation of reserves in power systems with high penetration of wind power. *IEEE Transactions on Power Systems*, 24(2):900–910, 2009.

[9] A. Tuohy, P. Meibom, E. Denny, and M. O'Malley. Unit commitment for systems with significant wind penetration. *IEEE Transactions on Power Systems*, 24(2):592–601, 2009.

[10] C. D. Jonghe, B. F. Hobbs, and R. Belmans. Value of price responsive load for wind integration in unit commitment. *IEEE Transactions on Power Systems*, 29(2):675–685, 2014.

[11] L. Wu, M. Shahidehpour, and Z. Li. Comparison of scenario-based and interval optimization approaches to stochastic SCUC. *IEEE Transactions on Power Systems*, 27(2):913–921, 2012.

[12] R. Jiang, J. Wang, and Y. Guan. Robust unit commitment with wind power and pumped storage hydro. *IEEE Transactions on Power Systems*, 27(2):800–810, 2012.

[13] C. Zhao, J. Wang, J. P. Watson, and Y. Guan. Multi-stage robust unit commitment considering wind and demand response uncertainties. *IEEE Transactions on Power Systems*, 28(3):2708–2717, 2013.

[14] A. Kalantari, J. F. Restrepo, and F. D. Galiana. Security-constrained unit commitment with uncertain wind generation: The loadability set approach. *IEEE Transactions on Power Systems*, 28(2):1787–1796, 2013.

[15] Y. Wang, Q. Xia, and C. Kang. Unit commitment with volatile node injections by using interval optimization. *IEEE Transactions on Power Systems*, 26(3):1705–1713, 2011.

[16] R. Jiang, J. Wang, M. Zhang, and Y. Guan. Two-stage minimax regret robust unit commitment. *IEEE Transactions on Power Systems*, 28(3):2271–2282, 2013.

[17] C. Zhao and Y. Guan. Unified stochastic and robust unit commitment. *IEEE Transactions on Power Systems*, 28(3):3353–3361, 2013.

[18] Q. Wang, Y. Guan, and J. Wang. A chance-constrained two-stage stochastic program for unit commitment with uncertain wind power output. *IEEE Transactions on Power Systems*, 27(1):206–215, 2012.

[19] D. Pozo and J. Contreras. A chance-constrained unit commitment with an $n - k$ security criterion and significant wind generation. *IEEE Transactions on Power Systems*, 28(3):2842–2851, 2013.

[20] M. Vrakopoulou, K. Margellos, J. Lygeros, and G. Andersson. A probabilistic framework for reserve scheduling and $n - 1$ security assessment of systems with high wind power penetration. *IEEE Transactions on Power Systems*, 28(4):3885–3896, 2013.

[21] D. He, Z. Tan, and R. G. Harley. Chance constrained unit commitment with wind generation and superconducting magnetic energy storages. *Power and Energy Society General Meeting*, pages 1–6, 2012.

[22] J. Hetzer, D. C. Yu, and K. Bhattarai. An economic dispatch model incorporating wind power. *IEEE Transactions on Energy Conversion*, 23(2):603–611, 2008.

[23] B. Venkatesh, P. Yu, H. B. Gooi, and D. Choling. Fuzzy MILP unit commitment incorporating wind generators. *IEEE Transactions on Power Systems*, 23(4):1738–1746, 2008.

[24] W. Zhou, H. Sun, and Y. Peng. Risk reserve constrained economic dispatch model with wind power penetration. *Energies*, 3(12):1880–1894, 2010.

[25] G. Liu and K. Tomsovic. Quantifying spinning reserve in systems with significant wind power penetration. *IEEE Transactions on Power Systems*, 27(4):2385–2393, 2012.

[26] R. Entriken, P. Varaiya, F. Wu, J. Bialek, C. Dent, A. Tuohy, and R. Rajagopal. Risk limiting dispatch. *Power and Energy Society General Meeting*, pages 1–5, 2012.

[27] K. Liu and J. Zhong. Generation dispatch considering wind energy and system reliability. *2010 IEEE PES General Meeting*, pages 1–7, 2010.

[28] Z. Zhou and A. Botterud. Dynamic scheduling of operating reserves in co-optimized electricity markets with wind power. *IEEE Transactions on Power Systems*, 29(1):160–171, 2014.

[29] P. Pinson, C. Chevallier, and G. N. Kariniotakis. Trading wind generation from short-term probabilistic forecasts of wind power. *IEEE Transactions on Power Systems*, 22(3):1148–1156, 2007.

[30] N. Zhang, C. Kang, Q. Xia, and J. Liang. Modeling conditional forecast error for wind power in generation scheduling. *IEEE Transactions on Power Systems*, 29(3):1316–1324, 2014.

[31] J. Usaola. Probabilistic load flow in systems with wind generation. *IET Generation Transmission and Distribution*, 3(12):1031–1041, 2009.

[32] R. D. Zimmerman, C. E. Murillo-Sanchez, and R. J. Thomas. Matpower's extensible optimal power flow architecture. *IEEE Power and Energy Society General Meeting, 2009. PES '09.*, pages 1–7, 2009.

[33] W. W. Hogan. Electricity scarcity pricing through operating reserves. *Economics of Energy and Environmental Policy*, 2(2), 2014.

[34] P. M Subcommittee. IEEE reliability test system. *IEEE Transactions on Power Apparatus and Systems*, PAS-98(6):2047–2054, 1979.

[35] A. Botterud, Z. Zhou, J. Wang, J. Sumaili, H. Keko, J. Mendes, R. J. Bessa, and V. Miranda. Demand dispatch and probabilistic wind power forecasting in unit commitment and economic dispatch: A case study of Illinois. *IEEE Transactions on Sustainable Energy*, 4(1):250–261, 2012.

10

Managing Renewable Energy Uncertainty in Electricity Market

With the rapidly increasing penetration of wind power, wind producers are becoming increasingly responsible for the deviation of the wind power output from the forecast. Such uncertainty results in revenue losses to the wind power producers (WPPs) due to penalties in ex-post imbalance settlements. This chapter explores the opportunities available for WPPs if they can purchase or schedule some reserves to offset part of their deviation rather than being fully penalized in the real-time market. The revenue for WPPs under such mechanism is modeled. The optimal strategy for managing the uncertainty of wind power by purchasing reserves to maximize the WPP's revenue is analytically derived with rigorous optimality conditions. The amount of energy and reserves that should be bid in the market are explicitly quantified by the probabilistic forecast and the prices of the energy and reserves. A case study using the price data from ERCOT and wind power data from NREL is performed to verify the effectiveness of the derived optimal bidding strategy and the benefits of reserve purchasing. Additionally, the proposed bidding strategy can also reduce the risk of variations on WPP's revenue.

10.1 Overview

The stochastic nature of wind power brings about significant challenges to power system operation and electricity market management. Wind power requires extra flexibility from other conventional generating units to ensure the required level of security and reliability for the power system [1, 2] . The extra flexibility requirement is commonly fulfilled by additional operation reserves and frequency regulations. In general, these ancillary service supplies would increase the utility company's operating costs and therefore passed onto consumers. As the integration of renewable energy increases, these costs are suggested to be paid by the wind power producers (WPPs). Penalizing the deviations of wind power is an effective way of shifting such costs from the utility company (consumer) to WPPs. In some United States (U.S.) and European markets, WPPs are now requested to commit themselves to the dispatched levels and are obligated to pay monetary penalties for the actual wind power deviation from their schedules [3]. Each electricity pool has its own special rules for renewable resources. In the Nord Pool, WPP is qualified to trade on the spot market and is subjected to full deviation penalties on the imbalance market. When the system is short of generation, only under-generation is penalized, while excess-generation is settled at the clearing price of the imbalance market, and vice versa [4]. In the PJM power market, a WPP sells energy through a two-settlement electricity market, in the same manner as conventional units. The wind power generation deviation is settled at a combined price of the real-time (RT) marginal price and the operating reserve deviation charge [5]. The uncertainty of wind power is costly to WPPs. In Spain, the WPP's revenue loss from deviation penalties is estimated as being up to 10% of the total revenue [6].

The WPP's revenue loss associated with the uncertainty of wind power can be mitigated through strategically bidding in the market [7]. Morales, et al. [8] established a two-stage stochastic optimization model to numerically obtain optimal bids that control the risks of profit variability. Matevosyan, et al. [9] introduced the scenario generation technique to represent the uncertainty of both wind power and market prices and further solved the optimal bidding strategy using the stochastic optimization problem. Botterud, et al. [10] applied the CVaR index to model the risk preference of a WPP and compared the performances of different day-ahead bidding strategies. Based on the wind power probabilistic forecasts, Pinson, et al. [11] and Bitar, et al. [12] analytically derived the optimal day-ahead (DA) energy bidding strategy that maximizes the expectation of revenue. The formulation of revenue loss associated with forecast errors is also derived and quantified in [12]. Instead of using wind power probabilistic forecasts, Bathurst et al. [13] applied the Markov process to simulate the wind power generation and proposed the corresponding optimal bidding strategy. A more comprehensive optimal bidding strategy described in [14] considers the asymmetry of imbalance penalty prices. The strategies depicted in [15] consider the statistical correlation between the wind and imbalance prices. Another effective method to manage wind power uncertainty is to combine a wind plant and storage devices as a virtual power plant (VPP). The role of storage and the optimal joint bidding strategy for a virtual plant were analyzed in [16]. The issue of optimal energy storage sizing was discussed in [17] and [18]. In addition, considering the complementary effect of multiple wind farms, the combined bidding strategy was introduced to reduce uncertainty penalties by submitting the combined forecasts of multiple wind farms as one bid [6].

Most of the prior research has studied the optimal bidding of wind power in a conventional two-settlement energy trading structure including a DA forward market and an RT spot market. In such a market, the ISO is responsible for scheduling and deploying reserves for all the deviations from advance schedules, and the WPPs are required to submit energy offers in DA market and are assumed to be penalized for the deviation [19]. Furthermore, it is possible for WPPs to improve their revenues if they can use more inexpensive reserves to mitigate the uncertainty of the wind power [20, 21]. In fact, research exists regarding the association of a WPP with storage devices or other flexible resources to form a VPP with a quite stable and controllable overall output [20]. Dai et al. [21] introduced a new bilateral reserve market in which wind power producers are allowed to buy energy from conventional power producers to minimize the risk of losing money due to their generation uncertainty. Liang et al. [3] explored the benefit of allowing WPPs to participate in the regulation reserve market with lower deviation penalties compared with that of the energy market, showing it will increase the WPPs' revenue and improve the system security. These studies show that employing its own reserves could benefit a WPP compared with paying the imbalance prices in the market.

It is of great interest to determine how much reserves the WPP should purchase from the DA reserve market (if qualified to participate in) or schedule from other resources. Methodologies of determining the reserve needed by a wind power producer have been widely discussed. When considering wind power integration, it is widely accepted that the reserve capacity should be determined by probabilistic approaches rather than deterministic approaches. From the perspective of the system operator, Holttinen, et al. [22] proposed a 3σ criterion to calculate the reserve requirements based on the combined distribution of wind and load deviations. Doherty and O'Malley [23] introduced a reliability criterion defined as the number of load shedding incidents tolerated per year to determine the effect of increasing wind power penetration. Ortega-Vazquez and Kirschen [24] proposed a method of determining the reserve requirements by comparing the cost of providing the reserves and the benefits of deploying the reserves. A stochastic unit commitment was proposed in [25–27] to optimally schedule the reserves while committing the generators.

However, all of these reserve scheduling approaches focus on the reserve requirement from the overall system perspective. This chapter focuses on the wind power optimal bidding strategy under the circumstances that a WPP commits some of its own reserves to offset its deviations, instead of entirely being penalized in the ex-post imbalance market. In this paper, we build a general model that considers a WPP is able to purchase reserve and bid energy in the DA market. The optimal amount of DA reserve purchase is analytically derived together with the optimal energy bidding strategy.

10.2 Market Model for Wind Power

10.2.1 Market Model

Most electricity markets in the U.S. are operated by the independent system operator (ISO) or regional transmission organization (RTO), employing a two-settlement energy clearing mechanism based on locational marginal prices (LMP) [28]. Specifically, this common pool energy trading structure consists of two successive ex-ante markets: a DA forward market and an RT spot market. DA market allows suppliers to submit energy offers for the following operating day. Once the market is cleared, the energy transactions are financially bound and settled with the DA LMP. RT spot market is employed after the close of DA market and cleared several minutes before actual power delivery to ensure the real-time balance between generation and demand by allowing market participants to adjust their DA schedules. The deviation between the DA schedule and RT delivery is settled at the RT LMP, which is determined by the ex-post actual energy delivery results.

For the market like the Nord Pool and UK market, the market structure mainly consists of an ex-post spot market and ex-post regulation market run by transmission system operator (TSO) [29]. The revenue of a participant is composed of the income on the spot market and the cost of imbalances. The former is determined by the energy contract fixed in the spot market and settled by spot price, while the latter is determined by the deviation between the advance schedule and actually generation and settled by up-regulation and down-regulation prices for the energy shortage and excessing respectively. A general model for WPP participating electricity market is established in this paper for the purpose of capturing the common characteristics of handling the wind power uncertainty in different markets. The revenue of a WPP π with DA schedule P_w^{DA} and RT delivery P_w^{RT} can be expressed as the following.

$$
\pi = \lambda_{DA} P_w^{DA} + \begin{cases} \lambda_{RT}^+ \left(P_w^{RT} - P_w^{DA} \right) , & P_w^{DA} \geq P_w^{RT} \\ \lambda_{RT}^- \left(P_w^{RT} - P_w^{DA} \right) , & P_w^{DA} < P_w^{RT} \end{cases} \tag{10.1}
$$

where λ_{DA} is the DA clearing price, λ_{RT}^+ and λ_{RT}^- are imbalance prices for positive and negative deviations.

In this paper, a two-price balancing market is considered, which means that the RT deviations are priced differently depending on the imbalance sign. Only deviations opposite in sign from the system imbalance are exempted and settled at the DA spot price, while deviations with the same sign are settled at the clearing price of the RT balancing market. Such a design is common in European electricity markets. Compared with the one-price balancing mechanism (deviations settled at RT price in all conditions), there is no punishment for the deviation that contributes to restoring the system balance and no arbitrage opportunity for a stochastic producer. This makes a two-price mechanism more reasonable.

$$
\lambda_{RT}^+ = \begin{cases} \lambda_{RT}, & if \ \lambda_{RT} \geq \lambda_{DA} \\ \lambda_{DA}, & if \ \lambda_{RT} < \lambda_{DA} \end{cases} \tag{10.2}
$$

$$
\lambda_{RT}^- = \begin{cases} \lambda_{DA}, & if \ \lambda_{RT} \geq \lambda_{DA} \\ \lambda_{RT}, & if \ \lambda_{RT} < \lambda_{DA} \end{cases} \tag{10.3}
$$

In the above two equations, λ_{RT} is RT clearing price. $(\lambda_{RT}^+, \lambda_{RT}^-)$ and λ_{RT} are decided by the load-generation balance of the entire system and the network congestion, and thus are highly variable. From the WPP's perspective, they are stochastic parameters with uncertainty.

10.2.2 Assumptions in This Study

1. The WPP is assumed to participate in the DA forward market and submit energy offers for the next operating day. The submitted offers are fully cleared in the DA market.

2. The WPP is assumed to be a price-taker in the ex-ante DA market. Because market rules allow power producers to differentiate the offered quantity depending on the day-ahead (DA) clearing price, the DA market price λ_{DA} is assumed to be known when analytically deriving the optimal bidding strategy.

3. λ_{RT}^+ and λ_{RT}^- are very hard to be precisely forecasted so that their expectations are used in this study, namely $\bar{\lambda}_{RT}^+$ and $\bar{\lambda}_{RT}^-$.

4. The probabilistic forecasting result of wind power is available before bidding in the DA market. The uncertainty of the wind power deviations is considered in this study.

5. The RT market prices are assumed to be independent with the offering of an individual wind producer.

6. The WPP is risk-neutral, meaning that it aims at the maximization of the expected profits.

10.2.3 Trading Wind Energy in Electricity Market

Considering the RT price uncertainty, the conditional expected revenue of a WPP with the DA schedule P_w^{DA} and RT delivery P_w^{RT} can be formulated according to 10.1 as:

$$E(\pi \,|\, P_w^{RT}) = \lambda_{DA} P_w^{DA} + E(\pi_w^{RT} \,|\, P_w^{RT}) \tag{10.4}$$

where $E(\pi \,|\, P_w^{RT})$ denotes the conditional expected revenue from the electricity market through energy trading with a possible wind power realization P_w^{RT}, and $E(\pi_w^{RT} \,|\, P_w^{RT})$ denotes the conditional expected revenue in the real-time imbalance market. The conditional expected revenue $E(\pi_w^{RT} \,|\, P_w^{RT})$ from the ex-post imbalance market is given by

$$E(\pi_w^{RT} \,|\, P_w^{RT}) = \begin{cases} \bar{\lambda}_{RT}^+ (P_w^{RT} - P_w^{DA}) \,, & P_w^{RT} \leq P_w^{DA} \\ \bar{\lambda}_{RT}^- (P_w^{RT} - P_w^{DA}) \,, & P_w^{RT} \geq P_w^{DA} \end{cases} \tag{10.5}$$

It should be noted that the conditionally expected revenue $E(\pi \,|\, P_w^{RT})$ can also be rewritten as the following.

$$E(\pi \,|\, P_w^{RT}) = \lambda_{DA} P_w^{RT} - \begin{cases} \left(\bar{\lambda}_{RT}^+ - \lambda_{DA}\right)\left(P_w^{RT} - P_w^{DA}\right), P_w^{RT} \leq P_w^{DA} \\ \left(\lambda_{DA} - \bar{\lambda}_{RT}^-\right)\left(P_w^{RT} - P_w^{DA}\right), P_w^{RT} \geq P_w^{DA} \end{cases}$$

$$(10.6)$$

Thus, the equivalent expected imbalance penalty for each MWh of wind power deviation would be

$$\begin{cases} \lambda_{short} = \bar{\lambda}_{RT}^+ - \lambda_{DA}, \ P_w^{RT} \leq P_w^{DA} \\ \lambda_{excess} = \lambda_{DA} - \bar{\lambda}_{RT}^-, \ P_w^{RT} \geq P_w^{DA} \end{cases}$$

$$(10.7)$$

where λ_{short} and λ_{excess} are the expected cost of each MWh of short-generation and the excess-generation, respectively.

The level of $\bar{\lambda}_{RT}^+$ and $\bar{\lambda}_{RT}^-$ depends on the real balance of supply and demand. Generally, $\bar{\lambda}_{RT}^+$ is more likely to be higher than λ_{DA}, while $\bar{\lambda}_{RT}^-$ is more likely to be lower than λ_{DA}, reflecting the cost of operational reserves responding to the deviation. Therefore, λ_{short} and λ_{excess} are more likely positive values than negative ones.

10.2.4   Reserve Purchasing for a WPP

Considering the wind power generation uncertainty, equation 10.7 suggests that a WPP cannot fully harvest its revenue from its actual power output with the LMP due to the unavoidable inaccuracy of the forecast. Obviously improving the forecast could reduce the revenue losses. Alternatively, it is also possible to reduce the losses by reducing the imbalance prices. Therefore, we consider the situation where the WPP is able to purchase some reserve capacity to offset its deviations. The reserves can be procured from conventional generation, demand response, or self-owned storage devices. Such a mechanism will benefit the WPP if the cost of reserves is less than the imbalance prices. Although it is widely accepted that it is most cost-effective to schedule reserves from a system perspective, allowing the WPP to schedule its own reserves is also justified for three reasons: 1) the WPP may combine with other flexible sources, which could internalize the deviation cost; 2) the imbalance penalties in some market may be higher than the cost of deploying reserves; and 3) the mechanism helps to discover the reserves truly needed by wind power by providing incentives to managing its uncertainty.

We now derive the WPP's revenue under such a mechanism. Assume that the WPPs purchase up/down optional reserve with the amount of R_U and R_D (MW) at capacity prices γ_U, γ_D (\$/MWh) in the reserve market or from other resources. The subscript U and D denote up-reserve and down reserve, respectively. The WPPs are charged at reserve deploy prices λ_U and λ_D (\$/MWh) for RT reserve deploy to compensate RT deviation. The income of selling energy and the cost of purchasing the reserves can be written as:

$$\pi_w^R = \lambda_{DA} P_w^{DA} + \begin{cases} \lambda_{RT}^-(P_w^{RT} - P_w^{DA} + R_U) \,, P_w^{RT} \le P_w^{DA} - R_U \\ 0 \,, P_w^{DA} - R_U \le P_w^{RT} \le P_w^{DA} + R_D \\ \lambda_{RT}^+(P_w^{RT} - P_w^{DA} - R_D) \,, P_w^{DA} + R_D \le P_w^{RT} \end{cases}$$

$$c_w^R = \gamma_U R_U + \gamma_D R_D + \begin{cases} \lambda_U R_U \,, P_w^{RT} \le P_w^{DA} - R_U \\ \lambda_U (P_w^{DA} - P_w^{RT}) \,, P_w^{DA} - R_U \le P_w^{RT} \le P_w^{DA} \\ -\lambda_D (P_w^{RT} - P_w^{DA}) \,, P_w^{DA} \le P_w^{RT} \le P_w^{DA} + R_D \\ -\lambda_D R_D \,, P_w^{DA} + R_D \le P_w^{RT} \end{cases}$$

$$(10.8)$$

where π_w^R and c_w^R denote the income of selling energy and the cost of purchasing reserve respectively. When the actual wind power deviation is less than $R_U(R_D)$, the deviation can be completely absorbed by the scheduled reserve. The deployed reserve energy is up to $R_U(R_D)$ when the deviations are beyond the purchased reserve capacity. The exceeded deviations are then settled with $\lambda_{RT}^+(\lambda_{RT}^-)$.

The WPP can procure reserves from either a coupled storage devices or self-owned controllable generator, a bilateral reserve trading market, or even a centralized reserve market. When cooperating with a storage device or self-owned controllable generator, reserve prices represent the cost of investing and operating these sources. In bilateral reserve trading market, reserve prices will be determined by double auction. In DA reserve market, reserve capacity prices reflect the opportunity cost of generations. When calculating the bidding strategy, their prices can be estimated using the historical expected prices:

$$\begin{aligned} \bar{\lambda}_U &= E(\lambda_U) \\ \bar{\lambda}_D &= E(\lambda_D) \end{aligned}$$

$$(10.9)$$

10.2.5 WPP's Revenue Combining Energy Bidding and Reserve Purchasing

After adding a reserve purchasing mechanism, the ex-ante DA market allows WPPs to purchase reserves with the capacity of R_U and R_D besides the DA bidding. The ex-post imbalance power clearing mechanism settles the actual deployed reserve energy and penalizes the deviations exceeding purchased reserve capacity. Considering the RT price uncertainties, the WPP's conditional expected revenue Π with a possible wind power realization P_w^{RT} is written as the function of the bidding decision (P_w^{DA}, R_U, R_D) shown below.

$$\Pi(P_w^{DA}, R_U, R_D)\big|_{P_w^{RT}} = E(\pi_w^R \,|\, P_w^{RT}) - E(c_w^R \,|\, P_w^{RT})$$

$$= \lambda_{DA} P_w^{DA} - \gamma_U R_U - \gamma_D R_D$$

$$- \bar{\lambda}_U \min(R_U, \max(P_w^{DA} - P_w^{RT}, 0))$$

$$+ \bar{\lambda}_D \min(R_D, \max(P_w^{RT} - P_w^{DA}, 0)) \qquad (10.10)$$

$$- \bar{\lambda}_{RT}^+ \max(P_w^{DA} - P_w^{RT} - R_U, 0)$$

$$+ \bar{\lambda}_{RT}^- \max(P_w^{RT} - P_w^{DA} - R_D, 0)$$

The first row of the revenue function 10.10 represents the profit of the energy contract and the cost of the reserve capacity scheduled in the DA forward market. The next two rows denote the expected cost of the deployed reserve energy. The last two rows formulate the penalties when the deviations exceed the purchased reserve capacity. In this study, the price relationship in 10.11 is assumed to be satisfied so that the expected reserve deployment price is less expensive than the real-time expected imbalance price.

$$0 \leq \bar{\lambda}_{RT}^- \leq \bar{\lambda}_D \leq \lambda_{DA} \leq \bar{\lambda}_U \leq \bar{\lambda}_{RT}^+ \qquad (10.11)$$

Figure 10.1 compares the WPP's revenue with and without reserve purchasing using an illustrative revenue curve. The gray curve represents the revenue of the wind plant participating in the energy market only, while the red curve denotes the revenue of wind plant participating in both energy and reserve markets. Apparently, the reserve purchasing does not provide an advantage until the deviation is higher than a threshold. Therefore, the amount of reserves should be strategically chosen to maximize the revenue of the WPP.

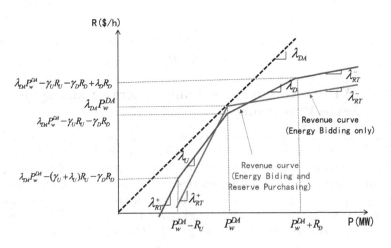

FIGURE 10.1 Comparison of revenue curve with respect to the deviation for WPP with and without reserve purchasing.

10.3   Optimal Wind Power Bidding Strategy

This section derives the optimal energy bidding reserve purchasing strategy based on the market model proposed above. The optimal strategy is analyzed considering the inherent uncertainty of wind power.

10.3.1   Uncertainty Model for Wind Power

The actual generation of wind power for each look-ahead time horizon is modeled as a random variable $w \in [0, W_{cap}]$, where W_{cap} denotes the installed capacity of the wind farm. The probabilistic density function (pdf) of the actual wind power output is denoted by $f(w)$, and the corresponding cumulative distribution function (cdf) is defined as $F(w) = \int_0^w f(x)dx$. Define $F^{-1} : [0,1] \rightarrow [0, W_{cap}]$ as the quantile function corresponding to the cdf $F(w)$. For $p \in [0, 1]$, the p-quantile value is calculated by

$$F^{-1}(p) = \sup\{w| \int_0^w f(x)dx \le p\} \qquad (10.12)$$

$f(w)$ and $F(w)$ can be obtained through the probabilistic forecasting of the wind power.

10.3.2   Expected Revenue for a WPP

Due to the limited accuracy of the wind power forecast, the real wind power output P_w^{RT} in equation 10.10 can be seen as a stochastic variable. The WPP's revenue also has uncertainty. Equation 10.13 provides the explicit analytical expression of the WPP's expected revenue under the energy bidding and reserve purchasing (P_w^{DA}, R_U, R_D).

$$J(P_w^{DA}, R_U, R_D) = \underset{P_w^{RT}}{E}\left[\Pi(P_w^{DA}, R_U, R_D)\right]$$

$$= \lambda_{DA} P_w^{DA} - \gamma_U R_U - \gamma_D R_D$$

$$- \bar{\lambda}_U \int\limits_{0}^{P_w^{DA}-R_U} R_U f(w)dw - \bar{\lambda}_U \int\limits_{P_w^{DA}-R_U}^{P_w^{DA}} (P_w^{DA}-w)f(w)dw$$

$$+ \bar{\lambda}_D \int\limits_{P_w^{DA}}^{P_w^{DA}+R_D} (w - P_w^{DA})f(w)dw + \bar{\lambda}_D \int\limits_{P_w^{DA}+R_D}^{W_{cap}} R_D f(w)dw \qquad (10.13)$$

$$- \bar{\lambda}_{RT}^{+} \int\limits_{0}^{P_w^{DA}-R_U} (P_w^{DA} - R_U - w)f(w)dw$$

$$+ \bar{\lambda}_{RT}^{-} \int\limits_{P_w^{DA}+R_D}^{W_{cap}} (w - P_w^{DA} - R_D)f(w)dw$$

The integration terms in formulation 10.13 can be explained using areas bounded by CDF of the wind power forecast $F(w)$ in Figure 10.2.

WPP's revenue can be re-written as:

$$J = \lambda_{DA} P_w^{DA} - \gamma_U R_U - \gamma_D R_D - \bar{\lambda}_{RT}^{+} U_1 - \bar{\lambda}_U U_2 + \bar{\lambda}_D D_1 + \bar{\lambda}_{RT}^{-} D_2 \qquad (10.14)$$

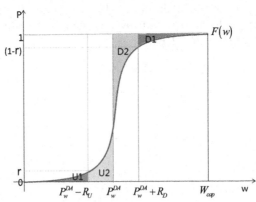

FIGURE 10.2 Graphical interpretation of a WPP's expected revenue.

From a graphical perspective, the areas U_1 and U_2 represent the uncertainty of wind power short-generation, while the areas D_1 and D_2 reflect the uncertainty of wind power over-generation. These areas determine the revenue loss by deviation penalty and decrease with the improvement of wind power forecast accuracy.

10.3.3 Value of Reserve Purchasing for a WPP

We first derive the conditional optimal reserve purchasing while assuming P_w^{DA} is constant. The optimal reserve $(R_{U|P_w^{DA}}^*, R_{D|P_w^{DA}}^*)$ is determined by maximizing the expected revenue $J(R_U, R_D)|_{P_w^{DA}}$:

$$(R_{U|P_w^{DA}}^*, R_{D|P_w^{DA}}^*) = arg \max_{\substack{R_U \in [0, P_w^{DA}] \\ R_D \in [P_w^{DA}, W_{cap} - P_w^{DA}]}} J(R_U, R_D)|_{P_w^{DA}} \qquad (10.15)$$

Theorem 10.1 *The solution to optimization problem 10.15 is:*

1. The optimal up-reserve contract is given by:

$$R_{U|P_w^{DA}}^* = \begin{cases} P_w^{DA} - F^{-1}\left(\dfrac{\gamma_U}{\bar{\lambda}_{RT}^+ - \bar{\lambda}_U}\right), & \dfrac{\gamma_U}{F(P_w^{DA})} + \bar{\lambda}_U \le \bar{\lambda}_{RT}^+ \\ 0, & otherwise \end{cases} \qquad (10.16)$$

2. The optimal down-reserve contract is given by:

$$R_{D|P_w^{DA}}^* = \begin{cases} F^{-1}\left(1 - \dfrac{\gamma_D}{\bar{\lambda}_D - \bar{\lambda}_{RT}^-}\right) - P_w^{DA}, & \dfrac{\gamma_D}{1 - F(P_w^{DA})} + \bar{\lambda}_{RT}^- \le \bar{\lambda}_D \\ 0, & otherwise \end{cases} \qquad (10.17)$$

3. The additional revenue compared with that of not purchasing reserves can be quantified by:

$$\Delta J^U = \left(\bar{\lambda}_{RT}^+ - \bar{\lambda}_U\right) \int_{P_w^{DA} - R_{U|P_w^{DA}}^*}^{P_w^{DA}} (P_w^{DA} - w) f(w) dw$$

$$\Delta J^D = \left(\bar{\lambda}_D - \bar{\lambda}_{RT}^-\right) \int_{P_w^{DA}}^{P_w^{DA} + R_{D|P_w^{DA}}^*} (w - P_w^{DA}) f(w) dw \qquad (10.18)$$

where $\Delta J^U, \Delta J^D$ represent the expected profit augment of purchasing up-reserve and down-reserve.

Proof *Assuming the energy contract P_w^{DA} is known, the expected revenue J formulated as 10.13 can be rewritten as*

$$J(R_U, R_D) = \lambda_{DA} P_w^{DA} + J^+(R_U) + J^-(R_D)$$

$$J^+(R_U) = -\gamma_U R_U - \bar{\lambda}_U \int_0^{P_w^{DA} - R_U} R_U f(w) dw - \bar{\lambda}_U \int_{P_w^{DA} - R_U}^{P_w^{DA}} (P_w^{DA} - w) f(w) dw$$

$$- \bar{\lambda}_{RT}^+ \int_0^{P_w^{DA} - R_U} (P_w^{DA} - R_U - w) f(w) dw$$

$$J^-(R_D) = -\gamma_D R_D + \bar{\lambda}_D \int_{P_w^{DA}}^{P_w^{DA} + R_D} R_D f(w) dw + \bar{\lambda}_D \int_{P_w^{DA} + R_D}^{W_{cap}} (P_w^{DA} - w) f(w) dw$$

$$+ \bar{\lambda}_{RT}^- \int_{P_w^{DA} + R_D}^{W_{cap}} (w - P_w^{DA} - R_D) f(w) dw$$

$$(10.19)$$

Apparently, $J^+(R_U)$ is independent of $J^-(R_D)$. The optimal up-reserve and down-reserve contracts are given by maximizing $J^+(R_U)$ and $J^-(R_D)$, respectively. Part a): For technical simplicity in the proof, the density function $f(w)$ is assumed to be continuous on $[0, W_{cap}]$. The first and second derivatives of $J^+(R_U)$ with respect to R_U are given by

$$\frac{\partial J^+(R_U)}{\partial R_U} = -\gamma_U + (\bar{\lambda}_{RT}^+ - \bar{\lambda}_U) \int_0^{P_w^{DA} - R_U} f(w) dw$$

$$= (\bar{\lambda}_{RT}^+ - \bar{\lambda}_U) F(P_w^{DA} - R_U) - \gamma_U \qquad (10.20)$$

$$\frac{\partial^2 J^+(R_U)}{\partial R_U^2} = (\bar{\lambda}_U - \bar{\lambda}_{RT}^+) f(P_w^{DA} - R^U)$$

As $\bar{\lambda}_U \le \bar{\lambda}_{RT}^+$, $J^+(R_U)$ is concave, which indicates that $R_{U|P_w^{DA}}^*$ is optimal if only if

$$(x - R_{U|P_w^{DA}}^*) \frac{\partial J^+(R_U)}{\partial R_U} \bigg|_{R_U = R_{U|P_w^{DA}}^*} \le 0 \ \forall x \in [0, P_w^{DA}] \qquad (10.21)$$

We now define the set:

$$M = \{ P_w^{DA} \in [0, W_{cap}] | \gamma_U / F\left(P_w^{DA}\right) + \bar{\lambda}_U \le \bar{\lambda}_{RT}^+ \} \qquad (10.22)$$

For $P_w^{DA} \in M$, the optimal point coincides with the stationary point given by

$$R_{U|P_w^{DA}}^* = P_w^{DA} - F^{-1}(\gamma_U / (\bar{\lambda}_{RT}^+ - \bar{\lambda}_U)) \qquad (10.23)$$

When $P_w^{DA} \notin M$, $\partial J^+(R_U) / \partial R_U$ is negative on $[0, P_w^{DA}]$, which yields $R_{U|P_w^{DA}}^* = 0$.

Notice that the optimal result is still achieved, even if $\bar{\lambda}_U > \bar{\lambda}_{RT}^+$.

Part b): The proof is analogous to that of part a).

Part c): The expected revenue of WPP, which only participate energy market without submitting reserve offers, is denoted as

$$
\begin{aligned}
\bar{J}(P_w^{DA}) &= P_w^{DA}\lambda_{DA} - \bar{\lambda}_{RT}^+ \int_0^{P_w^{DA}} (P_w^{DA} - w)f(w)dw \\
&+ \bar{\lambda}_{RT}^- \int_{P_w^{DA}}^{W_{cap}} (w - P_w^{DA})f(w)dw
\end{aligned}
\tag{10.24}
$$

Hence, the revenue addition is formulated as

$$
\begin{aligned}
\Delta J_{Add}^{reserve} &= J(P_w^{DA}, R_{U|P_w^{DA}}^*, R_{D|P_w^{DA}}^*) - \bar{J}(P_w^{DA}) \\
&= \left\{ J^+\left(R_{U|P_w^{DA}}^*\right) + \bar{\lambda}_{RT}^+ \int_0^{P_w^{DA}} (P_w^{DA} - w)f(w)dw \right\} \\
&+ \left\{ J^-\left(R_{U|P_w^{DA}}^*\right) - \bar{\lambda}_{RT}^- \int_{P_w^{DA}}^{W_{cap}} (w - P_w^{DA})f(w)dw \right\}
\end{aligned}
\tag{10.25}
$$

Define $\Delta J^U, \Delta J^D$ for separately representing the expected profit augment of purchasing the up-reserve and down-reserve offers. Then, $\Delta J^U, \Delta J^D$ can be analytically deduced as

$$
\begin{aligned}
\Delta J^U &= \bar{\lambda}_{RT}^+ \int_0^{P_w^{DA}} (P_w^{DA} - w)f(w)dw + J^+\left(R_{U|P_w^{DA}}^*\right) \\
&= (\bar{\lambda}_{RT}^+ - \bar{\lambda}_U) \int_{P_w^{DA}-R_{U|P_w^{DA}}^*}^{P_w^{DA}} (P_w^{DA} - w)f(w)dw \\
\Delta J^D &= \bar{\lambda}_{RT}^- \int_{P_w^{DA}}^{W_{cap}} (w - P_w^{DA})f(w)dw + J^-\left(R_{D|P_w^{DA}}^*\right) \\
&= (\bar{\lambda}_D - \bar{\lambda}_{RT}^-) \int_{P_w^{DA}}^{P_w^{DA}+R_{D|P_w^{DA}}^*} (w - P_w^{DA})f(w)dw
\end{aligned}
\tag{10.26}
$$

The theorem shows that when reserve deploy prices are forecasted to be lower than the deviation penalty imbalance prices, there are opportunities for a WPP to reduce the deviation penalty through strategic purchasing in the DA market. Conversely, with higher reserve deploy prices in the ex-ante reserve market, it is reasonable for wind producers to accepting all the deviation penalties. The optimal reserve purchase quantity is determined by a quantile value of the wind power distribution, which is strong relative to the reserve capacity price. In addition, because the reserve capacity price is generally much lower than the energy price, the quantile value is always very small, suggesting that an optimal reserve would be strongly sensitive to the tails of the probabilistic distribution of probabilistic forecast.

10.3.4 Optimal Bidding Strategy

The optimal contract $(\hat{P}_w^{DA}, \hat{R}_U, \hat{R}_D)$ is determined by maximizing the expected revenue $J(P_w^{DA}, R^U, R^D)$:

$$(\hat{P}_w^{DA}, \hat{R}_U, \hat{R}_D) = arg \max_{\substack{P_w^{DA} \in [0, W_{cap}] \\ R_U \in [0, P_w^{DA}] \\ R_D \in [P_w^{DA}, W_{cap} - P_w^{DA}]}} J(P_w^{DA}, R_U, R_D) \qquad (10.27)$$

According to Theorem 10.1, the optimal $R_{U|P_w^{DA}}^*, R_{D|P_w^{DA}}^*$ can be written analytically as 10.16 and 10.17. Equation 10.27 can thus be rewritten as

$$(\hat{P}_w^{DA}, \hat{R}_U, \hat{R}_D) = arg \max_{P_w^{DA} \in [0, W_{cap}]} J(P_w^{DA}, R_{U|P_w^{DA}}^*, R_{D|P_w^{DA}}^*) \qquad (10.28)$$

Theorem 10.2 *Proposition III.2: The optimal energy bidding together with the reserve purchasing for the problem 10.27 is derived as described in Table 10.1.*

Proof *The optimal P_w^{DA} is discussed under different cases, depending on whether $R_{U|P_w^{DA}}^*$ and $R_{D|P_w^{DA}}^*$ are non-zero.*

Case $R_{U|P_w^{DA}}^* > 0$ & $R_{D|P_w^{DA}}^* > 0$: The first and second derivatives of the optimal revenue $J(P_w^{DA}, R_{U|P_w^{DA}}^*, R_{D|P_w^{DA}}^*)$ with respect to P_w^{DA} is given by

$$\frac{\partial J(P_w^{DA}, R_{U|P_w^{DA}}^*, R_{D|P_w^{DA}}^*)}{\partial P_w^{DA}} = (\lambda_{DA} - \bar{\lambda}_D + \gamma_D - \gamma_U) - (\bar{\lambda}_U - \bar{\lambda}_D) F(P_w^{DA})$$

$$\frac{\partial^2 J(P_w^{DA}, R_{U|P_w^{DA}}^*, R_{D|P_w^{DA}}^*)}{\partial P_w^{DA2}} = -(\bar{\lambda}_U - \bar{\lambda}_D) f(P_w^{DA}) \leq 0$$

$$(10.29)$$

Apparently, $J(P_w^{DA}, R_{U|P_w^{DA}}^*, R_{D|P_w^{DA}}^*)$ is concave, indicating that the optimal point coincides with the stationary point given by

$$\hat{P}_w^{DA} = F^{-1}(\frac{\lambda_{DA} - \bar{\lambda}_D + \gamma_D - \gamma_U}{\bar{\lambda}_U - \bar{\lambda}_D}) \qquad (10.30)$$

Hence, the optimal reserve contracts are updated as

$$\hat{R}_U = \hat{P}_w^{DA} - F^{-1}(\frac{\gamma_U}{\bar{\lambda}_{RT}^+ - \bar{\lambda}_U})$$

$$\hat{R}_D = F^{-1}(1 - \frac{\gamma_D}{\bar{\lambda}_D - \bar{\lambda}_{RT}^-}) - \hat{P}_w^{DA} \qquad (10.31)$$

TABLE 10.1 Optimal bidding strategy for a wind producer participating in the combined energy and reserve market

Market prices condition	Optimal reserve purchasing
$\begin{cases} \bar{\lambda}_U \leq \bar{\lambda}_{RT}^+ - \dfrac{\gamma_U\left(\bar{\lambda}_{RT}^+ - \bar{\lambda}_D\right)}{\lambda_{DA} - \bar{\lambda}_D + \gamma_D} \\[4mm] \bar{\lambda}_D \geq \bar{\lambda}_{RT}^- + \dfrac{\gamma_D\left(\bar{\lambda}_U - \bar{\lambda}_{RT}^-\right)}{\bar{\lambda}_U - \lambda_{DA} + \gamma_U} \end{cases}$	$\begin{cases} \bar{\lambda}_U > \bar{\lambda}_{RT}^+ - \dfrac{\gamma_U\left(\bar{\lambda}_{RT}^+ - \bar{\lambda}_D\right)}{\lambda_{DA} - \bar{\lambda}_D + \gamma_D} \\[4mm] \bar{\lambda}_D \geq \bar{\lambda}_{RT}^- + \dfrac{\gamma_D\left(\bar{\lambda}_{RT}^+ - \bar{\lambda}_{RT}^-\right)}{\bar{\lambda}_{RT}^+ - \lambda_{DA}} \end{cases}$
$\begin{cases} \hat{P}_w^{DA} = F^{-1}\left(\dfrac{\lambda_{DA} - \bar{\lambda}_D + \gamma_D - \gamma_U}{\bar{\lambda}_U - \bar{\lambda}_D}\right) \\[4mm] \hat{R}_U = \hat{P}_w^{DA} - F^{-1}\left(\dfrac{\gamma_U}{\bar{\lambda}_{RT}^+ - \bar{\lambda}_U}\right) \\[4mm] \hat{R}_D = F^{-1}\left(1 - \dfrac{\gamma_D}{\bar{\lambda}_D - \bar{\lambda}_{RT}^-}\right) - \hat{P}_w^{DA} \end{cases}$	$\begin{cases} \hat{P}_w^{DA} = F^{-1}\left(\dfrac{\lambda_{DA} - \bar{\lambda}_D + \gamma_D}{\bar{\lambda}_{RT}^+ - \bar{\lambda}_D}\right) \\[4mm] \hat{R}_U = 0 \\[4mm] \hat{R}_D = F^{-1}\left(1 - \dfrac{\gamma_D}{\bar{\lambda}_D - \bar{\lambda}_{RT}^-}\right) - \hat{P}_w^{DA} \end{cases}$
$\begin{cases} \bar{\lambda}_U \leq \bar{\lambda}_{RT}^+ - \dfrac{\gamma_U\left(\bar{\lambda}_{RT}^+ - \bar{\lambda}_{RT}^-\right)}{\lambda_{DA} - \bar{\lambda}_{RT}^-} \\[4mm] \bar{\lambda}_D < \bar{\lambda}_{RT}^- + \dfrac{\gamma_D\left(\bar{\lambda}_U - \bar{\lambda}_{RT}^-\right)}{\bar{\lambda}_U - \lambda_{DA} + \gamma_U} \end{cases}$	$\begin{cases} \hat{P}_w^{DA} = F^{-1}\left(\dfrac{\lambda_{DA} - \bar{\lambda}_{RT}^- - \gamma_U}{\bar{\lambda}_U - \bar{\lambda}_{RT}^-}\right) \\[4mm] \hat{R}_U = \hat{P}_w^{DA} - F^{-1}\left(\dfrac{\gamma_U}{\bar{\lambda}_{RT}^+ - \bar{\lambda}_U}\right) \\[4mm] \hat{R}_D = 0 \end{cases}$
$\begin{cases} \bar{\lambda}_U > \bar{\lambda}_{RT}^+ - \dfrac{\gamma_U\left(\bar{\lambda}_{RT}^+ - \bar{\lambda}_{RT}^-\right)}{\lambda_{DA} - \bar{\lambda}_{RT}^-} \\[4mm] \bar{\lambda}_D < \bar{\lambda}_{RT}^- + \dfrac{\gamma_D\left(\bar{\lambda}_{RT}^+ - \bar{\lambda}_{RT}^-\right)}{\bar{\lambda}_{RT}^+ - \lambda_{DA}} \end{cases}$	$\begin{cases} \hat{P}_w^{DA} = F^{-1}\left(\dfrac{\lambda_{DA} - \bar{\lambda}_{RT}^-}{\bar{\lambda}_{RT}^+ - \bar{\lambda}_{RT}^-}\right) \\[4mm] \hat{R}_U = 0 \\[4mm] \hat{R}_D = 0 \end{cases}$

In addition, the condition $R_{U|P_w^{DA}}^* > 0$ & $R_{D|P_w^{DA}}^* > 0$ *is also equivalent as*

$$
\begin{cases} \dfrac{\gamma_U}{F\left(\hat{P}_w^{DA}\right)} + \bar{\lambda}_U \leq \bar{\lambda}_{RT}^+ \\[4mm] \dfrac{\gamma_D}{1 - F\left(\hat{P}_w^{DA}\right)} + \bar{\lambda}_{RT}^- \leq \bar{\lambda}_D \end{cases} \Rightarrow \begin{cases} \bar{\lambda}_U \leq \bar{\lambda}_{RT}^+ - \dfrac{\gamma_U\left(\bar{\lambda}_{RT}^+ - \bar{\lambda}_D\right)}{\lambda_{DA} - \bar{\lambda}_D + \gamma_D} \\[4mm] \bar{\lambda}_D \geq \bar{\lambda}_{RT}^- + \dfrac{\gamma_D\left(\bar{\lambda}_U - \bar{\lambda}_{RT}^-\right)}{\bar{\lambda}_U - \lambda_{DA} + \gamma_U} \end{cases} \tag{10.32}
$$

The proof of the other cases is analogous.

The results in Table I indicate that whether or not, and how much the WPP should purchase reserves is driven by the relationship of the prices of both energy and reserves (e.g. purchasing up reserve is more economic for the WPP only when the reserve deployment price λ_U is low enough compared with real-time penalty λ_{RT}^+. This criterion provides a sign that can tell whether wind power has the optimal decision in the market. Without reserve purchasing, the optimal bidding strategy is degenerated to the conclusions reported in [12].

10.3.5   Discussion

Though the derivation of optimal bidding strategy is based on a stylized market design, the proposed model and results provide some insights for the real market: 1) The deviation of wind power can be seen as an "accident" of wind power and the associated penalty is the loss. Purchasing reserves can be seen as a kind of "insurance" for such an accident. If the cost of the "insurance" is reasonable, the wind power can avoid the losses caused by deviation through purchasing such "insurance". 2) The cost-effectiveness of the "insurance" is determined by the relative value of the opportunity cost (reserve capacity price) and the saving from the difference between the deviation penalty and the reserve deployment cost. 3) The amount of "insurance" that the WPP should purchase is driven by the shape of the probability distribution of its forecast error, e.g. whether the distribution has a thick tail or thin tail. 4) The formulation can be used to justify the cost-effectiveness for the WPP to employ the self-owned storage or co-operate with a conventional generator to form a VPP to avoid being penalized.

Allowing a WPP to purchase its own reserves is also beneficial for the system security and reliability: 1) Such a mechanism will shift the extra cost brought by wind power uncertainty from ISO (consumer) to WPPs. It will also give incentives to WPPs to improve their forecast skills to reduce the uncertainty. 2) The system will benefit from facing fewer energy variations from wind plants, which will increase the security of the power system and reduce the price spikes. 3) The flexible resources can be more wisely used. For example, the reserves will be employed dynamically by WPPs since they seek to maximize their profits. The distributed flexible resources such as distributed storage and demand response which can not directly participate in the reserve market can be indirectly incorporated into the power system.

10.4   Case Study

Based on real market price signals and wind power data, the optimal bidding strategy for a WPP participating in the proposed combined energy and reserve market is calculated and compared with that of the WPP participating energy market only. The revenue comparison illustrates the advantage of reserve purchasing for handling wind power uncertainty. The sensitivities to both the market prices and wind power uncertainty are also analyzed.

10.4.1 Wind Power Probabilistic Forecasts and Market Prices

The wind power output and the forecast data used in this paper are from the Wind Integration Datasets of NREL [30]. Ten sites with an aggregated installed capacity of 1700 MW along the east coast of the U.S were selected as shown in [31] and were regarded as an aggregated wind plant. Day-ahead hourly forecasts and actual outputs are available from the dataset.

Probabilistic forecasts are obtained based on the point forecast given in the Dataset, using the copula based technique proposed in [31]. Figure 10.3 shows the modeled probabilistic forecasts for one day. The market prices for the energy and the reserve are extracted from the ERCOT nodal market in January 1st, 2015 and are shown in Figure 10.4. For both RT expected energy and reserve prices, we introduce reserve deploy price ratio γ and deviation penalty ratio ξ to evaluate the impacts of different cost of reserves on the optimal bidding strategy:

$$\begin{cases} \bar{\lambda}_U = (1+\gamma)\lambda_{DA} \\ \bar{\lambda}_D = (1-\gamma)\lambda_{DA} \end{cases} and \begin{cases} \bar{\lambda}_{RT}^+ = (1+\xi)\lambda_{DA} \\ \bar{\lambda}_{RT}^- = (1-\xi)\lambda_{DA} \end{cases} \qquad (10.33)$$

In this study, both ratios γ and ξ are set as 40% and 100% respectively. Different values of γ and ξ are further tested in the sensitivity analysis. It is noted $\xi = 100\%$ for $\bar{\lambda}_{RT}^-$ is equivalent to curtail wind power when there is energy spillage.

10.4.2 WPP's Optimal Bidding and its Benefit

To illustrate the application of the optimal bidding strategy and evaluate the revenue addition through strategic reserve purchasing, the optimal contracts under time-varying market prices during one day are calculated using Theorem 10.2, and the results are shown in Figure 10.3. The heights of the red and green bars represent the optimal down-reserve and up-reserve purchasing values, respectively, in the DA reserve market, while the level of the boundary between the red and green bins denotes the optimal energy bidding in the DA energy market. Some interesting observations are as follows. 1) When the up-reserve capacity price is expensive (e.g. during 18:00 20:00), the WPP will reduce the energy bids and increase the amount of down-reserve purchase. 2) When wind power has large ramps, it would more likely to have large deviation and is recommended for purchasing more reserve capacity. 3) Through strategic reserve purchasing, the expected revenue (considers all possible deviation) increases by 17.13%, compared with the corresponding revenue of participating energy market only. The detailed results are presented in Table 10.2. With comparing the expected revenue and real revenue for WPP, the results also indicate that WPPs can strategically purchase reserve to increase their revenue only when deviation penalty price ratio is larger than reserve deploy price ratio, namely$\xi > \gamma$. The larger difference exists between ξ andγ, the more revenue loss due to generation uncertainty can be compensated.

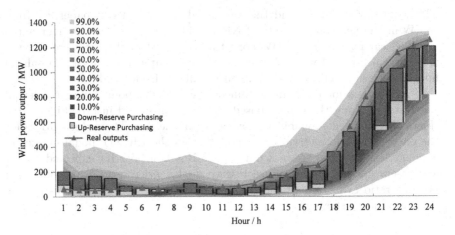

FIGURE 10.3 Probabilistic forecasts and the optimal bidding strategy in DA energy and reserve markets.

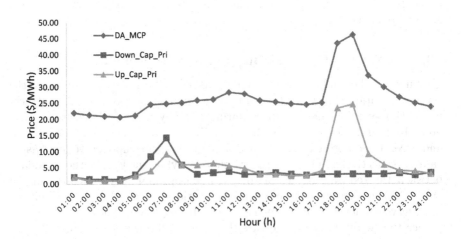

FIGURE 10.4 Examples of market prices for energy and reserve service in the ERCOT nodal market (January 1st, 2015).

TABLE 10.2 Revenue with and without reserve purchasing during one day

γ	ξ	Type	Reserve purchasing	Without reserve purchasing	Revenue increase
0%	0%	Expected	157391	157391	0%
		Real	248229	248229	0%
	40%	Expected	139674	134684	3.71%
		Real	208079	200257	3.91%
	100%	Expected	132556	94879	39.71%
		Real	185609	128299	44.67%
20%	0%	Expected	157391	157391	0%
		Real	248229	248229	0%
	40%	Expected	134897	134630	0.20%
		Real	200 387	200 257	0.07%
	100%	Expected	120201	94879	26.69%
		Real	164345	128299	28.10%
40%	0%	Expected	157391	157391	0%
		Real	248229	248229	0%
	40%	Expected	134684	134684	0%
		Real	200257	200257	0%
	100%	Expected	111133	94879	17.13%
		Real	147365	128299	14.86%

10.4.3 Sensitivity to Wind Power Uncertainty

Taking the prices of 1:00 AM as an example, the impact of the uncertainty of wind power on the optimal bids of both energy and reserve capacity are studied. The results for each output level of wind power are shown in Figure 10.5. The results suggest that: 1) the optimal energy bidding may not equal the point forecast. As suggested in Theorem 10.2, the optimal bidding is determined by a quantile of the probabilistic forecast, which is determined by both ex-ante and ex-post prices. In this case, the quantile is 50%. 2) The optimal reserve purchasing (ORP) quantity varies with wind power uncertainty level. As expected, the ORP of low or high output is relatively small, while the ORP of medium output is large. The optimal capacities of up-reserve and down-reserve are not identical because of the asymmetry of the probabilistic forecast.

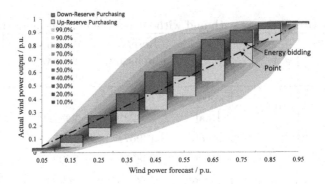

FIGURE 10.5 Optimal bidding strategy in energy and reserve markets with different point forecast level.

10.4.4 Sensitivity to Market Prices

Figure 10.6 depicts the sensitivity of optimal bidding strategies to the up-reserve deploy price when the point forecast is 45% of the capacity. The optimal amount of energy bidding decreases with increasing up-reserve price. A higher up-reserve deploy price also leads to a lower up-reserve capacity purchasing. It is interesting to find that the optimal amount of down-reserve is also affected by the up-reserve prices. The amount of down-reserve is suggested to increase as the up-reserve price increases until the up-reserve capacity purchase is suggested to be zero.

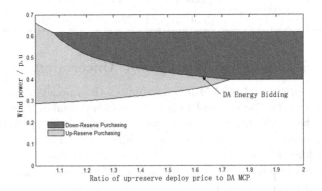

FIGURE 10.6 Optimal bidding strategy in energy and reserve markets with different up-reserve deploy prices.

10.4.5 Analysis of WPP's Revenue

Figure 10.7 demonstrates the expected revenue of a WPP with different reserve purchasing strategies when the point forecast is 45% of its capacity. The results suggest that purchasing reserves in the reserve market bring opportunities for WPPs to increase revenue compared with only participating in the energy market. Defining the revenue loss of wind power uncertainty as the revenue difference with and without deviation penalty, Figure 10.8 compares the ratio of expected revenue losses due to the uncertainty and expected reserve saved from purchasing the reserves, as the percentage of the revenue corresponding to the perfect forecast. The results show that the revenue loss takes a more significant share during lower output levels, making more room for the revenue saved through reserve purchasing. The ratio of revenue saved as the percentage of the revenue losses is approximately 20%.

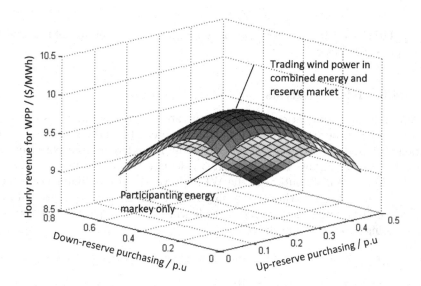

FIGURE 10.7 Expected revenues of a WPP with different reserve purchasing strategies.

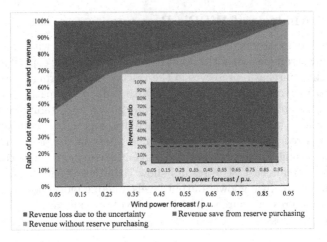

FIGURE 10.8 Ratio of revenue losses and saved in percentage of the revenue for perfect forecast.

10.4.6   Risk Analysis

A Monte-Carlo simulation is used to generate twenty thousand possible wind power outputs according to the probabilistic forecast and the revenue distribution with and without reserve purchasing under different wind power output level shown in Figure 10.9. Table 10.3 calculate the VaR and CVaR on the left side of the distribution of the WPP's revenue with a confidence level of 5%. The results show that purchasing reserve will shorten the left fat tail of revenue distribution, suggesting that allowing WPPs to purchase reserve can also bring down their market risks.

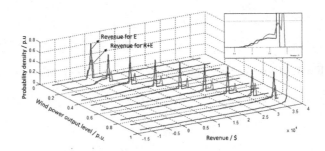

FIGURE 10.9 Probability density distribution for revenue with and without reserve purchasing under different point forecast levels.

TABLE 10.3 Risk index comparison with and without reserve purchase when the point forecast is 45% of its capacity

Unit($)	VaR (5%)	CVaR (5%)
Energy	2278	-449
Energy + Reserve	4063	1336

10.4.7   Discussions

Based on the numerical results, some generalized observations from the results of the case study can be drawn. 1) Strategically purchasing reserves can increase the expected revenue for a wind producer, and partly compensate the revenue loss associated with wind power generation uncertainty. The larger difference exists between reserve prices and deviation penalty prices, the more revenue loss due to generation uncertainty can be compensated. 2) The optimal quantity of reserve purchasing is sensitive to the tail of the probability distribution of wind power generation. More reserves are required if the distribution has a thick tail. 3) The costs of up-reserve and down-reserve products affect the optimal amount of energy bidding and reserve purchasing. For example, if the up-reserve price is expensive, WPPs tend to bid less power in DA energy market, increase the requirement of down-reserve and reduce the requirement of up-reserve. 4) The revenue risk of a wind producer can be efficiently reduced by strategically purchasing reserves.

It should be noted that though we verify that WPP is able to benefit from strategic reserve purchasing from a expect revenue point of view, it may not be the case for each of the scenarios. There exists the possibility that the WPP may lose profit from the reserves purchasing for some certain scenario. A simple illustrative example is carried out to demonstrate this argument. We assume the probability distribution of wind power P_w^{RT} as shown in Figure10.10 The probability distributions of RT market prices ($\lambda_{RT}^+, \lambda_{RT}^-$) is assumed to follow a normal distribution and is shown in Figure10.11 The expectation of these market prices are summarized in Table 10.4.

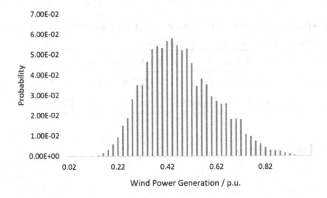

FIGURE 10.10 Probability distribution of wind power generation.

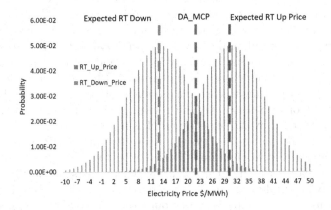

FIGURE 10.11 Probability distributions of market prices.

TABLE 10.4 Parameters of market prices

	λ_{DA}	$\bar{\lambda}_{RT}^{+}$	$\bar{\lambda}_{RT}^{-}$	$\bar{\lambda}_{U}$	$\bar{\lambda}_{D}$	γ_{U}	γ_{D}
Value($/MWh)	20.01	30.81	13.21	22.01	22.01	2.11	2.11

As P_w^{RT} and $(\lambda_{RT}^{+}, \lambda_{RT}^{-})$ are assumed having uncertainty in day-ahead market, Monte Carlo method is applied to generate 10000 scenarios $(P_w^{RT}, \lambda_{RT}^{+}, \lambda_{RT}^{-})$ according to their probability distributions, assuming they are independent with each other. Then, the probability distribution of real revenue can be acquired shown in Figure 10.12. The results are summarized in Table 10.5 (The capacity of the wind farm is set as 100MW).

The results validate that 1) Strategic reserve purchasing is able to increase the WPP's expected revenue. 2) The value of realized revenue has a variety of possibilities driven by the uncertainty of real-time prices and generation. There exists the possibility that the WPP may lose profit from the reserves purchasing. 3) It is practical to make decisions of whether to purchase reserve or not at the day-ahead market with the information of expected estimation of RT prices and probability distribution estimation of wind power generation.

TABLE 10.5 Revenue with and without reserve purchasing

Type	Expected Revenue	Optimal Biding (MW)
No Reserve	922.7	$\begin{cases} P_w^{RT} = 44.97 \\ \lambda_{RT}^+ = 0 \\ \lambda_{RT}^- = 0 \end{cases}$
Strategic Reserve Purchasing	943.8	$\begin{cases} P_w^{RT} = 44.97 \\ \lambda_{RT}^+ = 9.19 \\ \lambda_{RT}^- = 11.18 \end{cases}$

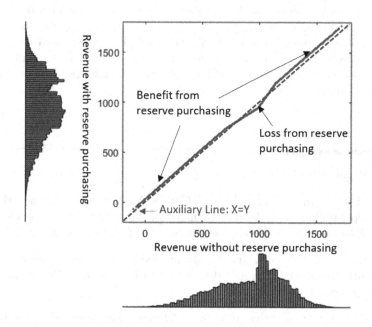

FIGURE 10.12 Probability distribution of real revenue.

10.5 Summary

This study provided insights on the problem of how the WPP could optimally bid for the energy and purchase reserve in the electricity market to reduce the revenue loss due to wind power generation uncertainty. A generalized market model is proposed that allows WPPs to purchase or schedule their own reserve to offset part of their deviation, instead of fully committing to the imbalance penalties of the ex-post market. The optimal bidding strategy under such a market model was analytically derived, which is determined by reserve price signals, imbalance clearing price signals, and the probabilistic forecast. The optimal bidding strategy provides an optimal way for WPPs to manage the uncertainty by strategically scheduling some reserve beforehand. The additional revenue through strategic reserve purchases is also quantified. A case study with price data from ERCOT and wind power data from NREL is presented to verify the optimal bidding strategy. The result shows that if the WPP could schedule the reserves with an expense of 40% of the cost of imbalance of penalty, 20% of the revenue loss associated with wind power uncertainty penalty could be saved by using its own reserve.

The derived optimal bidding and reserve scheduling provide a general means of managing the uncertainty in power systems using flexible resources with different quality. For example, the scheduling can be used to determine the optimal system-level reserves for wind power from an ISO point of view, with various reserve resources of different costs, availabilities, and probabilities of deployment.

Bibliography

[1] M. Milligan, E. Ela, D. Lew, D. Corbus, Y. Wan, B. Hodge, and B. Kirby. Operational analysis and methods for wind integration studies. *IEEE Transactions on Sustainable Energy*, 3(4):612–619, 2012.

[2] H. Xing, H. Cheng, and L. Zhang. Demand response based and wind farm integrated economic dispatch. *CSEE Journal of Power and Energy Systems*, 1(4):37–41, 2015.

[3] J. Liang, S. Grijalva, and R. G. Harley. Increased wind revenue and system security by trading wind power in energy and regulation reserve markets. *IEEE Transactions on Sustainable Energy*, 2(3):340–347, 2011.

[4] R. Madlener and M. Kaufmann. Power exchange spot market trading in Europe: theoretical considerations and empirical evidence. *OSCO-*

GEN (Optimisation of Cogeneration Systems in a Competitive Market Environment)-Project Deliverable, 5, 2002.

[5] M. Robinson. Role of balancing markets in wind integration. *2006 Power Systems Conference and Exposition*, pages 232–233, 2006.

[6] A. Fabbri, T. G. S. Roman, J. R. Abbad, and V. M. Quezada. Assessment of the cost associated with wind generation prediction errors in a liberalized electricity market. *IEEE Transactions on Power Systems*, 20(3):1440–1446, 2005.

[7] H. Huang and F. Li. Bidding strategy for wind generation considering conventional generation and transmission constraints. *Journal of Modern Power Systems and Clean Energy*, 3(1):51–62, 2015.

[8] J. M. Morales, A. J. Conejo, and J. Pérez-Ruiz. Short-term trading for a wind power producer. *IEEE Transactions on Power Systems*, 25(1):554–564, 2010.

[9] J. Matevosyan and L. Soder. Minimization of imbalance cost trading wind power on the short-term power market. *IEEE Transactions on Power Systems*, 21(3):1396–1404, 2006.

[10] A. Botterud, J. Wang, R. J. Bessa, H. Keko, and V. Miranda. Risk management and optimal bidding for a wind power producer. *Power and Energy Society General Meeting, 2010 IEEE*, pages 1–8, 2010.

[11] P. Pinson, C. Chevallier, and G. N. Kariniotakis. Trading wind generation from short-term probabilistic forecasts of wind power. *IEEE Transactions on Power Systems*, 22(3):1148–1156, 2007.

[12] E. Y. Bitar, R. Rajagopal, P. P. Khargonekar, K. Poolla, and P. Varaiya. Bringing wind energy to market. *IEEE Transactions on Power Systems*, 27(3):1225–1235, 2012.

[13] G. N. Bathurst, J. Weatherill, and G. Strbac. Trading wind generation in short term energy markets. *IEEE Transactions on Power Systems*, 17(3):782–789, 2002.

[14] M. S. Roulston, D. T. Kaplan, J. Hardenberg, and L. A. Smith. Using medium-range weather forcasts to improve the value of wind energy production. *Renewable Energy*, 28(4):585–602, 2003.

[15] C. J. Dent, J. W. Bialek, and B. F. Hobbs. Opportunity cost bidding by wind generators in forward markets: Analytical results. *IEEE Transactions on Power Systems*, 26(3):1600–1608, 2011.

[16] E. Bitar, R. Rajagopal, P. Khargonekar, and K. Poolla. The role of co-located storage for wind power producers in conventional electricity markets. *American Control Conference (ACC), 2011*, pages 3886–3891, 2011.

[17] P. Pinson, G. Papaefthymiou, B. Klockl, and J. Verboomen. Dynamic sizing of energy storage for hedging wind power forecast uncertainty. *Power & Energy Society General Meeting, 2009 IEEE*, pages 1–8, 2009.

[18] X. Y. Wang, D. M. Vilathgamuwa, and S. S. Choi. Determination of battery storage capacity in energy buffer for wind farm. *IEEE Transactions on Energy Conversion*, 23(3):868–878, 2008.

[19] A. Botterud, Z. Zhou, J. Wang, R. J. Bessa, H. Keko, J. Sumaili, and V. Miranda. Wind power trading under uncertainty in lmp markets. *IEEE Transactions on Power Systems*, 27(2):894–903, 2012.

[20] P. Denholm, E. Ela, B. Kirby, and M. Milligan. The role of energy storage with renewable electricity generation. *National Renewable Energy Laboratory*.

[21] T. Dai and W. Qiao. Trading wind power in a competitive electricity market using stochastic programing and game theory. *IEEE Transactions on Sustainable Energy*, 4(3):805–815, 2013.

[22] H. Holttinen, M. Milligan, B. Kirby, T. Acker, V. Neimane, and T. Molinski. Using standard deviation as a measure of increased operational reserve requirement for wind power. volume 32, pages 355–377. SAGE Publications Sage UK: London, England, 2008.

[23] R. Doherty and M. O'malley. A new approach to quantify reserve demand in systems with significant installed wind capacity. *IEEE Transactions on Power Systems*, 20(2):587–595, 2005.

[24] M. A. Ortega-Vazquez and D. S. Kirschen. Optimizing the spinning reserve requirements using a cost/benefit analysis. *IEEE Transactions on Power Systems*, 22(1):24–33, 2007.

[25] A. M. Da S., Leite and G. P. Alvarez. Operating reserve capacity requirements and pricing in deregulated markets using probabilistic techniques. *IET Generation, Transmission & Distribution*, 1(3):439–446, 2007.

[26] J. Wang, X. Wang, and Y. Wu. Operating reserve model in the power market. *IEEE Transactions on Power systems*, 20(1):223–229, 2005.

[27] J. M. Morales, A. J. Conejo, and J. Pérez-Ruiz. Economic valuation of reserves in power systems with high penetration of wind power. volume 24, pages 900–910. IEEE, 2009.

[28] A. Botterud, J. Wang, V. Miranda, and R. J. Bessa. Wind power forecasting in us electricity markets. *The Electricity Journal*, 23(3):71–82, 2010.

[29] P. Pinson, C. Chevallier, and G. Kariniotakis. Optimizing benefits from wind power participation in electricity markets using wind power forecasting embedded with uncertainty management tools. *European Wind Energy Conference, EWEC 2004*, 2004.

[30] M. Brower. Development of eastern regional wind resource and wind plant output datasets. *Rep. No. NREL/SR-550*, 46764, 2009.

[31] N. Zhang, C. Kang, Q. Xia, and J. Liang. Modeling conditional forecast error for wind power in generation scheduling. *IEEE Transactions on Power Systems*, 29(3):1316–1324, 2014.

11

Tie-Line Scheduling for Interconnected
Power Systems

A series of studies shows that larger balancing area is able to promote the
accommodation of renewable energy. However, different regions of the interconnected power system may be operated by different operators. So, how to
schedule the power exchanges considering the uncertainty of renewable energy
is an issue that needs to be addressed. This chapter proposes a day-ahead tie-
line power flow scheduling model for multi-area interconnected power systems
aiming to improve the wind power accommodation by allowing the exchange
of the flexibility among regional grids. A case of the IEEE 118-bus system is
carried out and the results show that the scheduling will greatly reduce the
wind power curtailment and improve the operation efficiency.

11.1   Overview

A larger balancing area facilitates the sharing of flexibility resources among a
wider area so that the renewable energy can be better accommodated. However, the situation in the real-life power system is that different regions of
the interconnected power system may be operated by different operators so
the system dispatching should be coordinated instead of monolithically dispatched. The uncertainty and intermittency of renewable energy require a
more strategic power exchange scheduling on the tie-lines. In the non-market
power system, such coordination gives rise to another issue that how much
responsibility of accommodating renewable energy should be taken by each
system since it usually results in additional cost to dispatch flexible resources.
Therefore, a quota system named Renewable Portfolio Standard (RPS) is usually proposed to address the responsibility problem. The policy will impose
wind power accommodation quota on each regional power grid, based on factors such as the load level, economic development, etc..

    This chapter addresses the issue of tie-line scheduling with renewable en-
ergy under the assumption that there is an RPS policy. Without renewable
energy integration or transaction, each regional power grid can operate independently. However, the integrated wind power will affect this circumstance

due to the inherent stochasticity and fluctuating nature. If there is only a fixed power curve interchange on tie-lines, sometimes it would be impossible for the grid with too much wind resource to accommodate its stochasticity and intermittency. Therefore, it is essential to introduce more strategic schemes to let the flexible sources exchange through the interconnected power grids, especially after implementing the RPS policy. However, most of the works focus on the reserve provision allocation among interconnected power systems [1–3], and none focus on the "wind accommodation quota" imposed by the RPS policy. M. F. Astaneh, *et al.* proposed an optimization framework that minimizes the cost of cross-border reserve procurement [1]. C. Chen, *et al.* conducted the multi-area reserve dispatch using a penalty function-hybrid direct search method considering network constraints [2]. A. Ahmadi-Khatir, *et al.* dispatched a multi-area reserve using a decomposition algorithm [3].

Based on the RPS policy, each regional grid has imposed a quota to accommodate a certain quantity of wind power. Then, the questions are how to deliver wind energy among interconnected power grids by tie-lines, how to quantify the wind energy amount delivered on tie-lines, and what kind of constraints should be considered when scheduling in a day-head time level? This chapter proposes a day-ahead tie-line power flow scheduling model considering system reserve, wind power forecast error, peak-shaving potential, load following capacity, transmission limits, and some special tie-line constraints. The proposed model aims to realize the minimum wind curtailment and maximum wind accommodation quota filling with a reasonable and feasible operating cost.

11.2 Problem Statement

11.2.1 Assumptions

Three basic assumptions are made during the modeling:

1. Each grid will fill its own wind accommodation quota as much as possible; the grid that does not fully meet the wind accommodation quota will pay a penalty.

2. Each grid will dispatch the power flow of tie-lines that deviate from long-term contracted power as little as possible, because vast fluctuation of tie-line power flow may threaten system security.

3. Each grid is supposed to supply sufficient generation capacity and system flexibility in case of wind power forecast error.

11.2.2   Model Framework

The framework of day-ahead tie-line power flow scheduling is shown in Figure 11.1.

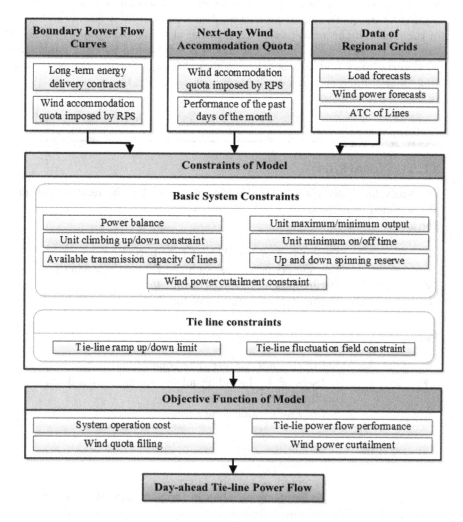

FIGURE 11.1 Framework of day-ahead tie-line power flow scheduling model

Firstly, the boundary tie-line power flow is determined by decomposing the energy delivery contract to each day, which should also consider the wind accommodation quota of each regional grid imposed by the RPS, which guarantees that the wind energy is properly allocated according to the RPS. This power flow is regarded as a "boundary power flow curve" in the proposed scheduling model. Secondly, the next-day quota of wind accommodation im-

posed for each regional grid is calculated based on the accumulative performance of the past day's in the same month. Since the wind accommodation
quota of each regional grid is imposed by month, grids are supposed to fully
meet the accommodation quota by the end of the month. The next day's accommodation quota should be increased to meet the total quota by the end
of the month if the wind power accommodated in the past days is relatively
less than the expected. Finally, the model is established towards minimizing
wind curtailment and meeting the accommodation quota as close as possible,
while maintaining a reasonable system operating cost.

11.3   Notations

To clearly introduce the proposed optimization model, the related notations
are provided as follows.

Indices

t	Parameters or variables at time period t
s	Parameters or variables on the sth day in a month
k	Parameters or variables of unit k
l	Parameters or variables of line l

Sets

Ψ_L^i	Set of tie-lines connected to zone i
Ψ_L^{i+}/Ψ_L^{i-}	Subsets of Ψ_L^i including all the tie-lines with flowing-in/flowing-out power for zone i
Ω_{D}^i	Set of buses inside zone i

Superscripts

i	Parameters or variables in zone i
r	Parameters or variables in the past days of the month
cur	Wind curtailment
ramp	Unit ramping rate

Functions

g	Unit operating cost function
h	Start-up and shut-down cost function

Parameters

Λ_1 to Λ_4	Weight coefficients for the objective function
β_m^i	Wind power accommodation quota of month m imposed by RPS for zone i
α_{nk}	Element of incidence matrix for bus n and unit k
γ_u/γ_d	Positive/negative reserve rate
$\gamma_{k,t}^{i+}/\gamma_{k,t}^{i-}$	Up/down reserve rate as response to the forecast error of $P_{k,t}^{w,i}$

Variable

$\omega_{l,t}^{i,u}/\omega_{l,t}^{i,d}$	Slack variables for the fluctuation up and down limit for tie-line l
$P_{l,t}^{i,b}$	Power flow on the tie-lines l in zone i fixed by long term contract
I	Binary decision variable: "1" if unit k is on at time t; "0" otherwise
u,v	Binary decision variables: "1" if unit k starts up /shuts down at time t; "0" otherwise

11.4 Model Formulation

The model is listed from Equations (11.1) to (11.17). It is a multi-objective model, and the objective function considers system operating cost, accommodation quota requirement, wind power curtailment, and tie-line power flow performance. The constraints are divided into two groups: basic system constraints (BSC) and tie-line constraints (TLC).

Objective function:

$$\min \quad \Lambda_1 \sum_i \left|Diff^i + Diff^{i,r}\right| + \Lambda_2 \sum_t \sum_i \sum_k P_{k,t}^{cur,i}$$
$$+ \Lambda_3 \sum_t \sum_i \sum_k \left[g(P_{k,t}^{c,i}) + h(P_{k,t}^{c,i})\right] \tag{11.1}$$
$$+ \Lambda_4 \sum_t \sum_i \sum_{l\in\Psi_L^i} \left(\omega_{l,t}^{i,u} + \omega_{l,t}^{i,d}\right)$$

where,

$$Diff^i = \sum_t \left[\sum_k (P_{k,t}^{\mathrm{w},i} - P_{k,t}^{\mathrm{cur},i}) + \sum_{l \in \Psi_{\mathrm{L}}^i} (P_{l,t}^i - P_{l,t}^{i,\mathrm{b}}) \right]$$
$$- \beta_m^i \sum_i \sum_t \sum_k P_{k,t}^{\mathrm{w},i} \tag{11.2}$$

$$Diff^{i,r} = \sum_{s=1}^{x-1} \left[\sum_t \left(\sum_k P_{k,s,t}^{\mathrm{w},i,r} + \sum_{l \in \Psi_{\mathrm{L}}^i} (P_{l,s,t}^{i,r} - P_{l,s,t}^{i,\mathrm{b}}) \right) \right.$$
$$\left. - \beta_m^i \sum_i \sum_t \sum_k P_{k,s,t}^{\mathrm{w},i,r} \right] \tag{11.3}$$

In Equation (11.1), $Diff^i$ is the value to judge the performance of wind quota filling in the following day, denoted in Equation (11.2) by the difference between actual accommodated quantity and the imposed wind accommodation quota for grid i. $Diff^{i,r}$ denoted in Equation (11.3) is the actual bias value of the wind quota filling in the past days of the month. The first part of the objective function in the form of absolute value is to maximize the quota filling, this part can be easily converted into a linearized form; the second part is to minimize wind power curtailment; the third part is the system operating cost composed of linearized unit operating cost function and start-up/shut-down cost function; the fourth part is the penalty function of slack variables for the tie-line fluctuation constraints.

Constraints of BSC group:

$$\sum_k P_{k,t}^{\mathrm{c},i} + \sum_k (P_{k,t}^{\mathrm{w},i} - P_{k,t}^{\mathrm{cur},i}) + \sum_{l \in \Psi_{\mathrm{L}}^{i+}} P_{l,t} = \sum_n D_{n,t}^i + \sum_{l \in \Psi_{\mathrm{L}}^{i-}} P_{l,t} \tag{11.4}$$

$$I_{k,t}^i \underline{P_k^{\mathrm{c},i}} \leqslant P_{k,t}^{\mathrm{c},i} \leqslant I_{k,t}^i \overline{P_k^i} \quad , \forall k \tag{11.5}$$

$$\underline{P_l} \leqslant P_{l,t} \leqslant \overline{P_l}, \forall l \tag{11.6}$$

$$P_{l,t} = \sum_i \sum_{n \in \Omega_{\mathrm{D}}^i} GSDF_{ln} \left[\sum_k a_{nk} P_{k,t}^{\mathrm{c},i} + \sum_k a_{nk}^{\mathrm{w}} (P_{k,t}^{\mathrm{w},i} - P_{k,t}^{\mathrm{cur},i}) - D_{n,t}^i \right], \forall l \tag{11.7}$$

$$\sum_k I_{k,t}^i \overline{P_k^{\mathrm{c},i}} + \sum_k P_{k,t}^{\mathrm{w},i} + \sum_{l \in \Psi_{\mathrm{L}}^{i+}} P_{l,t} \geqslant$$

$$(1 + \gamma_{\mathrm{u}}^i) \sum_{n \in \Omega_{\mathrm{D}}^i} D_{n,t}^i + \sum_k \gamma_{k,t}^{i+} P_{k,t}^{\mathrm{w},i} + \sum_{l \in \Psi_{\mathrm{L}}^{i-}} P_{l,t} \tag{11.8}$$

$$\sum_k I_{k,t}^i \underline{P_k^{c,i}} + \sum_{l \in \Psi_L^{i+}} P_{l,t} \leqslant (1 - \gamma_d^i) \sum_{n \in \Omega_D^i} D_{n,t}^i - \sum_k \gamma_{k,t}^{i-} P_{k,t}^{w,i} + \sum_{l \in \Psi_L^{i-}} P_{l,t}$$

(11.9)

$$P_{k,t+1}^{c,i} - P_{k,t}^{c,i} \leqslant (2 - I_{k,t}^i - I_{k,t+1}^i)\underline{P_k^{c,i}} + (1 + I_{k,t}^i - I_{k,t+1}^i)P_k^{ramp,u,i}, \ \forall k$$

(11.10)

$$P_{k,t}^{c,i} - P_{k,t+1}^{c,i} \leqslant (2 - I_{k,t}^i - I_{k,t+1}^i)\underline{P_k^{c,i}} + (1 - I_{k,t}^i + I_{k,t+1}^i)P_k^{ramp,d,i}, \ \forall k$$

(11.11)

$$\left(I_{k,t}^i - I_{k,t-1}^i\right) t_{k,on}^i + \sum_{j=t-t_{k,on}^i-1}^{t-1} I_{k,j}^i \geqslant 0, \ \forall k$$

(11.12)

$$\begin{cases} -v_{k,t}^i \leqslant I_{k,t}^i - I_{k,t-1}^i \leqslant u_{k,t}^i \\ \sum_t u_{k,t}^i \leqslant \overline{u_i}, \sum_t v_{k,t}^i \leqslant \overline{v_i} \\ 0 \leqslant v_{k,t}^i, u_{k,t}^i, I_{k,t}^i \leqslant 1 \end{cases}, \ \forall k$$

(11.13)

$$P_{k,t}^{cur,i} \leqslant P_{k,t}^{w,i}, \ \forall k$$

(11.14)

The BSC group consists of a power balance constraint (Equation (11.4)), unit output constraints (Equations (11.5), (11.6) and (11.14)), a network constraint ((11.7)), spinning reserve constraints (Equations (11.8) and (11.9)), unit ramping constraints (Equations (11.10) and (11.11)), a minimum on time constraint (Equation (11.12)) and a maximum start-up/shut-down times constraint (Equation (11.13)). The constraints in the BSC group form a regular security-constrained unit commitment (SCUC) model with the objective of minimum system operating cost. In Equations (11.8) and (11.9), $\sum_k \gamma_{k,t}^{i+} P_{k,t}^{cur,i}$ and $\sum_k \gamma_{k,t}^{i-} P_{k,t}^{cur,i}$ are the spinning reserve prepared to supply up and down generation capacity in response to the wind output forecast error, which is determined by the forecasted output level based on the conditional-distributed prediction errors [4]. Equations (11.10) and (11.11) not only restrict ramping up and down speed, but also restrict the unit power at the minimum output right after the starting up period and before the shutting down period. Equations (11.12) and (11.13) are the minimum on-time requirements, and the maximum start-up/shut-down times constraint, respectively. Equation (11.14) restricts that the wind curtailment cannot exceed the wind power output.

Constraints of TLC group:

$$-P_l^{ramp,tie} \leqslant P_{l,t+1} - P_{l,t} \leqslant P_l^{ramp,tie}, \ \forall l \in \Psi_L^i$$

(11.15)

$$-d_l^i - \omega_{l,t}^{i,d} \leqslant P_{l,t} - P_{l,t}^b \leqslant d_l^i + \omega_{l,t}^{i,u}, \ \forall l \in \Psi_L^i$$

(11.16)

$$\omega_{l,t}^{i,u}, \omega_{l,t}^{i,d} \geqslant 0, \ \forall l \in \Psi_L^i \ \forall t, i$$

(11.17)

In the above equations, d_l^i is the maximum deviation between the scheduled power flow and long-term contracted power flow of tie-line l. The TLC group consists of tie-line ramp up/down constraint (Equation (11.15)) and fluctuation field constraint (Equations (11.16) and (11.17)). Equation (11.15) is to restrict the ramp-up and ramp-down speed of tie-line power flow for system security. Equations (11.16) and (11.17) restrict the power flow on tie-lines to fluctuate in a certain window around the "boundary power flow curve". If the power flow on tie-lines exceeds this window, there will be a penalty in the fourth part of the objective function.

11.5   Case Study

11.5.1   IEEE 118-Bus System

This chapter takes the IEEE 118-bus system [5] shown in Figure 11.2 to conduct the case study. The system is divided into three areas, and here we take each area as a regional power grid. All parameters are extracted from Matpower 4.0 [6] and others are shown in Table 11.1. The model is solved by CPLEX. The total capacity of 54 conventional units is 9966 MW, and five of them are replaced by wind farms (similar studies can be done with other renewable energy). The actual power output of five wind farms located in north China is applied in this chapter as the input of the wind power. The wind farms are located at bus 10, 25, 54, 61, and 91 with the capacity of 100, 200, 100, 100, and 150 MW respectively. Therefore, Grid 1, 2, and 3 have 300 MW, 200 MW, and 150 MW wind power capacity, respectively. The total wind power capacity accounts for 7.0% of the total interconnected system capacity, and for the three grids, this number is 12.0%, 4.6%, and 6.3% respectively. Moreover, the year-round load curves come from the IEEE RTS-79 system [7], and the proportion of the load on each bus remains the same for the whole period of study. The power flow limit of each branch is set to be 30% of the original value. In addition, the boundary power flow curves derived from long-term energy contracts and RPS are assumed as 200 MW at peak-hour time periods (11th~13rd and 19th~21st time periods) and 100 MW at non-peak-hour time periods. $\gamma_{k,t}^{i+}$ and $\gamma_{k,t}^{i-}$ are set as in [4].

The wind accommodation quota of each grid is set to be in proportion to its monthly maximum forecasted load, so Grid 2 takes up about half of the total quota. Meanwhile, nearly half of wind power capacity is located at Grid 1 and only about 30% at Grid 2. Thus, only if Grid 1 delivers wind energy to Grid 2 and Grid 3, could the accommodation quota be fully filled by all grids.

TABLE 11.1 Parameters of the model for tie-line scheduling under uncertainty

Type	Value
$P_k^{\text{ramp,u},i}, P_k^{\text{ramp,d},i}$	100 (%/hour)
$t_{(k,on)}^i$	12 (hour)
$\overline{u_i}, \overline{v_i}$	1, 1
$\beta_m^1, \beta_m^2, \beta_m^3$	28%, 51%, 21%
$P_l^{\text{ramp,tie}}, d_l^i$	5%, 15% of the ATC
$\gamma_u^1, \gamma_u^2, \gamma_u^3$	5%, 5%, 7%
$\gamma_d^1, \gamma_d^2, \gamma_d^3$	4%, 4%, 5%
$\Lambda_1, \Lambda_2, \Lambda_3, \Lambda_4$	1000, 1, 1, 1

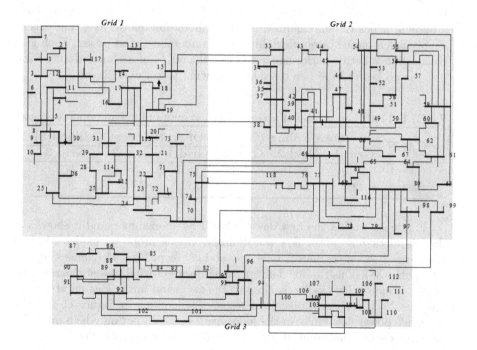

FIGURE 11.2 IEEE 118-bus system and the three grids

11.5.2 Case List

Two cases are studied for comparison: as shown in Table 11.2. Case 1 allows that wind energy can be delivered between two grids by tie-lines; Case 2 makes each grid accommodate wind power independently. For Case 2, $\Lambda_1 = 0$,

$\Lambda_4 = 10000$, and other settings of these two cases are identical as described above. A 30-day simulation of April is carried out.

TABLE 11.2 List of case studies for tie-line scheduling under uncertainty

Case	Day number	Transfer wind on tie-line
Case 1	30	Yes
Case 2	30	No

11.5.3   Results and Discussions

Table 11.3 compares the results of quota filling and average daily operating cost between two cases after the 30-day simulation. The results indicate that there will be less system operating cost with the proposed model except Grid 2 because it has the maximum wind accommodation quota (51%) and has to supply much more system flexibility to accommodate wind power. Each grid could achieve better performance to meet the quota of accommodating wind power, and both of the two cases have no wind power curtailment.

TABLE 11.3 Comparison of accommodation quota filling and operating cost

CASE	Quota Filling (%)			Operating cost (Million $)		
	Grid 1	Grid 2	Grid 3	Grid 1	Grid 2	Grid 3
Case 1	27.6	47.1	25.4	28.3	58	24.7
Case 2	42.5	34.8	22.7	31.2	57.4	26.5

Figure 11.3 takes the first day of Case 1 as an example, and it shows the day-ahead tie-line power flow scheduling from the proposed model. Most of the power flow on tie-lines is relatively stable except Line 54 (connecting bus 30 and bus 38). This tie-line delivers the majority of wind power generated by Grid 1 because this tie-line is close to the two wind farms located at bus 10 and bus 25. If each grid takes this result of scheduling in Figure 11.3 as the exchange power flow curve in the following day, it will not only help meet the accommodation quota but also help reduce the total operating cost.

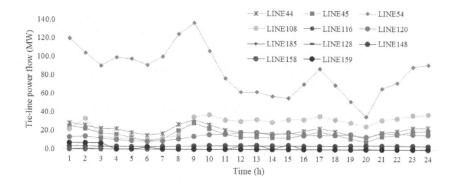

FIGURE 11.3 Day-ahead tie-line power flow scheduling of the first day in Case 1

Figure 11.4 shows the 30-day wind power accommodation performance of each grid. The grid in Case 1 takes responsibility to deliver wind energy to other interconnected grids and makes efforts to accommodate wind energy matching its accommodation quota, while the grid in Case 2 just accommodates the wind energy generated by itself. Figure 11.5 shows the statistical result of tie-line power flow deviation frequency, which is the bias between the dispatched tie-line power flow and the "long-term contracted power flow curve". Therefore, the deviation is brought about by wind power integration, and the results in Figure 11.5 are the different patterns of deviation in two cases. Though Case 2 does not allow deviations on tie-lines, there are still power flow deviations, indicating that if all power flow of tie-lines is controlled to exactly match the "long-term contracted power flow curve", there would be load shedding. The deviation in Case 1 is mostly located in a small-frequency triangle zone marked in Figure 11.5, except Line 54. During wind energy delivery, the tie-line closed to wind farms inside a sending-wind grid should thus be paid more attention. To sum up, the proposed model achieves a tie-line power flow scheduling with small power flow deviation and low deviation occurrence frequency, except for certain tie-lines located near wind farms.

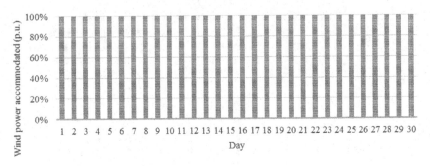

(a) Case 1

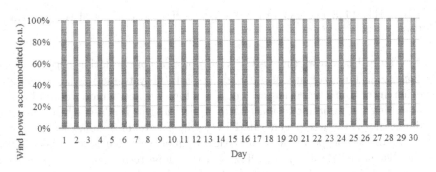

(b) Case 2

FIGURE 11.4 Wind power accommodation performance of each grid for 30 days

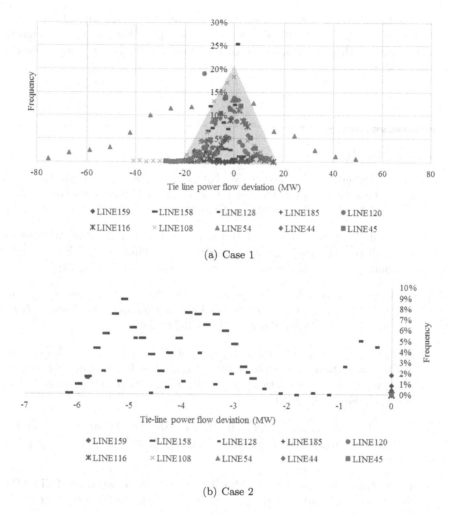

(a) Case 1

(b) Case 2

FIGURE 11.5 Frequency diagram of tie-line power flow deviation caused by wind power delivery

11.6   Summary

This chapter proposes a novel day-ahead tie-line power flow scheduling model that facilitates the exchanges of flexible resources between regional grids under RPS policy. The model schedules the tie-line power flow while minimizing the system operating cost, quota-filling deviation, wind power curtailment, and tie-line power flow deviations while considering the security of the power system. Test cases on the IEEE 118-bus system show that the proposed model could improve the accommodation of wind power while maintaining a reasonable operating cost.

Bibliography

[1] M. F. Astaneh, Z. H. Rather, W. Hu, and Z. Chen. Economic valuation of reserves on cross border interconnections: A Danish case study. *PES General Meeting—Conference & Exposition, 2014 IEEE*, pages 1–5, 2014.

[2] C. Chen, Z. Chen, and T. Lee. Multi-area economic generation and reserve dispatch considering large-scale integration of wind power. *International Journal of Electrical Power & Energy Systems*, 55:171–178, 2014.

[3] A. Ahmadi-Khatir, A. J. Conejo, and R. Cherkaoui. Multi-area energy and reserve dispatch under wind uncertainty and equipment failures. *IEEE Transactions on Power Systems*, 28(4):4373–4383, 2013.

[4] N. Zhang, C. Kang, C. Duan, X. Tang, J. Huang, Z. Lu, W. Wang, and J. Qi. Simulation methodology of multiple wind farms operation considering wind speed correlation. *International Journal of Power & Energy Systems*, 30(4):264, 2010.

[5] M. Shahidehpour, H. Yamin, and Z. Li. *Market Operations in Electric Power Systems: Forecasting, Scheduling, and Risk Management*. John Wiley & Sons, 2003.

[6] R. D. Zimmerman and C. E. Murillo-Sanchez. Matpower: A MATLAB® power system simulation package, user's manual, version 4.0. 2011.

[7] Probability Methods Subcommittee. IEEE reliability test system. *IEEE Transactions on Power Apparatus and Systems*, 6(PAS-98):2047–2054, 1979.

Part IV

Long-Term Planning Optimization

12

Power System Operation Simulation

In the power system planning problem, one of the most frequently asked questions is the "if-then" question, e.g. if installing energy storage in the power system, how much operating cost would be saved? The integration of renewable energy makes the problem more complex because of its stochastic nature. Under renewable energy integration, the study of power system planning requires the simulation of dispatch throughout the year, which is named "power system operation simulation". This chapter puts forward the framework and model of power system operation simulation and demonstrates its applications by answering three questions in power system planning: 1) How does the operating cost of thermal generation vary with wind power? 2) How much pumped hydro storage is needed for coordinating with wind power? 3) How much wind power can be accommodated by an existing power system?

12.1 Overview

Power system operation simulation denotes the reproduction of the scheduling and dispatch of the power system over the long time horizon. It includes the simulation of unit maintenance, long-term unit scheduling and short-term operation of the power system for a whole year on an hourly basis to study the overall operation configuration of the power system. Compared with the power system real-time simulation, power system operation simulation focuses on the steady-state operation of the power system instead of the transient process. Compared with the power system production analysis, power system operation simulation focuses more on economics and security of the normal operation situations instead of reliability aspects that consider the failure of generators and transmission lines. Power system operation simulation plays an important role in power system analysis, not only for the planning/investment studies but also for electricity market analysis. There are several tools currently used by power system planners all around the world, e.g. Wilmar, Multi Area Production Simulation Software (MAPS), GridView and Grid Optimization Planning Tool (GOPT).

The stochastic nature of renewable energy further gives rise to the necessity of power system operation simulation in power system planning study.

The number of possible operation modes of renewable energy is enormous so that it is hard to provide a comprehensive understanding of the economics and security of the power system when only analyzing the typical operating mode. The year-round power system operation simulation is able to accumulate the impact of renewable energy on the daily/hourly dispatch so that an overall view of the power system operation can be obtained. In this chapter, we will introduce the framework and model of power system operation simulation in the software GOPT developed by the authors' laboratory. We also demonstrate how it can be used in power system planning studies.

It is widely acknowledged that wind power requires more flexible capacity provided by conventional thermal units so that it increases the operating cost of conventional thermal units [1–3]. These additional costs cannot be recovered if there is no spot market or ancillary service payment. This is exactly the case of many power systems, especially in China [4–6], where wind power producers receive feed-in tariffs. Power system operation simulation is a useful tool for evaluating how much extra operating cost is caused by wind power from both the perspectives of the power system operation and the single generator level.

Energy Storage System (ESS) is another important issue for renewable energy integrated power systems. ESS helps to accommodate wind power or solar energy by providing ramping and reserve capacity at the expense of a large amount of investment cost. Several studies have already addressed the planning of storage in power systems with a large amount of wind power. In particular, A. Tuohy, et al. assessed the Irish power system with wind penetration from 17% to 80% [7]. B. C. Ummels, et al. explored the Dutch power system [8, 9]. P. Denholm, et al. evaluated the ERCOT system with 80% renewable energy [10]. All of these studies found that storage reduces the system operating cost and makes possible the integration of higher penetration of wind power. Pumped storage would also benefit the system by balancing wind power in a market [11] or in an isolated power system [12]. However, from a generation planning perspective, the economical pumped storage capacity is strongly influenced by its investment cost. Power system operation simulation provides a useful means of assessing the value of ESS to the wind power penetrated power system.

How much renewable energy can be economically integrated into the power system is a question that must be answered in power system planning. E.g. for wind power, the grid-accommodable wind power capacity (GWPC) is mainly affected by both wind power characteristics (wind power output distribution, diurnal/seasonal pattern, fluctuation characteristic and spatial correlation among wind farms), and load characteristics (peak-valley difference and ramping up/down speed). Both again directly influence five major types of flexibility in the power system, namely peak-valley regulation capability, frequency regulation capability, load-following capability, spinning reserve capacity and available transmission capacity (ATC). Power system operation simulation is able to capture all of the main factors that affect GWPC, making it an ideal tool for its assessment.

12.2 Power System Operation Simulation Model

12.2.1 Framework

The framework of power system simulation is shown in Figure 12.1, and the major modules in the platform are introduced in the following content.

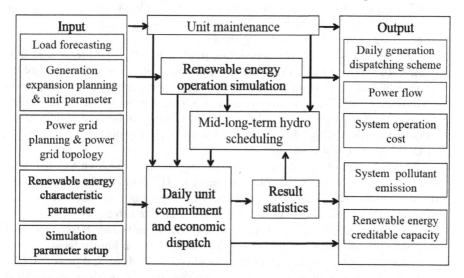

FIGURE 12.1 Framework of power system operation simulation platform

1) Unit maintenance module

The unit maintenance module handles the long-term generation dispatching. The classical unit maintenance model of equal risk target is adopted in the module, considering the maintenance continuity, exclusion, and fixed time constraints. A heuristic algorithm is used to solve this model.

2) Renewable energy operation simulation module

This module produces a chronological simulated output for the production of renewable energy, which is used as an input to the daily dispatch and scheduling module. This module consists of a wind farm power simulation module and a PV output simulation module.

Wind power output simulation: Firstly, according to the wind speed historical data of wind farms, parameters such as the probability distribution of wind speed, wind speed chronological correlation coefficient, multi-wind-farm wind speed correlation coefficient matrix, and wind farm monthly average wind speed curve, can be statistically calculated. Based on the above parameters, the stochastic differential equation model is used to generate wind farm simulated wind speed. Then, combined with the wind turbine output characteristic curve, the wind power output curve can be obtained. Finally, combined with the wake effect model, the total output curve of multiple wind farms can be obtained. A more detailed model is analyzed in Section 6.2.

PV output simulation: Firstly, according to the information of light resources and the installed capacity of PV power plants, chronological simulated output curves that meet the stochastic characteristics of PV can be obtained, considering the lighting conditions, the probability of different meteorological days, cloud occlusion, and other factors. Also, the statistical characteristics of PV output can be calculated and the typical PV output curves can be generated. A more detailed model is analyzed in Section 6.3.

3) Mid-long-term hydro scheduling module

The mid-long-term hydro scheduling module handles the problem of how to allocate long-term water inflow into daily scheduling. The problem is solved using a bilevel optimization model. The higher-level model is the long-term optimization model aiming to maximize the total power generation of the hydro system. The lower-level model is the mid-term optimization model aiming to maximize the utility of peak regulation and power generation.

4) Daily operation simulation module

This key part of the platform determines the daily dispatch for each unit using a security constrained unit commitment (UC) model. It schedules the daily operation of all units to minimize the operating cost while respecting various constraints. These units are divided into four categories:

Gas turbines and small thermal units: These units can be started and stopped daily and their on/off states are modeled on an hourly basis.

Nuclear plants, large thermal units, and combined heat and power (CHP) units: These units cannot or should not be cycled daily, so they are modeled as being in one state for an entire day.

Wind farm and PV stations: Their output can be reduced due to minimum load constraints or transmission constraints. The maximum reduction is equal to the forecast output obtained from the renewable energy operation simulation module.

Hydro and pumped storage units: These units have daily maximum energy constraints. In particular, the pumping loads of the pumped storage units are modeled using extra variables, and balancing constraints for their reservoirs are considered.

12.2.2 Detailed Model of Daily Operation Simulation Module

12.2.2.1 Notations

Indices

i	Index for generators
t	Index for time periods

Sets

Ω_C	Set of all conventional thermal generating units
Ω_{CHP}	Set of all CHP units
Ω_G	Set of all gas turbines
Ω_H	Set of all hydro units
Ω_{ESS}	Set of all energy storage systems
Ω_W	Set of all wind farms and PV stations
Ω	Set of all units

Parameters

$\underline{t}_i^{on}$ and $\underline{t}_i^{off}$	Minimum on and off time intervals of generator i
o_i^{on}	Start-up cost of generator i
H_t	Heat supply demand of period t

Variable

$s_{i,t}^{on}$	Start-up cost of generator unit during period t
$H_{i,t}^{CHP}$	Scheduled heat output of CHP during period t
$u_{i,t}$	On/off status of the units (except ESS) at time t
$u_{i,t}^{ESS}$	Discharging/charging state of ESS at time t

12.2.2.2 Model Formulation

1) Objective function

The objective function in Equation (12.1) aims to minimize the total generating cost and on/off cost of all the generators.

$$
\begin{aligned}
\text{minimize} \ \ &\sum_{t=1}^{T} \sum_{i \in \Omega_C, \Omega_G, \Omega_H} C_i(P_{i,t}) + \sum_{t=1}^{T} \sum_{i \in \Omega_{CHP}} C_i(P_{i,t}, H_{i,t}^{CHP}) \\
&+ \sum_{t=1}^{T} \sum_{i \in \Omega_C, \Omega_{CHP}, \Omega_G, \Omega_H} C_{i,t}^{on}
\end{aligned}
\tag{12.1}
$$

2) Constraints

$$
\text{s.t.} \ \sum_{i \in \Omega} P_{i,t} = D_t, \ t = 1, 2, ..., T
\tag{12.2}
$$

$$
\begin{aligned}
\sum_{i \in \{\Omega_C, \Omega_{CHP}, \Omega_G, \Omega_H, \Omega_{ESS}\}} u_{i,t} \overline{P}_i + \sum_{i \in \Omega_W} P_{i,t} &\geqslant D_t + \underline{R}_t^u, \\
\sum_{i \in \{\Omega_C, \Omega_{CHP}, \Omega_G, \Omega_H, \Omega_{ESS}\}} u_{i,t} \underline{P}_i + \sum_{i \in \Omega_W} P_{i,t} &\leqslant D_t - \underline{R}_t^d, \\
t &= 1, 2, ..., T
\end{aligned}
\tag{12.3}
$$

$$u_{i,t}\underline{P}_i \leqslant P_{i,t} \leqslant u_{i,t}\overline{P}_i, \ t = 1,2,...,T, \ i \in \{\Omega_{\mathrm{C}}, \Omega_{\mathrm{CHP}}, \Omega_{\mathrm{G}}, \Omega_{\mathrm{H}}\} \qquad (12.4)$$

$$\underline{P}_{i,t}^{\mathrm{CHP}} \leqslant P_{i,t} \leqslant \overline{P}_{i,t}^{\mathrm{CHP}}, \ t = 1,2,...,T, \ i \in \Omega_{\mathrm{CHP}} \qquad (12.5)$$

$$0 \leqslant P_{j,t} \leqslant \overline{P}_{j,t}^{\mathrm{W}}, \ t = 1,2,...,T, \ j \in \Omega_{\mathrm{W}} \qquad (12.6)$$

$$P_{i,t} - P_{i,t-1} + u_{i,t}\left(\overline{P}_i - \underline{P}_i\right) + u_{i,t-1}\left(\underline{P}_i - P_i^{\mathrm{ramp,u}}\right) \leqslant \overline{P}_i,$$
$$P_{i,t-1} - P_{i,t} + u_{i,t-1}\left(\overline{P}_i - \underline{P}_i\right) + u_{i,t}\left(\underline{P}_i - P_i^{\mathrm{ramp,d}}\right) \leqslant \overline{P}_i, \qquad (12.7)$$
$$t = 2,3,...,T, \ i \in \{\Omega_{\mathrm{C}}, \Omega_{\mathrm{CHP}}, \Omega_{\mathrm{G}}, \Omega_{\mathrm{H}}\}$$

$$(u_{i,t} - u_{i,t-1})\,\underline{t}_i^{\mathrm{on}} + \sum_{j=t-\underline{t}_i^{\mathrm{on}}-1}^{t-1} u_{i,j} \geqslant 0,$$
$$(u_{i,t-1} - u_{i,t})\,\underline{t}_i^{\mathrm{off}} + \sum_{j=t-\underline{t}_i^{\mathrm{off}}-1}^{t-1} (1 - u_{i,j}) \geqslant 0, \qquad (12.8)$$
$$t = 1,2,...,T, \ i \in \{\Omega_{\mathrm{C}}, \Omega_{\mathrm{CHP}}, \Omega_{\mathrm{G}}, \Omega_{\mathrm{H}}\}$$

$$C_{i,t}^{\mathrm{on}} \geqslant o_i^{\mathrm{on}}\left(u_{i,t} - u_{i,t-1}\right), C_{i,t}^{\mathrm{on}} \geqslant 0, \ t = 1,2,...,T, \ i \in \{\Omega_{\mathrm{C}}, \Omega_{\mathrm{CHP}}, \Omega_{\mathrm{G}}, \Omega_{\mathrm{H}}\} \qquad (12.9)$$

$$\eta_i^{\mathrm{ESS}} \sum_{t=1}^{T} P_{i,t}^{\mathrm{ESS}} = \sum_{t=1}^{T} P_{i,t},$$
$$\sum_{t=1}^{T} P_{i,t}^{\mathrm{ESS}} \leqslant Cap_i, \qquad i \in \Omega_{\mathrm{ESS}} \qquad (12.10)$$

$$P_{i,t} + u_{i,t}^{\mathrm{ESS}}\overline{P}_i \leqslant \overline{P}_i, \ P_{i,t} > 0, \ t = 1,2,...,T, \ i \in \Omega_{\mathrm{ESS}} \qquad (12.11)$$

In the above equations, Equation (12.2) is the generation-load balance equation, which ensures the aggregated output of all the generators to meet the load demand on an hourly basis. Equation (12.3) denotes the up/down spinning reserve constraint. Equation (12.4) limits the output of the generators by its state variable. The constraints are performed on all kinds of generators except for renewable energy and ESS. Equation (12.5) adds the operation zone constraints of electric output and heat output for the CHP plant. Equation (12.6) indicates that renewable energy should be scheduled less than or equal to their available power output. Equation (12.7) forms the ramping rate constraints, which guarantees that the generator is scheduled at the minimum output for the first time period after starting up and for the last time period before shutting down. Equation (12.8) guarantees the minimum on and off time periods for the generators. Equation (12.9) identifies the start-up costs of generators. Equation (12.10) ensures the balance between the consuming and generating energy of ESS considering cycling efficiency and the maximum energy storage capacity limited by the capacity. Equation (12.11) shows the mutually exclusive constraint of discharging and charging. In Equation (12.3), the system reserve requirement takes into account the forecast error on both the load and the output of renewable energy. The up and down reserve requirements from load are set at 3% and 2% of the load demand for each period, respectively. The amount of reserves for renewable energy varies with its forecast outputs.

Take wind power as an example. According to the study of wind power forecast uncertainties [13], the uncertainty interval of wind power output is conditional upon the point forecast. The forecast error for wind power in Jiangsu has been studied using the wind farms operation simulation module, assuming that the wind speed forecast errors have a Gaussian distribution with a standard deviation of 1.5 m/s. Conditional output intervals for each point forecast level have been calculated with a 90% confidence level. The up and down reserve requirements are then determined by comparing the conditional interval with the point forecast. Figure 12.2 shows that for higher forecasts, wind power requires more up reserve and less down reserve. The average up and down reserve requirements are 16.19% and 19.99% of total wind power capacity. The system reserve is further determined as the root mean square of the reserve requirement for load and wind power.

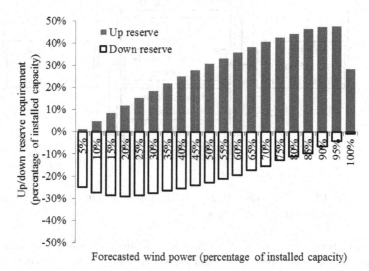

Forecasted wind power (percentage of installed capacity)

FIGURE 12.2 Conditional up and down reserve requirement from wind power forecast errors

12.3 Wind Power Impacts on Conventional Unit Operating Cost

12.3.1 Mechanism

The fundamental difference between wind and thermal generation is such that wind turbines cannot be controlled to serve loads in the same way as conventional units. It provides energy to the system but doesn't offer flexibility, which must thus be provided by conventional or energy storage units. The effect of wind power on the operating cost of thermal generation can be analyzed from two perspectives: the single unit perspective and the system-wide perspective. Figure 12.3 is a schematic representation of the effects of wind power integration.

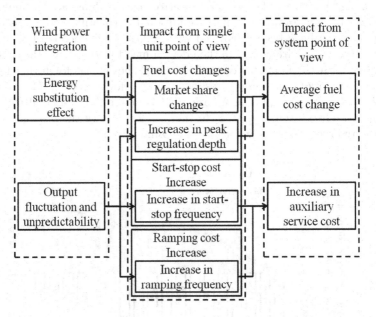

FIGURE 12.3 Schematic diagram of the effect of wind power integration on thermal generation operating cost

Wind power energy substitution effect means that when integrated into a power system, wind power is usually given priority in the dispatch and thus reduces the net output of thermal generation. The market share of thermal generation decreases, especially for the marginal units with high fuel costs. From a system perspective, the average fuel cost declines accordingly.

However, exceptions can be easily found: the substitution effect sometimes impacts low-cost units rather than high-cost marginal units because the high-cost units are better able to provide flexibility. Reference [2] found that in Denmark, the market share of units with low start-up and turn down performance and high minimum load level decreased more than the market share of units with high start-up and turn down performance or low minimum load level. The total cost to the power system would be underestimated by about 5% if this factor was neglected at a wind power penetration level of 24%.

Due to the stochastic nature of wind, the output of wind turbines fluctuates and is hard to predict with high accuracy. While great progress has been made in wind power forecasting, large fluctuations over periods of a few minutes or an hour are still unavoidable. These wind power fluctuations, combined with deviations in the load swings, must be balanced by conventional units. Thus, the additional fuel cost of thermal generation results in deeper peak load regulation, more frequent starts, and stops, and ramping.

The research described in [3] showed that the demand for peak load regulation capacity would increase significantly with large-scale wind power integration. The efficiency of thermal generation is always lower when the output power is below the rated value. When deeper peak-valley load difference is required, the thermal generation moves further away from its economic optimum. These increases in fuel cost are relatively small and are usually concealed by the wind power energy substitution effect.

The term ramping cost represents the extra cost incurred when the output of a thermal generating unit is adjusted above or below the static cost operating point. The ramp cost also includes the physical deterioration, control cost and maintenance cost caused by extra ramping service. Reference [14] analyzed ramp costs in the Ontario market. An additional ramp payment of $4.24 per MW/h is described as a conservative estimate. In this chapter, the concept of "fuel cost" only refers to the cost resulting from energy production. The cost related to ramping service is called "ramping cost" in order to distinguish it from the cost of peak load regulation.

12.3.2 Evaluation Metrics

As discussed in Section 12.3.1, the thermal generation extra operating cost caused by the wind output fluctuations mainly consists of three parts: extra fuel cost, extra ramping cost, and extra start-stop cost. These costs can be easily calculated through the results of the dispatching simulation presented in the previous section.

A new generation cost function, as shown in Equation (12.12), is used to evaluate the inefficiency effect resulting from deeper peak-valley load regulating and more frequent ramping. This cost function is composed of fuel cost C_{fi}, which is a classical quadratic curve, and ramp cost C_{ri}, which is in direct proportion to the absolute ramp rate.

$$C_i(P_{i,t}) = C_{fi}(P_{i,t}) + C_{ri}(P_{i,t})$$
$$= a_i(P_{i,t})^2 + b_i P_{i,t} + c_i + d_i \left| \mathrm{d}P_{i,t}/\mathrm{d}t \right| \tag{12.12}$$

The extra fuel cost of thermal generation could be concealed by the wind power energy substitution effect, so it is calculated through the average fuel cost difference, as described by Equation (12.13). The extra ramping cost can be calculated directly by considering the change in aggregated ramp cost, as shown in Equation (12.14). Finally, the extra start-stop cost can be calculated by counting the number of additional starts and stops and multiplying it by the cost of a single start and stop.

$$E_{fi} = \left(\frac{\sum\limits_{t \in \Gamma} C_{fi}(P'_{i,t})}{\sum\limits_{t \in \Gamma} P'_{i,t}} - \frac{\sum\limits_{t \in \Gamma} C_{fi}(P_{i,t})}{\sum\limits_{t \in \Gamma} P_{i,t}} \right) \times \sum_{t \in \Gamma} P'_{i,t} \tag{12.13}$$

$$E_{ri} = \sum_{t \in \Gamma} C_{ri}(P'_{i,t}) - \sum_{t \in \Gamma} C_{ri}(P_{i,t}) \tag{12.14}$$

12.3.3   Compensation Mechanism

It could be argued that this extra cost should be allocated to the wind farms, but this requires the design of a fair compensation mechanism. Figure 12.4 suggests a possible framework. The ancillary service cost for wind farms could be evaluated on a system-wide basis and allocated to the wind farms based on the amount of wind energy they produce, as shown in Equations (12.14) and (12.15). This would avoid cross-subsidization. This compensation can be seen as a kind of variable cost that wind farms incur because of their lack of flexibility. Including such a compensation mechanism in the feed-in tariffs would improve the fairness and efficiency of the whole market. Moreover, it also provides an incentive for conventional generators to maintain and provide greater flexibility.

$$\sigma = \frac{\sum\limits_{i \in \Omega_{con}} (E_{fi} + E_{ri} + E_{si})}{\sum\limits_{i \in \Omega_{wind}} \sum\limits_{t \in \Gamma} P'_{i,t}} \tag{12.15}$$

12.3.4   Case Study

Jiangsu provincial power system in China is used in this Subsection to investigate the concepts discussed above. The influence of wind power integration on thermal generation cost is evaluated based on actual data in order to study to what extent the thermal generation operating cost increases and should be

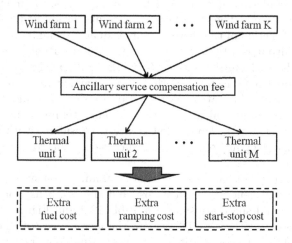

FIGURE 12.4 Framework of compensation mechanism for thermal generation extra cost

compensated. The value of pumped storage on thermal generation operating cost is also studied.

1) Basic data

The data used in this case comes from one of the electric power planning blueprints of Jiangsu province of China in 2020, in which more than 10 GW of wind power is supposed to be installed. The generation mix of conventional units in this plan is shown in Table 12.1, from which we can see that thermal plants make the major proportion. It is forecasted that in 2020 the whole system will have a load level of 125 GW and a total electric energy consumption of 661 TWh. In this section, the wind power integration scenarios are set as 50%, 100%, 150% and 200% of the original plan in order to investigate the impact of wind power on system cost.

TABLE 12.1 Jiangsu provincial power system conventional generation mix in 2020

Generation type	Capacity (MW)	Proportion (%)
Fossil-coal	116208.5	82.01
Combustion	5065	3.57
Combined heat and power	1901.9	1.34
Nuclear	8580	6.06
Hydro	4940	3.49
Pumped storage	5000	3.53
Total	141695.4	100

In the simulation, data related to wind speed is obtained from existing anemometer towers. The wind farm average utilization hour of a year is 2115 h. The cost function of the thermal units is quadratic and the efficiency at the minimum output is 10% lower than at the rated output. The ramping cost of all the thermal units is set at $5 (RMB33.33) per MW/h. Nuclear units are treated as fixed output units that do not participate in ancillary services. System positive and negative reserve ratios are set at 5% and 2% respectively. The power system is modeled as a two-region system with an interconnection capacity of 13000 MW. Unit maintenance is considered with a maintenance duration time ranging from 21 to 42 days according to the capacity of the unit.

2) Operating cost changes and wind farm impact on system operating cost

The thermal generation extra cost and the compensation fee that wind generators would share are shown in Table 12.2 and Figure 12.5. Unlike the results obtained with the RTS system, the average cost of the thermal generation in Jiangsu province increases when wind power capacity increases. This implies that when wind generation increases, more flexible generation with higher cost is dispatched for peak load regulation and ramping. This increases the average cost. From Figure 12.5 we can see that the compensation shared by each MWh of wind energy is also increasing (from $0.198/MWh to $0.711/MWh), especially the compensation caused by the extra fuel cost. This result means the system will adopt more expensive ways of peak load regulation when balancing more wind power. In the base scenario of 10380 MW wind power, the compensation fee is $0.412/MWh (RMB2.745/MWh), which is about 0.5% of the current wind energy price of $81/MWh.

TABLE 12.2 Thermal generation extra cost analysis of Jiangsu provincial power system in 2020 with different wind power installation

Wind power installation scenario (MW)	Thermal generation average cost ($/MWh)	$\sum\limits_{i\in\Omega_C} E_{fi}$ (million $)	$\sum\limits_{i\in\Omega_C} E_{ri}$ (million $)	$\sum\limits_{i\in\Omega_C} E_{si}$ (million $)
0	19.3241	0	0	0
5190	19.3277	0.5072	1.4417	0.2200
10380	19.3396	4.8390	3.5321	0.6711
15570	19.3602	13.4887	6.3908	0.7756
20760	19.3798	20.0834	9.8796	1.2702

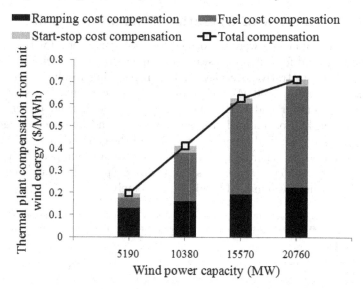

FIGURE 12.5 Compensation fee of each MWh of wind energy of Jiangsu provincial power system in 2020 with increasing amounts of installed wind power

12.4   Pumped Storage Planning with Wind Power Integration

Pumped storage can provide some of the flexibility that power system operators need to balance load and generation in an uncertain environment, and thus enhance a power system's ability to incorporate wind power. Since the process of balancing wind power involves various combinations of wind generation and loads, the amount of pumped storage capacity needed should be evaluated using a substantial number of operating scenarios.

This section uses the power system operation simulation to study the pumped storage capacity planning for the Jiangsu (China) provincial power system, where there is large-scale wind power integration. The year-round operation simulation is able to capture the intermittency of wind power. A detailed cost model for thermal generating units provides an accurate estimate of the benefits of pumped storage. Simulation results clearly show how much generation cost and wind power curtailment should be expected for different amounts of pumped storage capacity. A comparison between the operating and investment costs is then used to determine the optimal pumped storage capacity. The data used in this study are based on one of the electric power planning scenarios for Jiangsu in 2020.

12.4.1   Overview of Operation Simulation Results

Operation simulations were performed for a one-year period with a resolution of one hour and for different installed pumped storage capacities. Figure 12.6 provides an illustrative overview of the dispatching results for one week in January assuming 4500 MW of pumped storage capacity. Units are grouped as follows: nuclear and CHP, wind power generated, wind power curtailed, thermal and hydro, pumped storage (pumping) and pumped storage (generating). The system load is equal to the sum of the generation from nuclear and CHP, wind power generated, thermal and hydro and pumped storage (generating). The week chosen includes the Chinese Spring festival (January 26th and 27th) when system load would be extremely low.

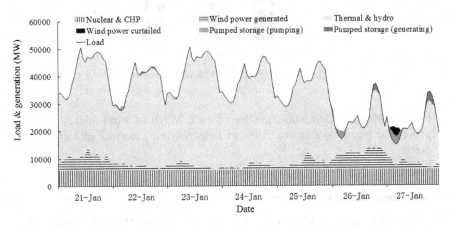

FIGURE 12.6 Dispatching results for one week in 2020 with 4500 MW of installed pumped storage capacity

According to the results, pumped storage units are not dispatched to cycle every day. They are only dispatched to reduce the peak-valley load difference on certain occasions, i.e. when wind farms cause a minimum load problem or when it is more economical to operate pumped storage than to run combined cycle gas turbines (CCGTs). Otherwise, they are dispatched to provide up/down reserves for the load and wind power. At certain moments (the early morning of January 27), the output of the nuclear, CHP, thermal and hydro units cannot be decreased further and exceeds the net load. The output of the wind farms must then be curtailed.

12.4.2   Wind Power Curtailment

Figure 12.7 shows that the amount of wind energy that has to be curtailed decreases rapidly as the pumped storage capacity increases. However, the curtailed wind energy represents a very small proportion of the total wind energy produced, i.e. 0.12% for 4500 MW of pumped storage.

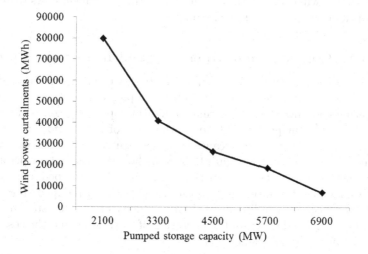

FIGURE 12.7 Wind power curtailment as a function of the installed pumped storage capacity

12.4.3 Savings in Thermal Generation Operating Costs

Table 12.3 shows the operating costs of thermal generation for different amounts of installed pumped storage capacity:

TABLE 12.3 Thermal generation cost analysis with different amounts of installed pumped storage capacity

Pumped storage capacity(MW)	Thermal generation average cost (RMB/MWh)	$\sum_{i \in \Omega_C} E_{fi}$ (billion RMB)	$\sum_{i \in \Omega_C} E_{ri}$ (billion RMB)	$\sum_{i \in \Omega_C} E_{si}$ (million RMB)
2100	279.54	160.82	1.25	34.01
3300	278.43	160.18	1.25	31.28
4500	277.36	159.64	1.21	24.67
5700	276.23	159.16	1.13	2.39
6900	275.54	158.96	1.04	0.00

The results show that the fuel, ramping and start-stop costs all decrease when the pumped storage capacity increases. This shows, as one would expect, that pumped storage is effective in bringing down the total operating cost, by reducing the peak-valley load difference or providing economical reserves. The next section will compare the operating and the investment costs to determine the optimal amount of pumped storage.

12.4.4   Comparison of Operating and Investment Costs

Figure 12.8 shows the total annual operating cost and the annualized investment cost as a function of the installed pumped storage capacity. The total operating cost includes the thermal generation operating costs and the wind power curtailment penalty. The operating cost of hydro and nuclear generation is not considered here because their operating costs do not vary with the pumped storage capacity. Figure 12.9 shows how the sum of the operating and investment costs varies as a function of the installed pumped storage capacity. It suggests that installing a total of 5700 MW of pumped storage is the optimal decision. Installing more than 5700 MW of pumped storage capacity would have a small effect on the operation of the system and the cost of these investments is not justified.

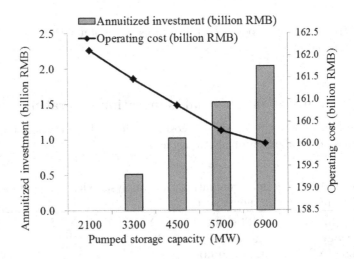

FIGURE 12.8 Operating and investment costs for different amounts of installed pumped storage capacity

One should note that building more pumped storage may substitute some investment in conventional generation from a system adequacy point of view. With this consideration, the equivalent investment of planning pumped storage would be much lower taking into account the avoided investment in conventional generation. On the other hand, in the Jiangsu power system, newly built conventional generation would be coal-fired and would have a low operating cost. Less investment in such a generation would increase the market share of higher cost generations and the operating costs of the entire system. The relative strength of these two factors requires a deeper analysis, which is outside the scope of this book.

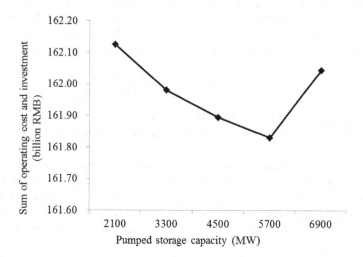

FIGURE 12.9 Sum of the operating and investment costs as a function of the pumped storage capacity

12.5 Grid-Accommodable Wind Power Capacity Evaluation

When conducting wind power planning, it is very important for electric power companies to evaluate the penetration limit of the grid-accommodable wind power. This section proposes a probabilistic method for determining GWPC based on power system operation simulation. Typical power system operation scenarios are generated from the combination of different wind power scenarios and demand scenarios.

12.5.1 Definitions of Grid-Accommodable Wind Power Capacity Evaluation Model

First, the wind power production scenario set Ψ and the system load scenario set Φ are established. In detail, a wind power production scenario is defined as a daily aggregated system wind power output, denoted as W_j, and a system load scenario denotes a daily system load profile, denoted as D_i. Therefore, the system operation scenario used in this section is a random combination of the elements in Ψ and Φ, represented by a pair of (D_i, W_j).

1) Curtailment ratio and accommodation state

The wind power curtailment is determined by both the system load scenario and the wind power production scenario. Thus, the wind power curtailment ratio can be expressed as a function of the system operation scenario (D_i, W_j). The curtailment ratio $\alpha(D_i, W_j)$ is defined as Equation (12.16):

$$\alpha(D_i, W_j) = \frac{Q^w(D_i, W_j)}{Q^w(D_i, W_j) + CUR^w(D_i, W_j)}. \tag{12.16}$$

In Equation (12.16), $Q^w(D_i, W_j)$ is the actual generated wind power electricity function. $CUR^w(D_i, W_j)$ is the wind power electricity curtailment function.

Each time a dispatch simulation is conducted, if the wind power curtailment ratio does not exceed the settled limited amount (i.e., if $\alpha(D_i, W_j)$ is no more than the given α_{max}), (D_i, W_j) is regarded as a success scenario, otherwise is defined as a failure scenario. An indicative function $u^{\alpha_{max}}(\alpha(D_i, W_j))$ is introduced to describe this accommodation state and it can be expressed as Equation (12.17):

$$u^{\alpha_{\max}}(\alpha(D_i, W_j)) = \begin{cases} 1, & \alpha(D_i, W_j) \leqslant \alpha_{\max} \\ 0, & \alpha(D_i, W_j) > \alpha_{\max} \end{cases}, \quad \begin{matrix} D_i \in \Phi \\ W_j \in \Psi \end{matrix}. \tag{12.17}$$

2) Probability of wind power accommodation

For the given Ψ and Φ, the probability of successful wind power accommodation under a given α_{max}, which is denoted by $\Theta^{\alpha_{max}}$, can be calculated. For a certain system scenario (D_i, W_j), the accommodation state can be successful or unsuccessful, and the corresponding $u^{\alpha_{max}}(\alpha(D_i, W_j))$ equals 1 or 0, respectively. $\Theta^{\alpha_{max}}$ can be determined after evaluating all the accommodation state of system scenarios (D_i, W_j). Because the unsuccessful accommodation state equals 0, $\Theta^{\alpha_{max}}$ equals to the expected value of $u^{\alpha_{max}}(\alpha(D_i, W_j))$, as shown in Equation (12.18). $\Theta^{\alpha_{max}}$ varies from 0 to 1, and a greater $\Theta^{\alpha_{max}}$ means that the wind power is more easily grid-accommodable.

$$\begin{aligned} \Theta^{\alpha_{\max}}(\Phi, \Psi) &= 1 \times p\left(u^{\alpha_{\max}}(\alpha(D_i, W_j)) = 1\right) + \\ &\quad 0 \times p\left(u^{\alpha_{\max}}(\alpha(D_i, W_j)) = 0\right) \\ &= 1 \times p\left(u^{\alpha_{\max}}(\alpha(D_i, W_j)) = 1\right) \\ &= E\left(u^{\alpha_{\max}}(\alpha(D_i, W_j))\right), \quad D_i \in \Phi, W_j \in \Psi \end{aligned} \tag{12.18}$$

3) Grid-accommodable wind power

With the increasing of the wind power penetration level, the occurrence probability of curtailment will be higher. $\Theta^{\alpha_{max}}$ will show a decreasing trend.

Therefore the maximum amount of grid-accommodable wind power $A_w^{\alpha_{max}}$ at the given α_{max}, is usually defined as the value at the pivotal point when $\Theta^{\alpha_{max}}$ starts to decrease, shown in Figure 12.10. In Figure 12.10, $\alpha_{max} = 0$, $\Theta^{\alpha_{max}} = 0$, and $A_w^{\alpha_{max}=0}$ is 400 MW at this time.

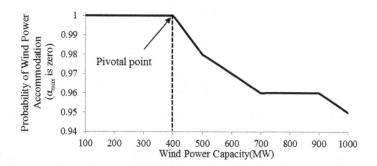

FIGURE 12.10 Schematic diagram for GWPC evaluation

12.5.2 Evaluation Methodology

The structure of the proposed GWPC evaluation method is illustrated in Figure 12.11. Firstly, the typical wind power production scenarios and system load scenarios are selected to compose the system operation scenario set used by GWPC evaluation. Then, the unit commitment and economic dispatch (UCED) simulation are run for each system operation scenario in the set. Finally, a probability model is applied here to conduct GWPC evaluation.
1) Probability-based scenario generation

In this case, two typical daily load curves per month over a one-year period are selected. Thus, there are a total of 24 system load scenarios in Φ. Half of these come from the workday load curve set and the remaining half are drawn from the weekend load curve set.

Given the established Φ, we will apply the law of total probability to calculate the $\Theta^{\alpha_{max}}$. Then, as a joint probability function, $\Theta^{\alpha_{max}}$ can be extended to a probability aggregation of the typical system load scenarios in Φ and is represented as Equation (12.19). The correlation between load and wind is neglected for simplicity as in [15].

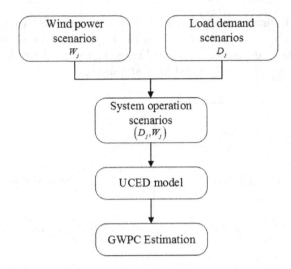

FIGURE 12.11 Overall framework of the GWPC evaluation method

$$\Theta^{\alpha\max} = E\left(u^{\alpha\max}\left(\alpha(D_i, W_j)\right)\right) = \sum_{D_i \in \Phi} \left[p(D_i) \cdot E\left(u^{\alpha\max}\left(\alpha(D_i, \Psi)\right)\mid D_i\right)\right]$$

$$= \sum_{D_i \in \Phi} \left\{p(D_i) \cdot \sum_{W_j \in \Psi} \left[p(W_j \mid D_i) \cdot u^{\alpha\max}\left(\alpha(D_i, W_j)\right)\right]\right\}$$

$$= \sum_{D_i \in \Phi} \left\{p(D_i) \cdot \sum_{W_j \in \Psi} \left[p(W_j) \cdot u^{\alpha\max}\left(\alpha(D_i, W_j)\right)\right]\right\}$$

$$(12.19)$$

Moreover, the wind power output simulation methodology proposed in [16] is applied here to specify Ψ. The simulator comprehensively considers multiple wind farm output curves so as to better match the given statistical indices. Wind speed seasonal rhythms and diurnal patterns, turbine output characteristics, spatial correlations of wind farms, the wake effect are also considered in the model. In this case, N_{max} wind power production scenarios are selected by the Monte Carlo method for each month. This number is decided by an analysis of the tradeoff between the accuracy and computation efforts.

It should be mentioned that the probability of occurrence of each scenario in Ψ is assumed to be the same. This assumption is acceptable because the equal probability scenario set is the fairest way to simulate wind power output when no accurate forecasting result exists. In addition, the probability of occurrence of each scenario in Φ that comes from the workday load curve set is 5/7, and 2/7 of that comes from the weekend load curve set.

2) GWPC evaluation model

With the introduction of scenario generation and system operation simulation, the whole GWPC evaluation model is presented in Figure 12.12. First, a system load scenario set is generated month by month so that its seasonal characteristics can be extracted. Second, wind scenarios are generated based on the selection of the system scenario. A total of N_{max} wind power production scenarios for each month are selected. The system operation simulation model is then run with the new system scenario (D_i, W_j) until all the system scenarios are considered. Finally, the yearly $\Theta^{\alpha max}$ is given.

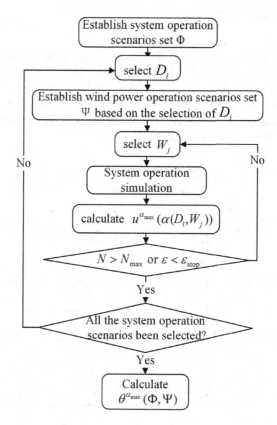

FIGURE 12.12 Flow chart of the calculation

With this model, several analyses with different objectives can be performed to study the GWPC in detail. The first analysis is performed by adjusting the integrated wind power penetration. The $\Theta^{\alpha max}$ will change with the different penetration percentages. This analysis will determine the maximum GWPC under the given requirement of α_{max}. The second analysis changes the constraints of the system operation simulation model so that the influences of each factor on the GWPC can be evaluated. In this section, a simulation with the five factors commonly used in China mentioned above is performed to act as a comparison sample. Based on the system operation simulation considering single-factor constraint, the results can then be analyzed to find out the short bar of the studied system when accommodating wind power. The third analysis changes the wind curtailment limitation α_{max} so that the tradeoff between wind farm profits and operating cost can be studied. The result offers the potential for a quantified cost analysis of wind power planning based on the curtailment requirement.

12.5.3   Case Study

1) IEEE 39-bus system

This subsection takes the IEEE 39-bus system shown in Figure 12.13 as an example to study how $\Theta^{\alpha_{max}}$ varies with wind power capacity and α_{max}. All parameters are taken from Matpower 4.0 [17]. Two wind farms are added to the system, one at bus 8 with a capacity of 804 MW and the other at bus 13 with a capacity of 933 MW. The total wind power capacity accounts for 19.1% of the total system capacity.

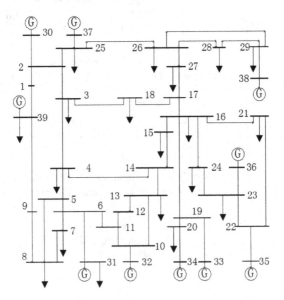

FIGURE 12.13 IEEE 39-bus system

Figure 12.14 presents $\Theta^{\alpha_{max}}$ for the modified IEEE 39-bus system. A higher α and lower amount of installed wind power will result in higher $\Theta^{\alpha_{max}}$. When α_{max} is 5%, $\Theta^{\alpha_{max}}$ would be no less than 85%, even when the wind power capacity reaches 1.4 p.u. This figure shows the quantified accommodation probability under different wind power capacities and α_{max}. Thus, according to the actual condition of the power system, the planner could perform accurate wind power planning with the expected α_{max} and $\Theta^{\alpha_{max}}$.

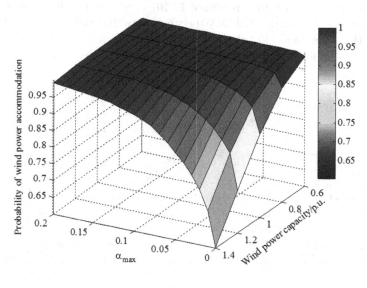

FIGURE 12.14 Relationship of $\Theta^{\alpha_{max}}$, α_{max} and wind power capacity

2) GWPC evaluation for a large power system in China

In this subsection, the GWPC evaluation method is applied to a real large power system planning in China for a certain year. The power system is divided into three subzones, and the two sections draw great focus because of serious power flow congestion, as shown in Figure 12.15. There are totally 11 wind resource areas, and they are dispersed at three subzones. The distribution proportion of wind power capacity for each wind resource area is fixed. The proposed method is used to evaluate the maximum GWPC for the whole power system when $\alpha_{max} = 0$.

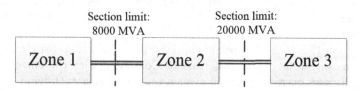

FIGURE 12.15 Three subzones of the real power system and the transmission limit of two sections

The GWPC evaluation result is shown in Figure 12.16. When wind power capacity is lower/greater than 16750 MW, peak-valley regulation constraints (PRC) / frequency regulation constraints (FRC) is the bottleneck factor limiting wind power accommodation. The accommodation probability considering all constraints (black line in Figure 12.16) decreases rapidly when the wind power capacity exceeds 12700 MW because more than two factors (PRC, FRC) limit the accommodation at this time. If the expected wind power accommodation probability is preset over 0.8, the maximum grid-accommodable wind power is 12150 MW.

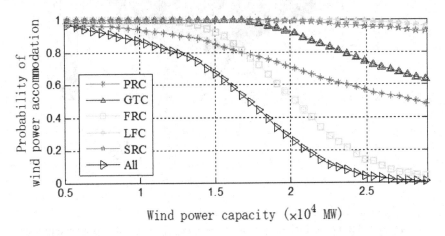

FIGURE 12.16 $\Theta^{\alpha_{max}=0}$ as a function of wind power capacity for a real large power grid in China

12.6 Summary

This chapter presents a general power system operation simulation model considering renewable energy integration and shows its application in the following three aspects:

Section 12.3 gives new insights into thermal generation cost changes caused by increasing wind power penetration. Compared with the current research which focuses on reductions on system operating cost with wind power penetration, the section proposes a model to evaluate the thermal generation market share changes and the extra cost of balancing wind power, including the efficiency loss at low load level, ramping cost, and start-stop cost. The simulation results show that wind power brings considerable cost increase on thermal generation. The results based on the Jiangsu provincial power system reveal that when wind power is integrated, the more flexible generation with higher cost is dispatched in peak load regulation, units start and stop more often and more money is spent on ramping costs, which increases the average cost. If these extra costs are borne by the wind energy producers, the compensation fee will be 0.5% of the wind energy price.

Section 12.4 describes a multi-scenario approach to determine the optimal pumped storage capacity to be installed in Jiangsu to accommodate the planned wind power generation. Results show that pumped storage brings down the operating cost and reduces the number and size of the wind power curtailments. Considering the sum of the operating and investment costs, installing 5700 MW of pumped storage capacity is the best option.

Section 12.5 proposes a probabilistic method for determining GWPC. The method considers the influences of related factors of the generation side, transmission side, and the load side. Utilizing a multi-factor system operation simulation, the method not only can quantify the technically grid-accommodable wind power for certain power system but also can identify the bottleneck factors that affect the accommodation situation. In a deregulated market, GWPC could help evaluate the capacity factor of wind farms, the value of which will directly affect the returns of investment for wind farms. In a monopolistic market, such as China, GWPC is one of the most important information required by the planners to make optimal wind development plans.

Bibliography

[1] B. C. Ummels, M. Gibescu, E. Pelgrum, W. L. Kling, and A. J. Brand. Impacts of wind power on thermal generation unit commitment and dispatch. *IEEE Transactions on energy conversion*, 22(1):44–51, 2007.

[2] L. Göransson and F. Johnsson. Dispatch modeling of a regional power generation system–integrating wind power. *Renewable Energy*, 34(4):1040–1049, 2009.

[3] N. Zhang, T. Zhou, C. Duan, X. Tang, J. Huang, Z. Lu, and C. Kang. Impact of large-scale wind farm connecting with power grid on peak load regulation demand. *Power System Technology*, 34(1):152–158, 2010.

[4] National Development and Reform Commission (NDRC) People's Republic of China. Renewable energy prices and the cost-sharing management pilot scheme. http://www.sdpc.gov.cn/jggl/zcfg/t20060120_57586.htm.

[5] T. D. Couture, K. Cory, C. Kreycik, and E. Williams. A policymaker's guide to feed-in tariff policy design. Technical report, National Renewable Energy Lab. (NREL), Golden, CO (United States), 2010.

[6] The impact of feed-in tariffs across Europe. `http://www.joneslanglasalle.com/Pages/`.

[7] A. Tuohy and M. O'Malley. Pumped storage in systems with very high wind penetration. *Energy policy*, 39(4):1965–1974, 2011.

[8] B. C. Ummels, E. Pelgrum, and W. L. Kling. Integration of large-scale wind power and use of energy storage in the Netherlands' electricity supply. *IET Renewable Power Generation*, 2(1):34–46, 2008.

[9] B. C. Ummels, M. Gibescu, E. Pelgrum, W. L. Kling, and A. J. Brand. Impacts of wind power on thermal generation unit commitment and dispatch. *IEEE Transactions on Energy Conversion*, 22(1):44–51, 2007.

[10] P. Denholm and M. Hand. Grid flexibility and storage required to achieve very high penetration of variable renewable electricity. *Energy Policy*, 39(3):1817–1830, 2011.

[11] J. Garcia-Gonzalez, R. de la Muela, L. M. Santos, and A. M. Gonzalez. Stochastic joint optimization of wind generation and pumped-storage units in an electricity market. *IEEE Transactions on Power Systems*, 23(2):460–468, 2008.

[12] P. D. Brown, JA Peças Lopes, and M. A. Matos. Optimization of pumped storage capacity in an isolated power system with large renewable penetration. *IEEE Transactions on Power Systems*, 23(2):523–531, 2008.

[13] P. Pinson and G. Kariniotakis. Conditional prediction intervals of wind power generation. *IEEE Transactions on Power Systems*, 25(4):1845–1856, 2010.

[14] C. W. Hamal and A. Sharma. Adopting a ramp charge to improve performance of the Ontario market. `http://www.ieso.ca/imoweb/pubs/consult/mep/MP_WG-20060707-ramp-cost.pdf`.

[15] N. Zhang, C. Kang, D. S. Kirschen, Q. Xia, W. Xi, J. Huang, and Q. Zhang. Planning pumped storage capacity for wind power integration. *IEEE Transactions on Sustainable Energy*, 4(2):393–401, 2013.

[16] N. Zhang, C. Kang, C. Duan, X. Tang, J. Huang, Z. Lu, W. Wang, and J. Qi. Simulation methodology of multiple wind farms operation considering wind speed correlation. *International Journal of Power & Energy Systems*, 30(4):264, 2010.

[17] R. D. Zimmerman and C. E. Murillo-Sanchez. Matpower: A MATLAB® power system simulation package, user's manual, version 4.0. 2011.

13

Capacity Credit of Renewable Energy

The capacity credit of renewable energy is defined as the ratio of the capacity it can serve to that of conventional generation under the same load and reliability level. How much capacity credit should be given to renewable energy in generation system adequacy analysis is a question of great interest worldwide. Both a theoretical analysis and an accurate evaluation of the renewable energy power capacity credit are essential for understanding its contribution to power system reliability. Taking wind power as an example, this chapter reviews the definitions and algorithms of the capacity credit and summarizes the factors that affect its value. A rigorous model based on the definition of the reliability function (RF) is also presented. Numerical tests demonstrate the accuracy of the proposed model and its potential applicability under different circumstances.

13.1 Introduction

The integration of large amounts of renewable generation creates a new set of reliability issues. Intermittent and stochastic forms of renewable generation, such as wind power, simply do not "serve" loads in the same way as conventional units. While they are clearly able to supply energy, what proportion of their nominal capacity can be counted in reliability calculations is an open question. In other words, what capacity credit should be given to the renewable energy?

The concept of a power system capacity credit was first proposed by Garver in 1966 [1] for the purpose of assessing the load-carrying capability of units with different random outage rates in the reliability sense. The calculation formula for the credible capacity of a single unit is given based on the two-state model. Edward Kahn [2], Haslett John [3] *et al.* first applied the concept of the capacity credit to the analysis of wind power in the late 1970s. In more than 40 years of research, domestic and foreign scholars have presented diverse definitions and calculation methods of the wind power capacity credit. The summary and classification of various definitions and methods are shown in Figure 13.1.

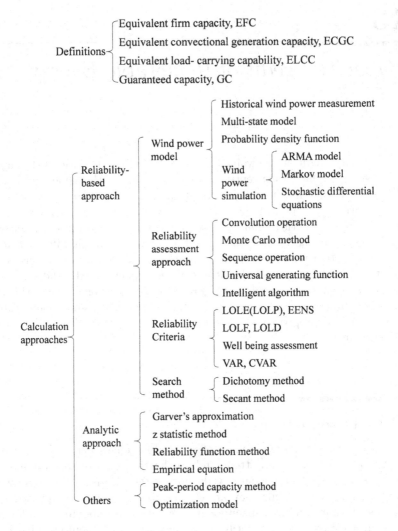

FIGURE 13.1 Classification of capacity credit definitions and calculation methods

The capacity credit of renewable energy represents its contribution to the generation adequacy of the system. Various definitions of capacity credit are possible [4]. The equivalent firm capacity (EFC) and effective load-carrying capability (ELCC) are the two most common definitions. In principle, these definitions are similar because they replace the fluctuating power output with an equivalent capacity that can be combined with the capacity of conventional generating units for estimating generation adequacy. The concept of capacity credit thus bridges the gap between the stochastic behavior of renewable energy and traditional system generation adequacy analysis. This concept has a long history and was first proposed to estimate the load-carrying capabilities of conventional generation in power system adequacy analysis [1]. The concept of the capacity credit of wind power was first introduced in approximately 1980 [2, 3]. Methods for evaluating the capacity credit can be divided into four categories: Monte Carlo methods [5, 6], peak-period capacity factors [7], convolution methods [8], [9] and analytical methods [10–12]. In addition, Patil and Ramakumar [13] proposed the weighted capacity credit method, and Zhu *et al.* [14] calculated the capacity credit based on a reliability constrained generation optimization model.

Based on a panel session on "Capacity Credit Value of Wind Power" held at its 2008 General Meeting, a Task Force of the IEEE Power and Energy Society recommended a method based on convolution [8]. This method relies on a capacity outage probability table (COPT) to evaluate the reliability of the system with and without wind power using the hourly load and wind power output. The load is continually increased progressively in the equivalent system without wind power until the reliability values of the two systems match. At that point, the increase in load represents the wind power capacity value. This method is applicable to a wide range of wind power penetrations and captures the possible correlation between load and wind power. However, the method requires numerical case studies and thus cannot support analytical analysis. The z-statistic method [11] takes the difference between available resources and load over peak demand hours as a random variable with an associated probability distribution. An analytical formula is derived in which the capacity credit is the capacity factor of the wind power minus a fractional function related to the z-statistic and the standard deviation of the distribution. Zachary and Dent [12] presented a similar analytical formula and showed that the capacity credit of wind power depends on its capacity factor and a correction factor proportional to its variability. These two analytical formulas provide insight into the factors that determine the credible capacity. However, they are both based on the assumption that the capacity of renewable energy is small relative to the load, and their applicability is questionable if this assumption is not satisfied.

Other authors have discussed the factors that determine the value of renewable energy capacity credit. For example, Ensslin *et al.* [15] presented results obtained at U.S. and European wind farms and showed that the credit depends on the statistical characteristics of the wind power and the load profile and that it decreases as the wind power penetration increases. Voorspools *et al.* [16] and Castro *et al.* [17] emphasized that the use of chronological data of load and wind power yields more accurate results. Vallee *et al.* [5] and Billinton *et al.* [18] showed that the capacity credit is highest when the production from different wind farms is uncorrelated and decreases as the correlation between the wind farms increases. Caralis and Leite [19] found that pumped storage generation tends to increase the capacity credit of wind power.

This chapter conducts in-depth research into the calculation methods and influential factors of the renewable energy capacity credit. First, the calculation method for the capacity credit is presented based on the principle of dependent probability sequence operation (DPSO) [11], which applies the sequential method to describe the randomness of conventional units, renewable energy units and loads. In addition, the DPSO is applied to implement the process of random production simulation that considers the renewable energy output, which enables fast calculation of the capacity credit by iteration. An analytical model of the capacity credit is presented to analyze its influential factors from the perspective of macroscale statistics and a microscale time sequence respectively; this model provides a theoretical basis for calculating the capacity credit by using the peak-load capacity factor method. Finally, the analytical-based calculation method for capacity credit is described, and the IEEE RTS-79 system is used to verify the effectiveness of the above models and methods.

13.2   DPSO-Based Capacity Credit Calculation

13.2.1   Definition and Calculation Framework

According to the EFC definition of the renewable energy capacity credit, the renewable energy capacity in the system is set to Cap_w, and the remaining conventional machine installed capacity is set to Cap_c. Cc_w represents the conventional machine installed capacity that is equivalent to renewable energy. The equivalent conventional unit here can be set to be 100% reliable or can be given a certain random outage rate. The equivalent conventional unit capacity for renewable energy units is simply referred to as the virtual unit below. $f(Cap_w)$ represents the reliability index of the system with diverse units. The credible capacity of the renewable energy units is defined by Cc_w and satisfies the following condition:

$$f(\{Cap_c, Cap_w\}) = f(\{Cap_c, Cc_w\}). \tag{13.1}$$

The capacity credit Cr_w is defined by the ratio of the credible capacity to the actual installed capacity of the wind turbine:

$$Cr_w = Cc_w/Cap_w. \tag{13.2}$$

It is evident that the concept of capacity credit transforms the unit capacity of different reliability levels into a unified level, which makes the adequacy assessment of the system more accurate and objective. The reliability index in the above calculation, in conjunction with the loss of load expectation (LOLE) or expected energy not supplied (EENS), can be selected as the calculation reference [4].

Based on the DPSO, the renewable energy capacity credit modeling framework is shown in Figure 13.2. The procedure is as follows: First, conduct the sequential modeling of conventional units, renewable energy units and loads. In the computational process, use the DPSO to carry out the stochastic production simulation of the actual system with wind farms and introduce an equivalent system. In the equivalent system, remove the wind turbine while adding the virtual unit with a certain forced outage rate (FOR). Perform random production simulations on the equivalent system, and conduct an iterative operation of the actual system reliability to obtain the conventional unit capacity that can be replaced by renewable energy, which is exactly the capacity credit of the wind turbine.

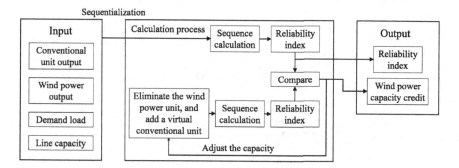

FIGURE 13.2 Capacity credit calculation modeling concept for renewable energy generation based on the DPSO

13.2.2   Overall Calculation Process

The calculation process for the renewable energy capacity credit is shown in Figure 13.3. First, apply DPSO theory to calculate the reliability index of the actual system with the renewable energy units. Then, calculate the reliability index EENS of the virtual system that includes the virtual unit without the wind turbine. Obtain the virtual unit capacity by means of the bisection method so that the reliability index of the actual system and the equivalent system is within a certain error range. At this point, the virtual unit capacity is the credible capacity of the wind turbine.

This chapter constructs the algorithm flow according to the first category of definition for the renewable energy capacity credit. However, since the capacity credit is calculated based on the system reliability regardless of the methods used, the proposed calculation method is also valid in calculating the second category of definition for the capacity credit.

At present, most studies on the capacity credit of renewable energy are developed under the framework of the power generation system reliability analysis. The reliability calculation considers the random outage of the power generation system only but the random outage of the transmission system is not considered. The reason is that renewable energy will have different capacity credits in different regions of the system when the reliability of the transmission component is considered, which is not conducive to an in-depth analysis of the variation pattern and influential factor of the capacity credit. In fact, whether the reliability of the transmission system is considered in the calculation of the renewable energy capacity credit does not change the overall calculation process. The only difference lies in the factors being considered in the calculation of the system reliability index. The renewable energy capacity credit calculation based on the power system reliability is described in detail in the following section. The reliability calculation method that considers the random outage of the transmission system is specified in [20].

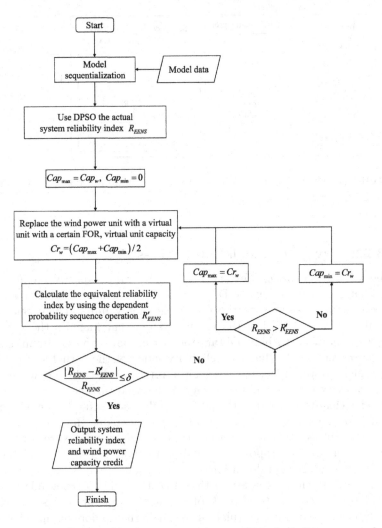

FIGURE 13.3 Renewable energy capacity credit calculation process based on the DPSO

13.2.3 Detailed Calculation Process

1) Reliability index calculation

There are m conventional units in the system, and the initial available margin sequence is $S_{si}^{(0)}(j), (i = 1, 2, ..., m)$. The initially available margin sequence of renewable energy is $S_w^{(0)}(j)$, which is the probability sequence corresponding to the total output of all wind farms in the system that can be obtained from the corresponding historical data. This parameter can also be calculated by DPSO on the sequence corresponding to each wind farm. The initial system load margin sequence is $S_d^{(0)}(j)$, and its modeling principle is the same as that of $S_w^{(0)}(j)$. In the capacity credit calculation, the modeling principle of the virtual unit sequence is the same as that of the conventional unit sequence. If the set virtual unit has a certain random outage rate, its sequential model and the conventional unit constitute a two-state model; otherwise, it is a single-state model, that is, its initial available margin is a unit sequence. The initial available margin sequence of the virtual unit is designated $S_v^{(0)}$; if there are several units, then they are designated $S_{v1}^{(0)}, S_{v2}^{(0)}, \cdots,$ $S_{vk}^{(0)}$, where k is the number of virtual units.

In the above algorithm flow, the key element in the calculation process is the calculation of the reliability index for the actual system that includes renewable energy and the equivalent system that contains the virtual unit. The application of single-node stochastic production simulation based on sequence operation [20] can implement fast calculation of the reliability index for the actual and equivalent systems. First, perform dependency subtraction-type convolution of the load and wind turbine:

$$S_d^{(1)} = S_d^{(0)} \overset{c}{\ominus} S_w^{(0)}. \tag{13.3}$$

Based on $S_d^{(1)}$, the conventional unit is incorporated for production simulation. Since there are a total of m conventional units, the final load residual margin sequence $S_d^{(m+1)}$ is the result of continuous subtractive-type convolution of the conventional unit margin sequence with respect to the initial load margin sequence:

$$S_d^{(m+1)} = S_d^{(1)} \ominus S_{s1}^{(0)} \ominus S_{s2}^{(0)} \ominus \cdots \ominus S_{sm}^{(0)}. \tag{13.4}$$

The system reliability indices R_{LOLE} and R_{EENS} can be obtained by the residual margin sequence $S_d^{(m+1)}$:

$$R_{LOLE} = 8760[1 - S_d^{(m+1)}(0)]$$
$$R_{EENS} = 8760\bar{C}\sum_{j=1}^{N_l}(jS_d^{(m+1)}(j)). \tag{13.5}$$

In the calculation of the equivalent system reliability index, according to the exchangeable order property of continuous subtraction-type convolution in sequence operation, first, perform continuous subtractive-type convolution on the initial load available margin $S_d^{(0)}$ along with the conventional unit; next, the subtraction-type-convolution calculation is conducted by using the result of continuous subtraction-type convolution $S'^{(m)}_d$ along with the virtual unit available margin sequence to obtain the final residual margin of the equivalent system $S'^{(m+1)}_d$:

$$S'^{(m)}_d = S_d^{(0)} \ominus S_{s1}^{(0)} \ominus S_{s2}^{(0)} \ominus \cdots \ominus S_{sm}^{(0)}$$
$$S'^{(m+1)}_d = S'^{(m)}_d \ominus S_v^{(0)}. \tag{13.6}$$

The reliability indices of the equivalent system R'_{LOLE} and R'_{EENS} are given by Equation (13.5).

A correlation may exist between the load and renewable energy output in the calculation stated above, so it is necessary to apply the DPSO. There is generally no correlation in outage between the conventional generator units, so the remaining operations are basic sequence operations.

2) Virtual unit capacity quick search

The virtual system reliability can be changed by adjusting the virtual unit capacity Cc_w. Through the bisection method, a virtual unit capacity Cc_w can be found when the reliability index difference between the virtual and actual system is less than δ. EENS is selected for the reliability index, which gives

$$\frac{|R_{EENS} - R'_{EENS}|}{R_{EENS}} \le \delta. \tag{13.7}$$

At this point, the reliability of the virtual system matches that of the equivalent system. In addition, the virtual unit capacity is the credible capacity of the wind turbine, and the ratio of the renewable energy credible capacity to the actual renewable energy capacity is then the capacity credit.

The calculation of the renewable energy capacity credit by sequence operation theory can reduce the number of repeated calculations, which greatly improves the calculation efficiency. In Equation (13.6), the virtual unit is not incorporated into the production simulation first; instead, the subtraction-type convolution between the load and conventional unit available margin is calculated. The reason is that the load residual margin sequence $S'^{(m)}_d$ needs to be calculated only once when following the said calculation order. In the bisection method, only the subtraction-type-convolution process between $S'^{(m)}_d$ and the initial virtual available margin needs to be calculated to obtain $S'^{(m+1)}_d$ and then $S'^{(m+1)}_d$, followed by the calculation of the system reliability index, which greatly reduces the calculation amount and thereby improves the calculation speed. By the nature of the sequence operation, it is evident that the subtraction-type-convolution result is not affected by a change of order in the main subtraction sequence. Hence, the reliability index results of the above calculation process are the same as that obtained by incorporating virtual units into the production simulation first.

13.3 Analytical Model of the Renewable Energy Capacity Credit

13.3.1 Power System RF

The key concept in modeling capacity credit is the chosen power system RF. As seen from the supply side of the power system, the total available capacity x is a stochastic variable, and its distribution $P(x)$ can be calculated using iterative discrete convolution of each generator's capacity and FOR [8, 21]. The RF is then defined as the expected value of the reliability index for different load levels. For instance, the RFs with reliability indices LOLE and EENS are given by Equations (13.8) and (13.9), respectively:

$$LOLE: \ f_{RF}(d) = \int_0^d P(x)dx \tag{13.8}$$

$$EENS: \ f_{RF}(d) = \int_0^d (d-x)P(x)dx \tag{13.9}$$

where $f_{RF}(d)$ represents the RF at load level d. An illustration of $P(x)$ and $f_{RF}(d)$ based on EENS is given in Figure 13.4 for the IEEE RTS-79 system [22].

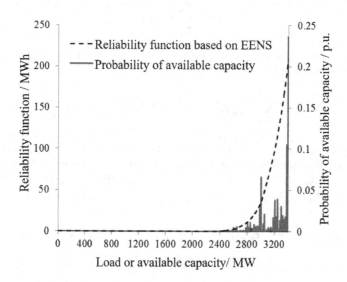

FIGURE 13.4 Probability distribution function of the available capacity and corresponding RF of the IEEE RTS-79 system

The RF in Equations (13.8) and (13.9) is associated with the reliability at a single instant in time. The reliability index for an extended period of time (i.e., a year) during which the load varies is then given by

$$R = \frac{1}{N} \sum_{i=1}^{N} f_{RF}(d_i). \tag{13.10}$$

If a non-chronological method is adopted, then d_i is the sampled system load, and N is the size of the sample.

13.3.2 Analytical Model of the Renewable Energy Capacity Credit Based on the RF

The idea starts with the formulation of the capacity credit evaluation method proposed and recommended by the IEEE Task Force [8]. Suppose that $f_{RF}(\cdot)$ is the RF for the conventional units only. When renewable energy is integrated into the system, conventional units need to supply only $(d_i - P_i)$ instead of d_i at each time period. The system reliability index is then

$$R = \frac{1}{N} \sum_{i=1}^{N} f_{RF}(d_i - P_i). \tag{13.11}$$

If Cc_w is the credible capacity of the renewable energy as specified by Equation (13.11), then the system reliability index can also be expressed as

$$R = \frac{1}{N} \sum_{i=1}^{N} f_{RF}(d_i - Cc_w). \tag{13.12}$$

According to the definition of credible capacity, combining Equations (13.11) and (13.12), we obtain

$$\frac{1}{N} \sum_{i=1}^{N} f_{RF}(d_i - P_i) = \frac{1}{N} \sum_{i=1}^{N} f_{RF}(d_i - Cc_w). \tag{13.13}$$

Each $f_{RF}(d_i - P_i)$ term on the left-hand side of this equation can be expressed using another value of $f_{RF}(\cdot)$ and the slope of $f_{RF}(\cdot)$ in a "point and slope" form. If we choose $(d_i - Cc_w)$ as this arbitrary point, for each $f_{RF}(d_i - P_i)$, we have

$$\begin{aligned} f_{RF}(d_i - P_i) &= f_{RF}(d_i - Cc_w) + k_i\left[(d_i - P_i) - (d_i - Cc_w)\right] \\ &= f_{RF}(d_i - Cc_w) + k_i(Cc_w - P_i) \end{aligned} \tag{13.14}$$

where k_i is the slope of $f_{RF}(\cdot)$ between $(d_i - P_i)$ and $(d_i - Cc_w)$:

$$k_i = \frac{f_{RF}(d_i - P_i) - f_{RF}(d_i - Cc_w)}{[(d_i - P_i) - (d_i - Cc_w)]} = \frac{f_{RF}(d_i - P_i) - f_{RF}(d_i - Cc_w)}{Cc_w - P_i}.$$

(13.15)

Substituting Equation (13.14) into each item on the left-hand side of Equation (13.13), we have

$$\frac{1}{N} \sum_{i=1}^{N} [f_{RF}(d_i - Cc_w) + k_i(Cc_w - P_i)] = \frac{1}{N} \sum_{i=1}^{N} f_{RF}(d_i - Cc_w).$$

(13.16)

Simplifying Equation (13.16), we obtain

$$\sum_{i=1}^{N} [k_i(Cc_w - P_i)] = 0$$

(13.17)

$$Cc_w = \sum_{i=1}^{N} \left(\frac{k_i}{\sum_{i=1}^{N} k_i} P_i \right).$$

(13.18)

The capacity credit Cr_w is then

$$Cr_w = \sum_{i=1}^{N} \left(\frac{k_i}{\sum_{i=1}^{N} k_i} \frac{P_i}{Cap_w} \right).$$

(13.19)

Equation (13.19) shows that the renewable energy capacity credit is a linear combination of the renewable energy output percentage (P_i/Cap_w) of each period with a weighting coefficient $\alpha_i = k_i / \sum_{i=1}^{N} k_i$. This weighting coefficient is further determined by the slope of the RF between ($d_i - P_i$) and ($d_i - Cc_w$) at time i proportional to the summation among all the time periods. The periods with a higher weighting coefficient have more significance in determining the value of the capacity credit, which indicates that the temporal correspondence between the wind and load is thus a key factor of the capacity credit.

Note that Equation (13.19) requires no assumption or approximation and is an accurate expression for the renewable energy capacity credit, irrespective of the level of renewable energy penetration and the specific dependence structure that exists between the load and the wind generation. However, k_i depends on Cc_w, which is not available before the capacity credit has been determined. Subsection 13.3.5 describes how this model can be calculated.

13.3.3   Formulation Based on the ELCC

The model above is based on the definition of the EFC. A similar analytical model can be derived for the ELCC. The Cc_w based on the ELCC is

$$\frac{1}{N}\sum_{i=1}^{N} f_{RF}(d_i) = \frac{1}{N}\sum_{i=1}^{N} f_{RF}(d_i - P_i + Cc_w). \qquad (13.20)$$

The left-hand side of Equation (13.20) represents the system reliability level without renewable energy, while the right-hand side takes into account the renewable energy and the increase in load. A derivation similar to the one given in the previous subsection leads to an equation identical to Equation (13.19) but with k_i given by

$$k_i = \frac{f_{RF}(d_i - P_i + Cc_w) - f_{RF}(d_i)}{[(d_i - P_i + Cc_w) - d_i]} = \frac{f_{RF}(d_i - P_i + Cc_w) - f_{RF}(d_i)}{Cc_w - P_i}. \qquad (13.21)$$

A comparison of Equation (13.21) and Equation (13.15) shows that the structures of the formulas are the same, but the numerator of the weighting coefficient is $f_{RF}(d_i - P_i + Cc_w) - f_{RF}(d_i)$ instead of $f_{RF}(d_i - P_i) - f_{RF}(d_i - Cc_w)$. This difference arises because the EFC definition assumes a constant load level, while the ELCC definition assumes a constant reliability level. When the renewable energy penetration is low (i.e., when $Cc_w << d_i$), the two definitions of the capacity credit give essentially the same value. The difference may become substantial as the renewable energy penetration increases. The EFC definition is thus better for comparing different wind integration scenarios for a given year, while the ELCC definition is better for comparisons over different time horizons.

13.3.4   Integrating the Value of the Capacity Credit from Different Definitions and RF Values

Choosing different definitions of capacity credit and RF results in different values of the capacity credit. It is possible to find a reasonable capacity credit value that can cover multiple metrics. The derivation in Subsection 13.3.3 shows that the factor that differentiates the capacity credit values is the weighting coefficients calculated by different metrics. The values of capacity credit can be integrated by combining the weighting coefficients calculated using different metrics for each time period, i.e., the average of the weighting coefficients. According to Equation (13.18), the integrated capacity credit $\tilde{C}c_w$ from M metrics can be formulated as

$$\tilde{C}c_w = \sum_{i=1}^{N} (\tilde{\alpha}_i P_i) \qquad (13.22)$$

where $\alpha_i^j, j = 1, 2, ..., M$ is the coefficient derived from metric j on time period i and $\tilde{\alpha}_i = (\alpha_i^1 + \alpha_i^2 + ... + \alpha_i^M)/M$ is the integrated weighting coefficients of time period i. $\tilde{C}c_w$ can be rewritten as

$$\tilde{C}c_w = \sum_{i=1}^{N} \left[\frac{1}{M} \sum_{j=1}^{M} (\alpha_i^j P_i) \right] = \frac{1}{M} \sum_{j=1}^{M} \left[\sum_{i=1}^{N} (\alpha_i^j P_i) \right] = \frac{1}{M} \sum_{j=1}^{M} P_{cc}^j \quad (13.23)$$

where Cc_w^j is the capacity credit value obtained using different metrics. Equation (13.23) shows that choosing such a set of integrated weighting coefficients is equivalent to directly averaging the values of capacity credit. Other integrating metrics are possible; however, their discussion is beyond the scope of this work.

13.3.5   Rigorous Method for Calculating the Capacity Credit

The basis of the proposed model is the RF (e.g., LOLE or EENS) that can be calculated using the well-known concept of the COPT, which associates the available capacities and their probabilities using one of several efficient computational techniques [21]. The RFs can be different over different time periods because different units are being maintained and are thus not included in the calculation of the COPT. Maintenance thus influences the value of the capacity credit because it dominates the RFs and hence the weighting of each time period in Equation (13.19).

The slope of the RF k_i cannot be directly calculated because Cc_w in Equations (13.15) and (13.21) is itself the credible capacity that this model calculates and is thus not known a priori. This quantity, therefore, has to be approximated. In the next section, we show how the errors caused by this approximation are eliminated using an iterative procedure.

The easiest approximation consists of replacing Cc_w by the average renewable energy output $E(P_w)$ in the calculation of k_i. Various authors [14], [7] have shown that the capacity factor (average output ratio) of renewable energy gives an acceptable approximation of the capacity credit. However, this approximation only works when the renewable energy penetration is low and the renewable energy and load profile are independent.

A more sophisticated approach consists of using the derivative of the RF to approximate each k_i. Since k_i is the slope of RF between $(d_i - P_i)$ and $(d_i - Cc_w)$, its derivative, i.e., $f'_{RF}(d_i - P_i)$, provides a reasonable approximation of the slope. It can be demonstrated that this approximation results in an error of the higher-order term of $(P_i - Cc_w)$. Taking the Taylor series expansion of each $f_{RF}(d_i - Cc_w)$ term around each $(d_i - P_i)$ in Equation (13.13), we obtain

$$
\begin{aligned}
f_{RF}(d_i - Cc_w) &= f_{RF}(d_i - P_i) + f'_{RF}(d_i - P_i)\left[(d_i - Cc_w) - (d_i - P_i)\right] \\
&\quad + o\left[(d_i - Cc_w) - (d_i - P_i)\right] \\
&= f_{RF}(d_i - P_i) + f'_{RF}(d_i - P_i)(P_i - Cc_w) + o(P_i - Cc_w)
\end{aligned}
\tag{13.24}
$$

where $f'_{RF}(d_i - P_i)$ is the derivative of $f_{RF}(\cdot)$ at $(d_i - P_i)$ and $o(P_i - Cc_w)$ represents the higher-order infinitesimal term of $(P_i - Cc_w)$, i.e., the Peano remainder term of Taylor's formula. Since $(P_i - Cc_w)$ is small relative to the load d_i, which dominates the value of $f'_{RF}(d_i - P_i)$, we can omit the infinitesimal $o(P_i - Cc_w)$ and substitute Equation (13.24) into each item on the right-hand side of Equation (13.23):

$$
\frac{1}{N}\sum_{i=1}^{N} f_{RF}(d_i - P_i) = \frac{1}{N}\sum_{i=1}^{N}\left[f_{RF}(d_i - P_i) + f'_{RF}(d_i - P_i)(P_i - Cc_w)\right].
\tag{13.25}
$$

Through simplification of Equation (13.25), we obtain

$$
Cc_w = \sum_{i=1}^{N}\left[\frac{f'_{RF}(d_i - P_i)}{\sum\limits_{i=1}^{N} f'_{RF}(d_i - P_i)} P_i\right].
\tag{13.26}
$$

The two ways of calculating the slope of the RF both give suitable approximations of k_i most of the time. However, they both require assumptions. In circumstances where the renewable energy and load profile are not independent (i.e., Cc_w is not close to $E(P_w)$) and the share of renewable energy is large ($(P_i - Cc_w)$ is not very small relative to the load), these approximations fail to provide satisfactory results. This is also a shortcoming of other models of the capacity credit [11], [12]. In the next subsection, we propose an iterative approach to minimize this error under all circumstances.

Figure 13.5 illustrates the proposed rigorous yet simple and practical method for calculating the capacity credit. The method takes $E(P_w)$ as a first approximation of the credible capacity and improves on this approximation of Cc_w using an iterative approach. Although in each iteration, k_i and Cc_w are inaccurate, Equation (13.19) itself does not involve any approximations; Cc_w is renewed through each iteration and converges to its true value through iteration. The essential difference between this model and the other models is that the only error in this model arises from an insufficient number of iterations. Therefore, the proposed method is valid for a wide range of renewable energy penetration and non-independent circumstances.

The convergence of this method is fast even for high wind energy penetration levels, not only because RF is usually a monotonic function but also because the approximation errors on the values of k_i mostly cancel each other

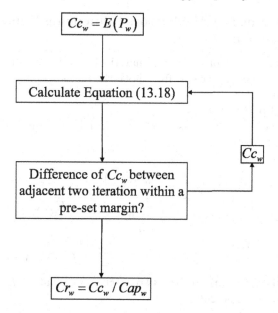

FIGURE 13.5 Flow chart of the iterative approach for calculating renewable energy capacity credit.

through the massive linear combination in the calculation of λ. This finding is confirmed by the case studies.

3) Discussion

Before evaluating the capacity credit, the RF can be calculated at each load level using a COPT and stored in a table. The capacity credit is then calculated directly using the values of the renewable energy output and the load. Since the iterative algorithm involves only analytical expressions, the calculation is very fast, and the algorithm is straightforward to implement.

The proposed method is also widely applicable for the following reasons:

1. It applies to both chronological and non-chronological conditions. In a chronological study, the terms d_i and P_i in Equation (13.15) represent the load and renewable energy output at a given time, respectively. In a non-chronological study, these variables represent samples of the load and renewable energy output.

2. The RF in the model can be replaced by other reliability indices associated with power system adequacy.

3. The method applies to other kinds of generation in addition to renewable energy, in particular, the types of the generation where temporal correspondences of the generation and load are critical in determining the capacity credit.

13.3.6 Analysis of the Factors Influencing Renewable Energy Capacity Credit

The proposed model provides an analytical way to explore the factors that determine the capacity credit, from both the statistical and chronological perspectives [23].

1) Statistical perspective

If we treat k_i and P_i as stochastic variables, Equation (13.19) can be rewritten as follows:

$$Cr_w = \frac{\frac{1}{N}\sum\limits_{i=1}^{N}\left(k_i \frac{P_i}{Cap_w}\right)}{\frac{1}{N}\sum\limits_{i=1}^{N}k_i} = \frac{E(KP_w)}{E(K)Cap_w} \tag{13.27}$$

where $E(K) = \sum\limits_{i=1}^{N} k_i/N$ and $E(KP_w) = \sum\limits_{i=1}^{N} (k_iP_i)/N$. Supposing $Cov(K, P_w)$ represents the covariance of the two stochastic variables K and P_w, the following relation holds [24]:

$$Cov(K, P_w) = E(KP_w) - E(K)E(P_w). \tag{13.28}$$

Combining Equations (13.27) and (13.28), leads to:

$$Cr_w = \frac{E(P_w)}{Cap_w} + \frac{Cov(K, P_w)}{E(K)C_w}. \tag{13.29}$$

The linear correlation coefficient [24] (also known as the Pearson correlation coefficient) between K and P_w is defined as:

$$c(K, P_w) = \frac{\sum\limits_{i=1}^{N}(k_i - E(K))\,(P_i - E(P_w))}{\sqrt{\sum\limits_{i=1}^{N}(k_i - E(K))}\sqrt{\sum\limits_{i=1}^{N}(P_i - E(P_w))}}. \tag{13.30}$$

The following relation holds:

$$c(K, P_w) = Cov(K, P_w)/[\sigma(K)\sigma(P_w)]. \tag{13.31}$$

From Equations (13.29) and (13.31), we get:

$$Cr_w = \frac{E(P_w)}{Cap_w} + \frac{\sigma(K)\sigma(P_w)c(K, P_w)}{E(K)Cap_w} \tag{13.32}$$
$$= \eta_w + V(K)\underline{\sigma}(P_w)c(K, P_w)$$

where $\eta_w = E(P_w)/Cap_w$ is the capacity factor of the renewable energy, $V(K) = \sigma(K)/E(K)$ is the standard deviation of k_i normalized by its mean value, and $\underline{\sigma}(P_w) = \sigma(P_w)/Cap_w$ is the standard deviation of the renewable energy normalized by its capacity.

Equation (13.32) shows that the renewable energy capacity credit is the sum of two terms. The first term is the capacity factor of the renewable energy while the second is a bias component that involves the standard deviations of the slope of RF and of the renewable energy as well as the linear correlation between these variables.

Since RF is usually a monotonically increasing function of the system load, the correlation coefficient $c(K, P_w)$ depends on the relation between load and wind. When the renewable energy is independent of the system load (i.e. $c(K, P_w) = 0$), the capacity factor of the renewable energy gives a good approximation of the capacity credit. On the other hand, when the renewable energy is positively correlated with the load, the capacity credit will be higher than the capacity factor. These observations confirm the empirical research that highlighted the effects that the correlation between wind and load has on the capacity credit. Equation (13.32) also indicates that even though the dependence structure between renewable energy and load might be complex, the capacity credit only depends on the linear correlation coefficient between K and P_w. Meanwhile, the renewable energy P_i also affects the K and further affects the dependence between K and P_w when the renewable energy is not small compared to the load. This also indicates that the capacity factor approximation is valid only for small wind capacities. Numerical results illustrating the effect of these factors are presented in the next section.

It should be noted that Equation (13.32) looks similar to the analytical formula proposed by Zachary and Dent [9], However, they are based on different assumptions and are significantly different.

2) Chronological perspective

As mentioned above, k_i determines the weight of each P_i in the capacity credit calculation. Since the RF always has a strong nonlinear relationship with the load level, there will be considerable differences between the weighting coefficients for different time periods. As argued in Subsection 13.3.2 f'_{RF} provides a good approximation of k_i. Using the IEEE RTS-79 system [22], Figures 13.6, 13.7 and 13.8 show the relationship between f'_{RF} and d_i and illustrate the variation of the weight under different load levels, assuming $P_i \ll d_i$. The derivative of EENS (DEENS) was calculated for different system load levels by differentiating the EENS between adjacent load levels. Figures 13.6 and 13.7 show that the relationship between DEENS and the load is significantly nonlinear. In Figure 13.8 the hourly load and the corresponding DEENS are arranged in descending order and show that DEENS is non-negligible for only the highest 20% values of the load. These observations mean that in Equation (13.19) only the period corresponding to the 20% highest loads are significant in determining the capacity credit. The contribution of the other 80% is small enough to be neglected. This conclusion is consistent with the work of Rui and Luís [17] and supports the use of the "peak-period capacity factors" method for evaluating the capacity credit.

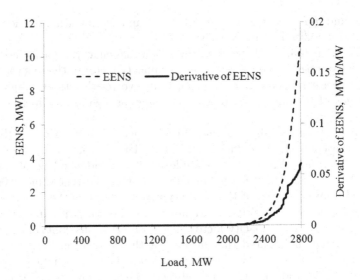

FIGURE 13.6 EENS and derivative of EENS for different load level in the IEEE RTS-79 system.

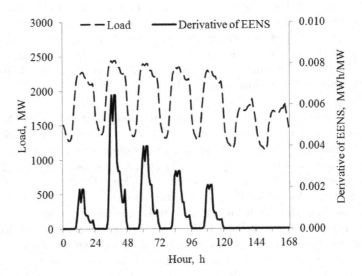

FIGURE 13.7 System load and derivative of EENS for the 21st week in the IEEE RTS-79 system.

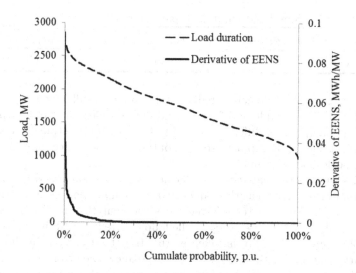

FIGURE 13.8 System load and corresponding DEENS arranged in descending order for the IEEE RTS-79 system.

13.4   Case Study

13.4.1   Basis Data

In this example, the IEEE RTS-79 system is selected for analysis. The system consists of 32 conventional units with a total installed capacity of 3,405 MW and a maximum system load of 2,850 MW. The load characteristics of this system and the unit reliability parameters are reported elsewhere [22].

In this example, a certain capacity of wind power is added to the original installed structure of the IEEE RTS-79 system, and the wind farm output is given by means of the wind farm operation simulation method presented in Chapter 6. The simulation assumes that the system contains five identical wind farms, as shown in Table 13.1. c and k are the scale and shape parameters of the Weibull distribution. θ is the exponential attenuation coefficient of the autocorrelation. η is the wake effect factor. If not otherwise stated, the example assumes that the outputs of these five wind farms are independent of each other, and daily and seasonal characteristics of the wind power are not taken into consideration. The probability distribution of the total output for the five wind farms is shown in Figure 13.9; the wind power simulation output average is 25.66%.

TABLE 13.1 Simulated wind farm parameters

Index	c/p.u.	k/p.u.	θ/h^{-1}	v_{in}/m/s	v_{rate}/m/s	v_{off}/m/s	η/p.u.
Value	8	3	0.05	3.5	12	25	0.95

To compare the capacity credit measurement of different wind power capacities, in the example of different wind power scales, the total output is obtained by multiplying different ratios by the same total output time sequence. Therefore, the total output factor and output time sequence for different wind power scales are the same.

The wind power capacity credit calculation method based on the DPSO (hereinafter referred to as the sequence operation method) and the analytical-based method (hereinafter referred to as the analytical method) are applied to the analysis of the case study in the following text. EENS is selected for the reliability index, 0.1 MW is taken as the discretization step in the sequence operation method, 0.01% is selected for the allowable error of the dichotomy iteration, and the iteration number of the analytical method is set to 5.

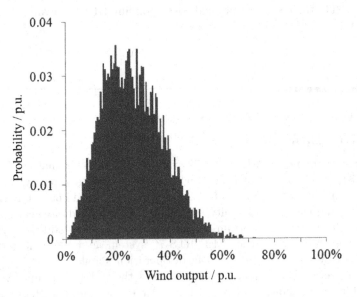

FIGURE 13.9 Total wind farm output probability distribution

13.4.2   Comparison of Different Methods

This section compares wind power capacity credit values calculated by different methods through a case study. The study selects wind power capacity from 35 MW increasing to 1700 MW, which occupies 1.23% to 59.65% of the maximum load, as shown in Table 13.2.

TABLE 13.2 Wind power capacity setting

Wind power capacity/MW	35	70	170	340	680	1700
Proportion of wind power capacity at maximum load/%	1.23	2.46	5.96	11.93	23.86	59.65
Wind power penetration rate/%	0.51	1.03	2.49	4.98	9.97	24.92

In terms of the calculation method, in addition to the sequence operation method and analytical-based method proposed in this section, the Monte Carlo simulation method [5, 25], and the peak load capacity factor method [7, 13, 17] are selected for comparison. The result of the Monte Carlo method is used as a standard to evaluate the computational error of other calculation methods. For this purpose, the number of iterations for the Monte Carlo method is chosen to be 10^8. This number of iterations already far exceeds that needed for the convergence of the Monte Carlo method and therefore, is regarded to obtain the true value of the capacity credit.

Since the wind power output is randomly generated, although the simulation does not consider the daily and seasonal characteristics of wind power, the spatial correlation inevitably exists between the simulated wind power output and the load. The analytical-based method can take this correlation into account because its calculations correspond to the time sequence. The sequence operation method does not consider time sequence information and can only use dependency probability sequence modeling to consider the correlation between the wind power output and load. For comparison of the case studies in this section and in subsequent sections, the wind power output and the load are assumed to be mutually independent; thus, the calculation results differ slightly from the results of the analytical-based method. To compare the calculation accuracies of the sequence operation method and analytical method, two types of Monte Carlo methods are introduced. The first type samples the load and the wind power output at the corresponding time simultaneously (termed the sequential sampling method), and the results are compared with the calculation results of the analytical-based method. The second type ignores the time sequence correlation between the wind power and the load (termed non-sequential corresponding sampling), and the results are compared with the calculation results of the sequence operation method. The wind power credible capacity results calculated by different methods are shown in Table 13.3. Figure 13.10 shows the variation trend of the wind power capacity credit under different installed capacities, and Figure 13.11 shows the calculation error of the different methods.

TABLE 13.3 Wind power credible capacity results calculated by different methods

	Sequential sampling	Analytical method	Non-sequential sampling	Sequence operation method	Average peak load output method
35	9.21	9.2	9.03	9.02	8.93
70	18.32	18.27	18.2	18.05	17.86
170	43.83	43.75	42.95	42.5	43.38
340	85.33	85.3	82.03	82.34	86.76
680	159.87	159.81	153.98	152.73	173.53
1700	323.52	323.55	305.01	306.71	433.82

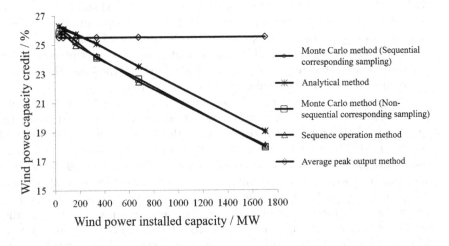

FIGURE 13.10 Wind power capacity credit results as calculated by different methods

The calculation results reveal that the outcome of the two methods presented in this chapter deviates slightly from that of the corresponding Monte Carlo method. Regardless of the proportion of wind power installed capacity, accurate calculation results can be obtained for the wind power capacity credit. In the case of low wind power penetration, the credible capacity is approximately equal to the average wind power output (when the wind power installed capacity is 35 MW, its capacity credit is 25.78%, which is equivalent to an average output of 25.66%). As the power capacity increases, the capacity credit decreases linearly as the wind power penetration increases. A comparison of the results given by the peak load capacity factor method indicates that this method is more accurate only when the wind power ratio is small.

When the proportion of wind power capacity to the highest load exceeds 10%, the calculation results are no longer accurate.

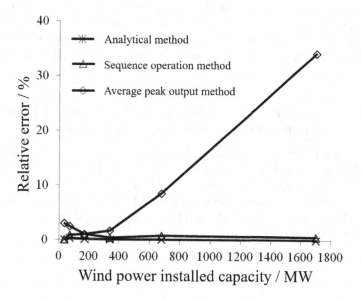

FIGURE 13.11 Error analysis of the wind power capacity credit calculated by different methods

TABLE 13.4 Analysis of influential factors for the wind power capacity credit

Wind power capacity/MW	η_w	$V(K)$	$\underline{\sigma}(P_w)$	$c(K, P_w)$
35	0.2592	4.0002	0.1163	0.0078
70	0.2592	4.0434	0.1163	0.0037
170	0.2592	4.1345	0.1163	−0.0039
340	0.2592	4.2908	0.1163	−0.0167
680	0.2592	4.5598	0.1163	−0.0456
1700	0.2592	5.3509	0.1163	−0.1107

To throughly analyze the variation mechanism of the capacity credit that decreases as the installed capacity increases, Table 13.4 calculates the value of each parameter for different wind power installed capacities in the analytical model of capacity credit Equation (13.32). The wind power capacity factor η_w and the standard deviation of the output $\underline{\sigma}(P_w)$ are equal for different wind power installed capacities. The standard deviation $V(K)$ of the weight coefficient is not large; it increases from 4.0002 p.u. at 35 MW for wind power to 5.3509 p.u at 1700 MW. The source of the decline in the credible capacity is the correlation coefficient $c(K, P_w)$ between the weight coefficient and the wind power output. When the wind power installed capacity increases, although the power output and the load are mutually independent, the negative correlation between the net loads gradually increases, resulting in negative correlation enhancement between weighting factors, which causes the offset $V(K)\underline{\sigma}(P_w)c(K, P_w)$ in Equation (13.32) to decrease and leads to a further reduction in capacity credit.

13.4.3   Influence of Spatial Correlation of Wind Farm Output on the Wind Power Capacity Credit

The spatial correlation of different power outputs has a fundamental impact on the total output of the wind farms. Based on the simulation parameters of the wind farm presented in Subsection 13.4.2, this subsection sets the correlation coefficient between the wind speeds to 0.7 for the five wind farms; the probability distribution of the wind farm total output given by simulation is shown in Figure 13.12.

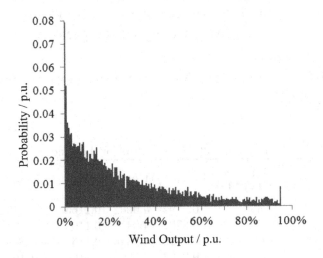

FIGURE 13.12 Probability distribution of the total wind farm output (for a strong correlation in wind farm output)

A comparison of Figure 13.12 with Figure 13.9 indicates that the probability distribution of the wind-power plant total output varies greatly with different spatial correlations. Taking the wind power with strong correlation as an example, the wind power credible capacity is calculated under different wind power installed capacities by means of the analytical method, and the corresponding parameters are calculated using Equation (13.32); the results are shown in Table 13.5.

A comparison of the calculation results of this subsection with that of

Subsection 13.4.2 indicates that in the case of strong correlation of wind farm output, the credible capacity is much lower than in the case of mutual independence. Furthermore, the wind power capacity credit decreases as the installed capacity increases: the capacity credit is reduced to 8.5% when the wind power installed capacity is 1700 MW.

The intrinsic factor for this variation can be analyzed using the calculation results of the corresponding parameters in Equation (13.32). A comparison of η_w, $V(K)$, $\underline{\sigma}(P_w)$, and $c(K, P_w)$ between Table 13.4 and Table 13.5 indicates that the spatial correlation of wind-power plant output affects the standard deviation $\underline{\sigma}(P_w)$ of the total wind power output. The standard deviation of the total wind power output is 0.1163 p.u. in the case of mutual independence of the wind farms. In the situation with a strong correlation of wind speed, the output is increased to 0.2353 p.u., corresponding to the increase of the offset $V(K)\underline{\sigma}(P_w)c(K, P_w)$ in Equation (13.32), which consequently reduces the wind power credible capacity.

TABLE 13.5 Analysis of influential factors for the wind power capacity credit under strong correlation of wind power output

Wind power capacity /MW	Credible capacity /MW	Capacity credit /%	η_w	$V(K)$	$\underline{\sigma}(P_w)$	$c(K, P_w)$
35	7.66	21.88	0.2549	4.0026	0.2353	−0.0383
70	14.85	21.22	0.2549	4.0715	0.2353	−0.0446
170	33.11	19.48	0.2549	4.2282	0.2353	−0.0605
340	57.22	16.83	0.2549	4.5182	0.2353	−0.0815
680	89.42	13.15	0.2549	5.0889	0.2353	−0.1031
1700	136.77	8.05	0.2549	6.5465	0.2353	−0.1133

13.4.4 Influence of Spatial Correlation between the Load and Power Output on the Wind Power Capacity Credit

The spatial correlation between the load and wind power output is mainly reflected in the corresponding temporal relationship. To this end, this subsection simulates the wind power output that is positively and negatively correlated with the seasonal characteristics of the load for the IEEE RTS-79 system. The example that analyzes the variation pattern of credible capacity is constructed in the case of non-independence between the load and wind power output. The seasonal characteristics of the wind speed in the two examples are shown in Figure 13.13.

In the examples, the total wind power capacity is 170 MW. We take the wind power installed capacity of 170 MW as described in Subsection 13.4.2 as the comparative example to carry out calculation by means of the sequence

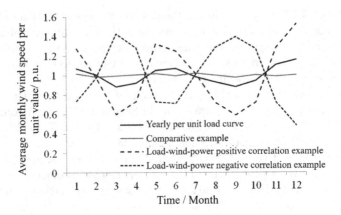

FIGURE 13.13 Monthly average wind speed per unit curve setting of the load and wind power in the non-independent examples

TABLE 13.6 Analysis of correlation between the load and wind power

Example	Wind power and load rank correlation coefficient
Comparative example	−0.0084
Load-wind-power positive correlation example	0.0993
Load-wind-power negative correlation example	−0.1491

operation method and analytical method. In the sequence operation method, first, the measurement of correlation between the wind power output and loads is statistically analyzed; then, the credible capacity is calculated by applying the DPSO. Table 13.6 calculates the rank correlation coefficient of the wind power output and load in each example. Figure 13.14 shows the joint distribution of the load and wind power output in the case of negative correlation along with the corresponding copula function scatter plot.

Table 13.7 shows the calculation results of the sequence operation method and the analytical method. The table also shows the corresponding parameters of different examples calculated in Equation (13.32). The calculation results reveal that the results given by the two calculation methods are similar. Compared with the case with independent load and wind power output, the results indicate that the wind power credible capacity increases in the case where the load is positively correlated with the wind power output and decreases when the load is negatively correlated. The spatial correlation between the load and wind power output mainly affects the parameter $c(K, P_w)$ in Equation (13.32),

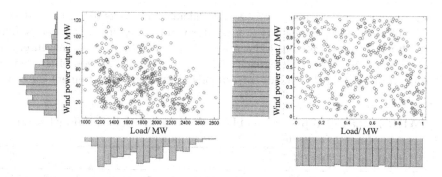

FIGURE 13.14 Seasonal characteristics of the load and wind power output joint probability distribution scatter plot (left) and copula function scatter plot (right) for the negative correlation example

and the wind power credible capacity measurement can be further determined by altering the offset $V(K)\underline{\sigma}(P_w)c(K, P_w)$.

TABLE 13.7 Wind power capacity credit and its influential factors in the scenario of non-independent load and wind power

Example	Wind power capacity credit/%		η_w	$V(K)$	$\underline{\sigma}(P_w)$	$c(K, P_w)$
	Sequence operation method	Analytical model method				
Benchmark	25	25.73	0.2592	4.1345	0.1163	−0.0039
Positive correlation	32.4	33.48	0.2601	4.0983	0.1445	0.1262
Negative correlation	16.22	16.87	0.2628	4.1424	0.145	−0.1566

13.4.5 Effect of the Virtual Unit Setting

For comparison with the capacity of conventional units, the wind power credible capacity is often equated to the virtual unit with a certain random outage rate (FOR) in the calculation process. In fact, the virtual unit can be interpreted as a "ruler" for measuring the credible capacity of wind power. The FOR setting affects the calculated wind power credible capacity.

On the basis of a wind power installed capacity of 680 MW as described in Subsection 13.4.2, the virtual unit FOR is altered to 2%, 4%, and 6%. The sequence operation method was applied in the calculation of the capacity credit;

the results are shown in Table 13.8. The calculation results reveal that a larger virtual unit FOR corresponds to a greater wind power credible capacity. The result can be interpreted as follows: a smaller "ruler" unit in the measurement of the wind power credible capacity (i.e., a smaller contribution of the virtual unit to the system reliability) corresponds to a higher measurement outcome.

TABLE 13.8 Wind power credible capacity and its influential factors in the scenario of non-independent load and wind power

Virtual unit FOR/%	0	2	4	6
Wind power credible capacity /MW	152.73	159.04	166.02	173.65

In the calculation of the wind power credible capacity, the wind farm is typically equated to a single conventional unit. When the capacity of the wind farm is large, the capacity of this equivalent conventional unit is much greater than the actual unit capacity of the system. At this point, we equate the wind farm to multiple conventional units, so it is more applicable to the actual unit of equal measurement. Based on the system reliability, it is found that as the virtual unit FOR becomes zero and the total capacities become equal, the contribution to the system reliability coming from either a single unit or multiple units stays the same. However, when FOR is not zero, the contribution to the reliability varies for both cases.

This subsection analyzes variations in the capacity of each single virtual unit and their impact on the calculation results of the wind power credible capacity. The calculation results are shown in Table 13.9. As the capacity of one single virtual unit decreases, the number of units increases, and the wind power capacity credit declines. Under the premise of equal total wind power capacity, multiple units make a greater contribution to the system reliability than single ones. In other words, the unit range of the "ruler" becomes larger, and the measured value decreases. The above calculation results indicate that when the wind power penetration is high and the wind power credible capacity is much greater than that of one single conventional unit, the error becomes quite large equating the total wind power to one virtual unit. Therefore, under these conditions, multiple virtual units should be used for the credible capacity calculation.

TABLE 13.9 Analysis of the impact of virtual single unit capacity on the wind power capacity credit

Capacity of each virtual unit/MW	200	100	50	25
Wind power credible capacity/MW	166.02	162.03	160.7	159.87

13.5 Summary

Renewable energy is unable to provide the power generation capacity equivalent to its designed capacity due to its strong outage randomness, volatility, and challenges in controlling output. In power system planning, it is necessary to accurately assess the contribution of renewable energy to system reliability, that is, the capacity credit of renewable energy. The capacity credit of renewable energy has practical significance in power planning and plays an important role in understanding the impact of the renewable energy mechanism on system reliability.

To this end, this chapter starts with three aspects, calculation methods, research objects and an operation mechanism, to conduct in-depth research into the capacity credit of renewable energy. Based on DPSO theory, we first present the calculation method to successfully implement the fast calculation of the capacity credit of renewable energy. According to the definition of capacity credit, a widely applicable analytical model for capacity credit is presented for the first time. This model analyzes the factors affecting the capacity credit from the perspectives of macroscale statistics and a microscale time sequence. The theoretical basis for calculating the renewable energy capacity credit by means of the peak-load capacity factor method is provided by this model. Furthermore, a renewable energy capacity credit calculation method based on the analytical model is proposed.

The two methods presented in this chapter that calculate the capacity credit are based on different principles and modeling assumptions. Therefore they are appropriate for different areas of applications. Because the calculation method based on the DPSO is flexible, the random outage of the transmission system and a virtual unit can be taken into consideration. This method is applicable to the calculation of the capacity credit under the condition that the power output and load probability distribution are known. The renewable energy capacity credit calculation method based on the analytical model has a defined physical meaning and is suitable for conditions where the renewable energy output and load time sequence data are known.

Bibliography

[1] L. L. Garver. Effective load carrying capability of generating units. *IEEE Transactions on Power Apparatus and Systems*, (8):910–919, 1966.

[2] E. Kahn. The reliability of distributed wind generators. *Electric Power Systems Research*, 2(1):1–14, 1979.

[3] J. Haslett and M. Diesendorf. The capacity credit of wind power: A theoretical analysis. *Solar Energy*, 26(5):391–401, 1981.

[4] M. Amelin and L. Soder. Taking credit. *IEEE Power and Energy Magazine*, 8(5):47–52, 2010.

[5] F. Vallee, J. Lobry, and O. Deblecker. System reliability assessment method for wind power integration. *IEEE Transactions on Power Systems*, 23(3):1288–1297, 2008.

[6] A. S. Dobakhshari and M. Fotuhi-Firuzabad. A reliability model of large wind farms for power system adequacy studies. *IEEE Transactions on Energy Conversion*, 24(3):792–801, 2009.

[7] C. J. Dent, A. Keane, and J. W. Bialek. Simplified methods for renewable generation capacity credit calculation: A critical review. *Power and Energy Society General Meeting, 2010 IEEE*, pages 1–8, 2010.

[8] A. Keane, M. Milligan, C. J. Dent, B. Hasche, C. D'Annunzio, K. Dragoon, H. Holttinen, N. Samaan, L. Soder, and M. O'Malley. Capacity value of wind power. *IEEE Transactions on Power Systems*, 26(2):564–572, 2011.

[9] B. Hasche, A. Keane, and M. O'Malley. Capacity value of wind power, calculation, and data requirements: the Irish power system case. *IEEE Transactions on Power Systems*, 26(1):420–430, 2011.

[10] K. R. Voorspools and W. D. D'haeseleer. An analytical formula for the capacity credit of wind power. *Renewable Energy*, 31(1):45–54, 2006.

[11] K. Dragoon and V. Dvortsov. Z-method for power system resource adequacy applications. *IEEE Transactions on Power Systems*, 21(2):982–988, 2006.

[12] S. Zachary and C. J. Dent. Probability theory of capacity value of additional generation. *Proceedings of the Institution of Mechanical Engineers, Part O: Journal of Risk and Reliability*, 226(1):33–43, 2012.

[13] S. Patil and R. Ramakumar. A study of the parameters influencing the capacity credit of WECS: a simplified approach. *Power and Energy Society General Meeting, 2010 IEEE*, pages 1–7, 2010.

[14] S. Zhu, Y. Zhang, and A. A. Chowdhury. Capacity credit of wind generation based on minimum resource adequacy procurement. *IEEE Transactions on Industry Applications*, 48(2):730–735, 2012.

[15] M. Milligan and K. Porter. Determining the capacity value of wind: A survey of methods and implementation; preprint. *National Renewable Energy Laboratory*, 2005.

[16] K. R. Voorspools and W. D. D'haeseleer. Critical evaluation of methods for wind-power appraisal. *Renewable and Sustainable Energy Reviews*, 11(1):78–97, 2007.

[17] R. M. Castro and L. A. Ferreira. A comparison between chronological and probabilistic methods to estimate wind power capacity credit. *IEEE Transactions on Power Systems*, 16(4):904–909, 2001.

[18] R. Billinton, D. Huang, and B. Karki. Wind power planning and operating capacity credit assessment. *Probabilistic Methods Applied to Power Systems (PMAPS), 2010 IEEE 11th International Conference on*, pages 814–819, 2010.

[19] G. Caralis and A. Zervos. Value of wind energy on the reliability of autonomous power systems. *IET Renewable Power Generation*, 4(2):186–197, 2010.

[20] C. Kang, Q. Xia, and N. Xiang. Sequence operation theory and its application in power system reliability evaluation. *Reliability Engineering & System Safety*, 78(2):101–109, 2002.

[21] R. N. Allan et al. *Reliability Evaluation of Power Systems*. Springer Science & Business Media, 2013.

[22] Probability Methods Subcommittee. IEEE reliability test system. *IEEE Transactions on Power Apparatus and Systems*, 6(PAS-98):2047–2054, 1979.

[23] N. Zhang, C. Kang, D. S. Kirschen, and Q. Xia. Rigorous model for evaluating wind power capacity credit. *IET Renewable Power Generation*, 7(5):504–513, 2013.

[24] J. A. Rice. *Mathematical Statistics and Data Analysis*. China Machine Press Beijing, 2003.

[25] Wang Haichao, Lu Zongxiang, Zhou Shuangxi, et al. Research on the capacity credit of wind energy resources. *Proceedings of the CSEE*, 25(10):103–106, 2005.

14

Sequential Renewable Energy Planning

Renewable energy planning aims to locate and size wind farms or PV sites optimally. Take wind power as an example. Wind power planners tend to choose the wind farms with the richest wind resources to maximize the energy benefit. However, the capacity benefit of wind power should also be considered in large-scale clustered wind farm planning because the correlation among the wind farms exerts an obvious influence on the capacity benefit brought about by the combined wind power. This chapter proposes a planning model considering both the energy and the capacity benefit of the wind farms. The capacity benefit is evaluated by the wind power capacity credit. The Ordinal Optimization (OO) Theory, capable of handling problems with non-analytical forms, is applied to address the model. To verify the feasibility and advantages of the model, the proposed model is compared with a widely used genetic algorithm (GA) via a modified IEEE RTS-79 system and the real-world case of Ningxia, China. The results show that the diversity of the wind farm enhances the capacity credit of wind power.

14.1 Introduction

Renewable energy planning aims to optimize the location and capacity of wind farms / PV according to the wind/radiation resources and grid connections. Currently, the economics of renewable energy may not be competitive compared with conventional generators. Therefore, during the planning, the overall capacity of renewable energy is usually determined beforehand, e.g. by the government, according to the resource endowment and grid accommodation capacity. The planning problem is then transformed into the capacity allocation problem.

The optimization of renewable energy planning has been widely researched. Taking wind power as an example, Xu and Zhuan [1] proposed a wind farm planning model to minimize the power system operation cost, to reserve costs, to experience the value of load shedding, and to take advantage of the cost of the opportunity for wind curtailments. The planning problem is modeled using a chance-constrained programming approach and solved in combination with a Monte Carlo approach, neural network, and genetic algorithm. Baringo and

Conejo [2] presented a risk-constrained two-stage wind farm planning model, which considered the planning decision in the first stage and the market clearing of different wind power output scenarios in the second stage. The model has transformed into a large scale Mixed Integer Linear Programming (MILP) problem and is solved using CPLEX. The same method was used in a transmission planning problem considering wind power output [3]. Nick *et al.* [4] established an optimal wind farm allocation model considering the capacity limits of the transmission networks. The problem is decomposed into a master problem determining wind farm allocation and a sub-problem simulating annual unit commitment accompanied by optimal DC power flow, which is solved with Benders decomposition. Liu *et al.* [5] proposed a robust optimization planning model to investigate the benefit of wind farm diversification for power system operation and reliability. The model is transformed into a coupled problem composed of a linear programming problem and a conic quadratic programming problem to be solved.

Traditionally, the optimization of wind farm planning tends to choose the wind farm with the best wind resources to maximize the energy benefit. However, the capacity benefit of wind power should also be considered in the case of large-scale clustered wind farm planning because the clustered wind farms can act as a firm load-carrier and benefit the power system in terms of reducing the investment in traditional plants [6], [7]. Previous research has shown that diversifying the wind farm could smooth out the fluctuations in wind power generation and improve the load carrying capability of wind power. Lu *et al.* [8] studied the optimal integration of offshore wind farms in China and showed that installing offshore wind farms in the dispersed regions of Bohai Bay, the Yangtze River Delta, and the Pearl River Delta could significantly smooth out the variability of wind power so that 28% of the overall offshore wind potential could be deployed as base-load power. Placing all the wind farms in some areas with rich resources might not be the best choice because a diversified distribution of wind farms would obtain a higher credit capacity.

This chapter focuses on the planning of wind farms while maximizing the energy benefit and capacity benefit. The capacity benefit is measured by the capacity credit for the wind power, which denotes the ratio of equivalent conventional generation capacity that the wind power can exhibit in a power system on a power system adequacy basis. The planning optimization model is proposed and solved using the Ordinal Optimization (OO) Theory.

14.2   Wind Power Capacity Credit

Properly evaluating the wind power capacity credit is challenging because the value of the capacity credit is affected by various factors. Normally, a better wind resource would result in a higher capacity credit. Almost all the studies have shown that the capacity credit of wind power decreases as the wind power penetration increases. The wind farms with more dispersed outputs are found to result in a higher capacity credit than the same capacity of a wind farm with highly correlated output [9].

Evaluation methods for capacity credit can be roughly divided into the Monte Carlo approach [10], the convolution approach [11] and the analytical approximation approach [12]. The Monte Carlo approach and convolution approach are more accurate but relatively time-consuming. However, the analytical approximation approach is more efficient but may not be as accurate and can be used only in rough estimations. An analytical form of wind power capacity credit in Chapter 13 is used for the evaluation of the wind power credit:

$$Cr_w = \sum_{t=1}^{T} (\frac{k_t}{\sum_{t=1}^{T} k_t} \frac{p_{wt}}{Cap_w})$$

$$\text{where } k_t = \frac{f_{RF}(d_t - p_{wt} + Cr_w Cap_w) - f_{RF}(d_t)}{Cr_w Cap_w - p_{wt}}$$

(14.1)

where $f_{RF}(d_t)$ is the reliability function with the load of d_t; p_{wt} represents the output of wind power for hour t. Equation (14.1) shows that the wind power capacity credit is a linear combination of the wind power output ratio. While k_t has the term of Cr_w which is not known to be a priority, the analytical form of the wind power capacity credit $Cr_{w,a}(x_{1,a}, ..., x_{N,a})$ is not an explicit expression and requires an evaluation approach.

14.3 Problem Formulation

14.3.1 Assumptions and Notations

In this section, we try to demonstrate how the capacity credit of wind power can be maximized in wind power planning. To discuss this issue more clearly, we simplify the current wind power planning model with three assumptions, as follows:

1. The capacity requirement for wind power is set before the planning. Because the proper capacity of wind power is determined by the power system flexibility, the cost-effectiveness of wind power and other policy issues are not the focus of this chapter. The constraints related to the operation of the power system are therefore omitted in the model.

2. The information on the wind resources of each candidate wind farm is available before planning. The hourly wind power can be simulated according to the characteristics of a wind resource such as the

distribution of wind speeds and the correlation between the wind speeds. The simulated hourly wind power output will be used in the planning model as the basis for calculating the capacity credit of wind power.

3. The planning of generations has been determined, and the hourly load forecasting or simulation result is available because the capacity of other generations and level of load have a great impact on the evaluation of the wind power capacity credit.

The long-term planning of wind power over several years is considered in the model to optimize the investment potential of wind farms. Therefore, the net present value (NPV) of the capacity benefit over all the years studied is considered in the model.

14.3.2   Notations

Indices

i	Index for generators
a	Index for operating years

Parameters

r	Interest rate used in calculating the annualized investment cost.
K_w	Payback period of wind farm investment.
K_c	Payback period of conventional thermal generator.
$(P/A, r, K)$	Coefficient of equivalent annual investment with the interest rate r and payback period K. $(P/A, r, K) = \left[(1+r)^K - 1 \right] / r (1+r)^K$
$I_{w,i}$	Unit investment cost of wind farm i (\$/kW).
$I_{c,i}$	Unit investment cost of conventional thermal generator i (\$/kW).
O_i	Unit yearly operation and maintenance (O & M) cost for wind farm i (\$/kW).
ρ_i	Energy price of wind farm i (\$/MWh).
$C_{\text{Plan},a}$	Requirement of wind power installation for the year a (MW).
N	Number of candidate wind farms considered in the planning model.
Y	Number of continuous target years considered in the planning model.

Variables

$x_{i,a}$	State variable for the investment decision of wind farm i during the year a (equals 1 if the wind farm is installed in the year a; equals 0 if the wind farm is installed before or after the year a).
$B_{Energy,i,a}$	Energy benefit of wind farm i during the year a (\$).
$B_{Capacity,a}$	Capacity benefit of wind farm i during the year a (\$).
$F_{i,a}$	Annualized capital cost, operation and maintenance cost of wind farm i during the year a (\$).
$p_{i,a,t}$	Simulated power output of wind farm i for hour t during the year a (\$).
$C_{w,i}$	The installed capacity of wind farm i (MW).
$Cr_{w,a}(x_{1,a},...,x_{N,a})$	The capacity credit of all the wind farms during the year a under the investment decision of $x_{1,a},...,x_{N,a}$.

14.3.3 Objective Function

The object of the planning is to minimize the net cost (maximize the net benefit) from the social perspective, which includes the capital cost and the O & M cost of the wind farm, while the benefit includes the energy benefit and the capacity benefit. Both the considered cost and the benefit are associated with the target years in the planning problem. To simplify, the cost and benefit in different years are aggregated as the NPV of the first target years of the planning model.

The capital cost and O & M cost are annualized considering the interest rate and its payback period.

$$F_{i,a}=(1+r)^{-a}\left[(P/A,r,K_w)I_{w,i}+O_i\right]C_{w,i}x_{i,a} \tag{14.2}$$

The energy benefit of the wind farm is modeled using the simulated hourly output multiplied by its energy price:

$$B_{Energy,i,a}=(1+r)^{-a}\rho\sum_{t=1}^{T}p_{i,a,t}x_{i,a}. \tag{14.3}$$

The capacity benefit of the wind farm is modeled using the credible capacity of wind power multiplied by the investment of conventional thermal generation:

$$B_{Capacity,a}=(P/A,r,K_c)I_{w,i}Cr_{w,a}(x_{1,a},...,x_{N,a})\sum_{i=1}^{N}C_{w,i}x_{i,a}. \tag{14.4}$$

Successively, the objective function of the planning model can be written as

$$\min \sum_{i=1}^{N}\sum_{a=1}^{Y}F_{i,a}-\sum_{i=1}^{N}\sum_{a=1}^{Y}B_{Energy,i,a}-\sum_{a=1}^{Y}B_{Capacity,a}. \tag{14.5}$$

14.3.4   Constraints

The capacity requirement of wind power for each year should be achieved in the target years:

$$\sum_{i=1}^{N}\sum_{a=1}^{\tilde{a}} x_{i,a}C_{w,i} \geq C_{\text{Plan},a}. \tag{14.6}$$

Each wind farm can be invested only once, or not be invested:

$$\sum_{a=1}^{\tilde{a}} x_{i,a} \leq 1. \tag{14.7}$$

Other constraints for the wind farm planning model are also available. For example, the earliest installed year of each wind farm can be set as a direct constraint on $x_{i,a}$. The associated power system operational constraints can also be added. However, these constraints can all be constructed in the MILP form and do not change the mathematical characteristics of this problem.

14.4   OO Theory Based Approach

14.4.1   OO Theory

Because the term of capacity credit $Cr_{w,a}\left(x_{1,a},...,x_{N,a}\right)$ is not an explicit expression, the proposed model cannot be solved directly. The OO theory is a novel approach to cope with the complex and difficult optimization problem and can significantly reduce the number of search samples of the huge feasible solution space formed by all discrete control variables [13] [14]. The OO theory aims to obtain sufficiently good solutions with a high probability instead of finding the perfect optimal solutions to avoid time-consuming calculation. The OO theory has been widely used in many aspects such as bidding strategy for electric power suppliers [15], optimal power flow with discrete variables [16], transmission system enhancement [17], reactive optimization in distribution systems [18], etc.

The principle of OO is that the random selection of feasible solutions would cover some sufficiently good solutions, and the crude model evaluation gives a rough ordering of the solutions. If we pick some certain number of solutions properly, we would be confident that the solution selected would include one or more sufficiently good solutions. Because the crude evaluation model is much faster than the accurate evaluation, this approach can improve the overall solving efficiency of the complex problem. OO has been applied to optimization problems in the power system. Figure 14.1 shows a visual relationship between the set of solutions defined in the OO approach. The details of this approach can be found in [14].

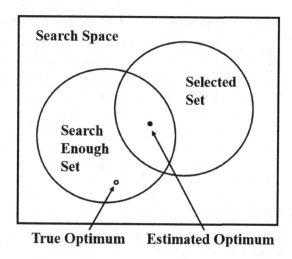

FIGURE 14.1 Schematic diagram of the set of solutions in OO.

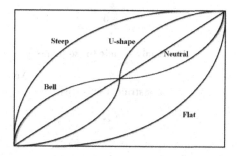

FIGURE 14.2 Five types of ordered performance curves (OPCs).

The procedure of OO is to first generate a large number of feasible solutions and use the crude model to rapidly evaluate the performance of each solution. Then, the solution is sorted by the evaluated results of the crude model, and a certain number of top performance solutions are selected to be evaluated by the accurate model. The sufficiently good solution is selected according to the accurate evaluation results.

As shown in Figure 14.2, five types of order performance curves (OPCs) can be roughly categorized in OO: Flat (many good designs), Steep (many bad designs), Bell (many mediocre designs), U-shape (either good or bad designs) and Neutral (uniformly distributed designs). The shape of the OPCs determines the number in the set to perform the accurate evaluation [13].

14.4.2   Crude Evaluation Model Used for Capacity Credit

According to Equation 14.1, the capacity credit of wind power Cr_w is neither linear nor analytical and thus is difficult to compute because the weight of different hours of wind power output k_t is a function of Cr_w. Here, we use the average output of wind power to approximate Cc_w:

$$k_t = \frac{f_{RF}(d_t - p_t + \sum\limits_{t=1}^{T} p_{wt}) - f_{RF}(d_t)}{\sum\limits_{t=1}^{T} p_{wt} - p_{wt}} \tag{14.8}$$

14.4.3   Procedures

The procedure for solving the proposed wind farm planning problem based on OO theory is shown in Figure 14.3.

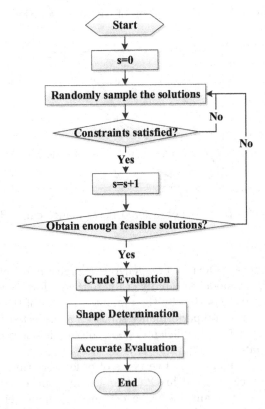

FIGURE 14.3 OO based procedures of wind farm planning theory.

Detailed procedures are listed below:

- Step I: Sample the value of each $x_{i,a}$ and test if the solution meets the constraints. If the solution does meet the constraints, put the decision into the feasible solution set. Get enough feasible solutions (i.e., 10000), and then go to step II.

- Step II: Evaluate the feasible solution set using the crude model in Equation (14.8). Sort the feasible solutions by the objective value of the crude model.

- Step III: According to the shape of the accumulated curve of the objective values (ordered performance curves, OPCs), decide the number in the set for which to perform the accurate evaluation. The shape of the OPC consists of a Bell shape (as the Case Study will show) so that the formulation of the solution is $M_{k,g}$.

- Step IV: Pick the top $M_{k,g}$ solutions in the crude model and evaluate each solution using the accurate model (Equation (14.1)). Choose the best solution as the final solution.

14.5 Illustrative Example

14.5.1 Data and Settings

The proposed model was tested using a modified IEEE RTS-79 system. Generator data and load were extracted from [19]. Twenty candidate wind farms with 100 MW each were introduced in the planning model and were divided into groups of five. The wind capacity factors in different groups were different and are shown in Figure 14.4. The wind farms #1-#5 (in zone 1) had the highest capacity factor while wind farms #16-#20 (in zone 4) had the lowest capacity factor. The correlation coefficients of wind speed among the different zones were set as shown in Table 14.1. The wind power outputs in the same zone were assumed to be highly correlated with a coefficient of 0.99. In the actual situation, the spatial correlation among the different wind power zones could be calculated according to the historical wind speed measured by anemometer towers. The yearly outputs of each wind farm were simulated using the method in [20]. The parameters in the planning model were set as shown in Table 14.2.

In the case study, we consider a five-year wind farm planning decision. The capacity requirement is 300 MW in the first year and increases by 300 MW for each year. The total wind farm capacity requirement is 1500 MW in the 5th year.

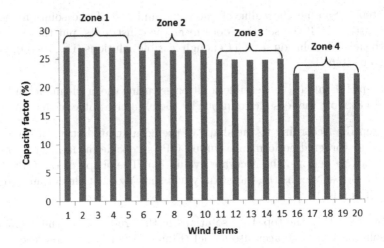

FIGURE 14.4 The capacity factor for candidate wind farms.

TABLE 14.1 Linear correlation coefficient among different zones.

	Zone 1	Zone 2	Zone 3	Zone 4
Zone 1	1	0.85	0.7	0.16
Zone 2	0.85	1	0.81	0.18
Zone 3	0.7	0.81	1	0.22
Zone 4	0.16	0.18	0.22	1

TABLE 14.2 Settings for the parameters in the case study

Parameter	Value
r	8%
K_w	15 years
K_c	25 years
$(P/\mathrm{A}, r, K_\mathrm{w})$	0.1082
$(P/\mathrm{A}, r, K_\mathrm{c})$	0.0867
$I_{\mathrm{w},i}, I = 1, 2, ..., N$	1193 \$/kW
$I_{c,i}, I = 1, 2, ..., N$	645 \$/kW
$O_i, I = 1, 2, ..., N$	24.03 \$
$\rho_i, I = 1, 2, ..., N$	48.4 \$/MWh
$C_{\mathrm{w},i}, I = 1, 2, ..., N$	100 MW
$M_{k,g}$	139

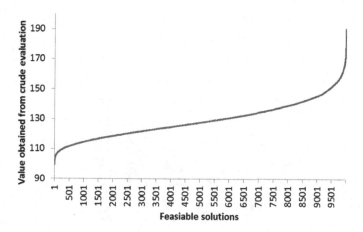

FIGURE 14.5 The OPC of the proposed planning model.

14.5.2 OPC Shape Determination

Figure 14.5 shows the OPC obtained in step II of the proposed approach. According to the classification of the OPCs in the OO theory [26], the OPC shown in Figure 14.5 belongs to the Bell-shaped OPCs.

14.5.3 Number of Solutions for Accurate Evaluation

The formulation of $M_{k,g}$ is therefore chosen under the Bell OPC category.

$$M_{k,g} = \left[e^{8.1998}k^{1.9164}g^{-2.0250} + 10\right] = 139$$

Here $[\cdot]$ denotes the rounding of the number, and g and k are the parameters of confidence and optimality. We chose $g = 10, k = 1$, meaning that we wished to find one or more decisions that ranked within the best 10 decisions in the feasible solution set, with the confidence level of 95%.

14.5.4 Solutions

To demonstrate how the capacity benefit of wind power impacts the planning, three cases were conducted as follows, and the optimal planning result as listed in Figure 14.6.

Case 1: Considering only the energy benefit.

Case 2: Considering only the capacity benefit.

Case 3: Considering both capacity and energy benefit.

In Figure 14.6, the planned wind power capacity in each zone for each year is compared to illustrate the difference in planning decisions under different objectives. The capacity factor and the capacity credit associated with each planning decision are listed in Table 14.3.

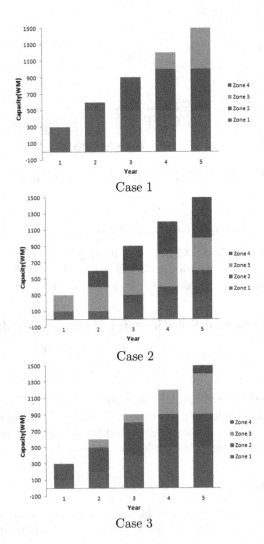

FIGURE 14.6 The planned wind farm capacity in each zone for the three cases.

TABLE 14.3 Average capacity factor and capacity credit for different cases.

	Year	Result (%)		
		Case 1	Case 2	Case 3
Capacity Factor	1	27.07%	25.48%	26.65%
	2	26.92%	24.65%	26.32%
	3	26.76%	24.56%	26.49%
	4	26.39%	24.73%	26.23%
	5	26.05%	24.82%	25.76%
Capacity Factor	1	21.29%	22.47%	21.26%
	2	15.76%	17.80%	16.55%
	3	12.24%	14.81%	12.62%
	4	10.38%	12.45%	10.61%
	5	8.99%	10.72%	9.54%

In Case 1, when considering only the energy benefit as most planning models do, the planning decision tends to give priority to the wind farms with higher capacity factors.

In Case 2, when considering only the capacity benefit, the planning decision tends to disperse the wind farm investment to obtain a higher capacity credit. In this case, the wind farms that have a higher capacity factor (in zone 1 and zone 2) are not fully planned because the correlation between these two zones is strong and would result in a smaller capacity credit. By contrast, the wind farms in zone 4 that have the lowest capacity factors are fully planned because the correlation with the wind farms in other zones is weak.

Case 3 has a neutralized result of case 1 and 2, where the wind farms are planned to be dispersed while giving the priority to the wind farm with the higher capacity factor. The results suggest that the proposed model can effectively take into consideration the benefit of the capacity value of wind farm in planning decisions.

The capital and O & M costs, energy benefit, and capacity benefit for different cases are shown in Table 14.4. The total cost can be minimized by considering both the energy benefit and the capacity benefit.

TABLE 14.4 Costs and benefits for different cases ($).

	Capital and O&M Costs	Energy Benefit	Capacity Benefit	Total
Case 1	56.319	41.33	1.552	13.437
Case 2	56.319	38.418	2.922	14.98
Case 3	56.319	40.858	2.569	12.892

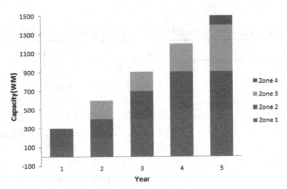

FIGURE 14.7 Optimal solution for GA.

14.5.5   Comparison with Genetic Algorithm (GA)

A GA method proposed by the Genetic Algorithm Solver in MATLAB® is employed to solve the same wind farm planning optimization model as Case 3. The key parameters of the GA model are listed below:

Population Size: 300
Generations: 100
Crossover Probability: 0.80
Mutation Probability: 0.02

Figure 14.7 shows the optimal solution for GA. Compared with the results shown in Fig. 6 for Case 3, there is little difference in the second and third planning years. The optimal value of GA is $12.994 million, which is a little larger than the optimal value of OO ($12.892 million). Additionally, the convergence of GA is not well performed because each simulation gets different results.

14.6   Application to Ningxia Provincial Power Grid

14.6.1   Wind Power Planning in Ningxia Province

In this section, an application of the proposed approach is conducted in Ningxia Province, China, which is rich in wind power.

There are several wind farms distributed in six zones, named Qingtongxia (QT), Taiyangshan (TY), Mahuangshan (MH), Xiangshan (XS), Jiucaishan (JC), and Magaozhuang (MG). Through the analysis of historical data in Ningxia, the Weibull parameters of different wind zones and correlation coefficients among them can be obtained (shown in Table 14.5 and Table 14.6, respectively). Table 15.4 shows that MH is the richest in wind resources, while

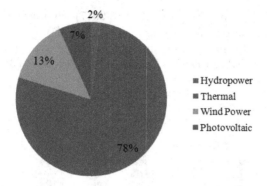

2%

7%

13%

78%

- Hydropower
- Thermal
- Wind Power
- Photovoltaic

FIGURE 14.8 Capacity of different forms of power of Ningxia in 2013.

TY is the poorest. The correlation among the six wind zones in Ningxia is at a relatively higher level.

TABLE 14.5 Weibull parameters of the six wind zones.

	c	k
QT	7.6636	2.2117
TY	7.5742	2.3222
MH	7.7237	2.1375
XS	7.6577	2.219
JC	7.6489	2.2299
MG	7.6291	2.2175

TABLE 14.6 Linear correlation coefficient among six zones.

	QT	TY	MH	XS	JC	MG
QT	1	0.6473	0.5754	0.6101	0.5121	0.592
TY	0.6473	1	0.7523	0.5761	0.5593	0.709
MH	0.5754	0.7523	1	0.5281	0.5521	0.7926
XS	0.6101	0.5761	0.5281	1	0.6187	0.6121
JC	0.5121	0.5593	0.5521	0.6187	1	0.6534
MG	0.592	0.709	0.7926	0.6121	0.6534	1

As shown in Figure 14.8, by the end of 2013, the installed capacities of hydropower, thermal power, wind power and photovoltaic power in Ningxia are 426 MW, 17313 MW, 3018 MW, and 1561 MW, respectively. The power consumption of the whole society in Ningxia is 8.112 billion kWh in 2013, and the maximum electricity load had reached 11260 MW.

We consider a four-year (from 2014 to 2017) wind farm planning decision. The capacity requirement in each year is 2334 MW, 4668 MW, 5742 MW, and 6815 MW. The maximum planning capacity of each of the six wind zones in Ningxia is 520 MW, 1950 MW, 1045 MW, 1872 MW, 2350 MW, and 2450 MW.

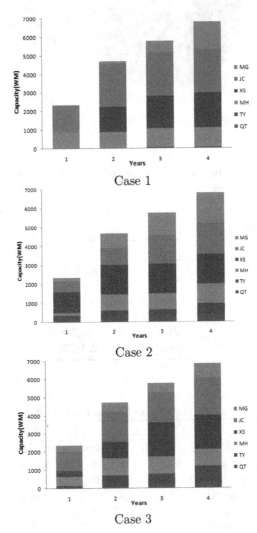

FIGURE 14.9 The planned wind farm capacity in Ningxia for the three cases.

14.6.2   Results and Discussion

The same three cases as shown in Figure 14.6 are used in the wind power planning in Ningxia, and the optimal solution is shown in Figure 14.9. Figure 14.9 illustrates that the zones with rich wind power resources (MH and JC) have the priority to be invested, while the zones with lower correlation (XS and MH) have the priority to be invested to maximize the capacity benefit as well.

14.7   Conclusions

Large-scale clustered wind farms could contribute to generation adequacy, so the capacity benefit of wind power should be considered in wind farm planning. This chapter proposes a wind farm planning model that takes into account the capacity benefit of wind farms. The OO technique is used to solve the problem. The widespread of wind farms with less correlation would result in a higher capacity credit. Considering only the energy benefit may mislead the investment and result in a lower wind power capacity credit. An illustrative example of a modified IEEE RTS 79 system and a case study in Ningxia shows that the capacity benefit of the wind power should be considered apart from the energy benefit in wind power planning.

Bibliography

[1] M. Xu and X. Zhuan. Optimal planning for wind power capacity in an electric power system. *Renewable Energy*, 53(9):280–286, 2013.

[2] L. Baringo and A. J. Conejo. Risk-constrained multi-stage wind power investment. *IEEE Transactions on Power Systems*, 28(1):401–411, 2013.

[3] L. Baringo and A. J. Conejo. Transmission and wind power investment. *IEEE Transactions on Power Systems*, 27(2):885–893, 2012.

[4] M. Nick, G. H. Riahy, S. H. Hosseinian, and F. Fallahi. Wind power optimal capacity allocation to remote areas taking into account transmission connection requirements. *IET Renewable Power Generation*, 5(5):347–355, 2011.

[5] S. Liu, J. Jian, Y. Wang, and J. Liang. A robust optimization approach to wind farm diversification. *International Journal of Electrical Power & Energy Systems*, 53(53):409–415, 2013.

[6] B. Hasche, A. Keane, and M. O'Malley. Capacity value of wind power, calculation, and data requirements: the Irish power system case. *IEEE Transactions on Power Systems*, 26(1):420–430, 2011.

[7] N. Zhang, Z. Chen, Y. Zhou, W. Xi, J. Huang, Q. Zhang, and C. Kang. Wind power credible capacity evaluation model based on sequence operation. *Proceedings of the CSEE*, 31(25):1–9, 2011.

[8] X. Lu, M. B. Mcelroy, C. P. Nielsen, X. Chen, and J. Huang. Optimal integration of offshore wind power for a steadier, environmentally friendlier, supply of electricity in china. *Energy Policy*, 62(7):131–138, 2013.

[9] N. Zhang, C. Kang, Q. Xu, C. Jiang, Z. Chen, and J. Liu. Modelling and simulating the spatio-temporal correlations of clustered wind power using copula. *Journal of Electrical Engineering & Technology*, 8(6):1615–1625, 2013.

[10] H. C. Wang, Z. X. Lu, and S. X. Zhou. Research on the capacity credit of wind energy resources. *Proceedings of the CSEE*, 25(10):103–106, 2005.

[11] R. Karki, Po Hu, and R. Billinton. A simplified wind power generation model for reliability evaluation. *IEEE Transactions on Energy Conversion*, 21(2):533–540, 2006.

[12] N. Zhang, C. Kang, D. S. Kirschen, and Q. Xia. Rigorous model for evaluating wind power capacity credit. *IET Renewable Power Generation*, 7(5):504–513, 2013.

[13] Y. C. Ho, Q. Zhao, and Q. Jia. *Ordinal Optimization: Soft Optimization for Hard Problems*. Springer Science & Business Media, 2008.

[14] X. Guan, Y. Ho, and F. Lai. An ordinal optimization-based bidding strategy for electric power suppliers in the daily energy market. *IEEE Transactions on Power Systems*, 21(9):64–64, 2001.

[15] X. Guan, Y. Ho, and F. Lai. An ordinal optimization-based bidding strategy for electric power suppliers in the daily energy market. *IEEE Transactions on Power Systems*, 21(9):64–64, 2001.

[16] S. Lin, Y. Ho, and C. Lin. An ordinal optimization theory-based algorithm for solving the optimal power flow problem with discrete control variables. *IEEE Transactions on Power Systems*, 19(1):276–286, 2004.

[17] Y. Chang, R. Chang, T. Hsiao, and C. Lu. Transmission system loadability enhancement study by ordinal optimization method. *IEEE Transactions on Power Systems*, 26(1):451–459, 2011.

[18] R. A. Jabr and B. C. Pal. Ordinal optimisation approach for locating and sizing of distributed generation. *IET Generation Transmission & Distribution*, 3(8):713–723, 2009.

[19] P. M Subcommittee. Ieee reliability test system. *IEEE Transactions on Power Apparatus & Systems*, PAS-98(6):2047–2054, 1979.

[20] N. Zhang, C. Kang, C. Duan, X. Tang, J. Huang, Z. Lu, W. Wang, and J. Qi. Simulation methodology of multiple wind farms operation considering wind speed correlation. *International Journal of Power & Energy Systems*, 30(30):264–173, 2010.

15

Generation Expansion Planning

The renewable energy output is obviously intermittent and stochastic. In future power generation planning, one of the main challenges is how the power system responds to various renewable energy generation scenarios. This chapter proposes a multi-region power generation planning model with a high penetration of wind power. The optimization objective minimizes the sum of system investment and operating costs. The uncertainty of wind power generation is captured using the scenario approach. Massive operating scenarios are generated through the cross combination of wind scenarios and typical load days. Power system operation constraints including peak and frequency regulation limits and multi-area line-transmission capacity limits are considered. Additionally, the group modeling approach is used to reduce the complexity of operational constraints. A case study on IEEE RTS-79 system is presented to verify the validation of the proposed planning model.

15.1   Overview

The stochastic and intermittent nature of renewable energy output requires the power system to provide more flexibility for renewable energy integration. Generally, generation expansion planning (GEP) researches can be categorized by its objective, including the expansion planning of conventional generation sources [1–4], renewable energy sources [5, 6] and the coordination with transmission system [7–9]. Planning methodologies can be categorized into two methods: establishing an integrated optimization model to solve the optimization planning program [5] and evaluating the performance of candidate planning schemes on reliability and economics using a production simulation approach [10, 11]. Additionally, GEP considering renewable energy integration often focuses on the topic of the coordination of renewable energy with thermal power, hydro-power [12], pumped storage [13] and battery devices [14, 15].

GEP programming is essentially a binary optimization model. The investment plan of each construction project is denoted by a binary variable and optimized to search the most economical planning scheme that satisfies various practical construction constraints and daily operational constraints. In traditional GEP models, the number of operational situations that must be considered is few, and the number of operational constraints for typical operating days is also low. However, the integration of renewable energy introduces many random factors in the planning model, and the traditional planning model is not capable of addressing this challenge.

The key point is how to model the uncertainty of renewable energy generation in the planning model. Generally, there are two methods mainly applied to many related studies. Taking wind power as an example, the first one is to convert wind farms into multi-state units, establishing corresponding constraints for each state. The objective function minimizes the expected total cost. The second one is based on the chance-constrained optimization theory. The basic idea is to make a trade-off between planning risk and economic cost. Renewable energy integration requires conventional thermal units to provide more operational flexibility. This may increase planning and operation costs. As a consequence, the trade-off between the penalty of renewable energy curtailment and the flexible generation power source investment needs to be considered in the optimization model. To effectively consider the uncertainty of renewable energy generation, a large number of renewable energy output scenarios should be considered in the planning model. This drives a significant increase in model complexity.

In summary, the main challenge of the GEP model considering renewable energy integration is the conflict between the stochastic of renewable energy generation and the model complexity. Considering too few renewable energy output scenarios fails to fully reflect the requirements of system peak regulation and transmission capacity, resulting in a potentially inaccurate estimate. On the contrary, considering too many renewable energy output scenarios greatly increases the model complexity, resulting in a significant or even unsatisfied requirement of computation ability. Therefore, the key to constructing GEP lies in reducing the model scale while considering the renewable energy output variability.

This chapter solves this problem in three aspects. (1) Find the key factors that influence the operation of the power system and thereby form vital operation constraints considering renewable energy integration. (2) Consider power balance constraints for extreme scenarios and energy balance constraints for typical scenarios. (3) Improve the computation efficiency by applying a grouping modeling method to reduce the number of decision variables and operation constraints variables.

15.2 Basic Ideas

The GEP model established in this chapter is an integrated planning programming model that simultaneously incorporates the upper-level investment sub-problem (investment) and the lower-level operation sub-problem. In the following text, wind power will be used as an example to explain the GEP model considering large-scale renewable energy integration.

15.2.1 Wind Power Output Modeling Method

The integration of wind power influences both the investment decision module and the operating simulation module. This chapter takes different approaches to these two issues.

First, an increase in wind power installation capacity does not mean the adequacy of power capacity, since the wind power generation is uncertain and unstable. Flexible thermal units are required to provided firm capacity to guarantee the system power capacity balance for all the time periods. This chapter applies the expectation method to determine the ratio of available firm capacity to installation capacity based on the average usage hour for wind power. The average usage hour is obtained from historical data.

Next, the incorporation of multiple wind power scenarios will undoubtedly make the power system operation simulation more accurate. However, it will also over-estimate the influence of wind power on the system if extreme scenarios are considered, and under-estimate the influence if extreme scenarios are ignored. Therefore, in this chapter, only the operating costs in typical scenarios are considered in the objective function. For extreme scenarios, the operating costs are not included, but the corresponding operational constraints are imposed in order to ensure the robustness of the solution.

In addition, in order to capture the peak-regulation constraint accurately, this chapter divides the generation units into two categories: start-stop and non-start-stop groups. A start-stop operation unit can adjust its output from 0 to the capacity, such as hydro-power or medium/small thermal power. The non-start-stop unit includes the units that cannot adjust their unit output, such as wind, PV, and CHP, and the units that must maintain a constant output after starting, such as nuclear power and large thermal power.

15.2.2 Group Modeling Method

This chapter utilizes a group modeling method to address the operation simulation problem. In reality, the online capacity and output for a set of generation units are discrete values. However, for a large region with a sufficient number of units, the online capacity and output of a set of generation units can be approximated as continuous variables.

This chapter classifies and combines generation units according to their regional location, unit type, and operating characteristics. The operating simulation is converted from a unit commitment problem into an economic dispatch problem. This aligns with the processing strategy of general medium- and long-term power planning problems. This processing method can significantly reduce the model scale and address the dimensionality disaster caused by introducing a binary variable for each unit. In operation simulation, only the overall output change of a type of units is concerned. This will benefit in modeling unit commitment and peak-regulation constraints, effectively reflecting the impact of wind power integration on the electric power system operation.

15.3 Model Formulation

15.3.1 Notations

Indices

y	Index for planning years
s	Index for operating scenarios
d	Index for operating typical days
t	Index for time periods, including two time periods (peak and valley) in each typical day
k	Index for unit types

Parameters

H_k	Annual usage hour of a k-type unit
N_d^{Day}	The total number of days at load level d within the year
N_T^{Time}	The number of periods at load level t during that day
$C_n^{Cutloss}$	Load cutting loss
$R_y^{Reserve}, D_y^{Max}$	The power generation reserve rate for year y
$R_y^{Positive}$	The positive reserve coefficient in planning year y
$R_y^{Negative}$	The negative reserve coefficient in planning year y
D_y^{Max}	The maximum load of the system for year y
$D_{y,s,d}^{Max}, D_{y,s,d}^{Min}$	The highest and lowest loads in a typical load day
$g_{l,n}$	Generation shift distribution factor
$F_{y,n}^{Min}$	The minimum frequency modulation capability in area n
$R_{n,k}$	The unit-hour ramp-up limit of a k-type unit

Investment Variables

$I_{y,u}$	if the u unit was put into operation in the y year, then the service state variable $I_{y,u}$ takes the value of 1; otherwise, the value is 0.

Operational Variables

$Cap_{n,k}^{y,s,d}$	the non-start-stop capacity of the unit
$Cap_{n,k}^{',y,s,d}$	the start-stop capacity of the unit
$P_{n,k}^{y,s,d,t}$	the non-start-stop output of the unit
$P_{n,k}^{',y,s,d,t}$	the start-stop output of the unit
$D_{CUT,n}^{y,s,d,t}$	the overall elimination load value

15.3.2 Objective Function

The objective function minimizes the sum of investment cost and operation cost during the planning years.

$$\min_{PlanYear} C^{TOTAL} = C^{INV} + C^{OPE} \tag{15.1}$$

The investment cost C^{INV} should consider the discount rate. When the discount rate is i, the total number of planning years is N, and the static investment cost is C_u^{INV}, we obtain:

$$C^{INV} = \sum_{y\in\Omega_Y} \sum_{u\in\Omega_U} (1+i)^{-y}(C_y^{GEN}+C_y^{STA}+C_y^{CUT}). \tag{15.2}$$

The operation cost C^{OPE} is the sum of the power generation cost, start-stop cost, and load shedding loss, and discount rate should also be considered.

$$C^{OPE} = \sum_{y\in\Omega_Y} (1+i)^{-y}(C_y^{GEN}+C_y^{STA}+C_y^{CUT}) \tag{15.3}$$

For planning year y, the power generation cost C_y^{GEN} is represented by the sum of the product of unit cost C_k^{GEN} and power generation amount $Q_{y,n,k}^{GEN}$ for all types of units.

$$C_y^{GEN} = \sum_{n\in\Omega_N} \sum_{k\in\Omega_K} C_k^{GEN} Q_{y,n,k}^{GEN} \tag{15.4}$$

The power generation amount $Q_{y,n,k}^{GEN}$ involves two modeling concepts according to the difference of the unit type. When the annual usage hours for the unit is given, the power generation amount is determined by the product of the installation capacity Cap_u and the annual usage hour H_k of a k-type unit, giving:

$$Q_{y,n,k}^{GEN} = \sum_{u\in\Omega_U} Z_{u,n}I_{y,u}Cap_uH_k. \tag{15.5}$$

For a specific typical day, d, N_d^{Day} represents the total number of days at a load level d within the year. For a typical time period t, N_T^{Time} represents the number of periods at load level t during that day. When the annual usage hour for the unit is uncertain, the annual power generation amount is the sum of power generation from start-stop units and non-start-stop units.

$$Q_{y,n,k}^{GEN} = \sum_{d\in\Omega_D} \sum_{t\in\Omega_T} N_d^{Day} N_T^{Time} (P_{n,k}^{',y,s,d,t} + P_{n,k}^{y,s,d,t}) \qquad (15.6)$$

For planning year y, the system start-stop cost C_y^{STA} is represented by the product of unit start-stop cost C_k^{STA} and the online capacity.

$$C_y^{STA} = \sum_{n\in\Omega_N} \sum_{k\in\Omega_K} \sum_{d\in\Omega_D} \sum_{t\in\Omega_T} N_d^{Day} C_k^{STA} Cap_{n,k}^{',y,s,d} \qquad (15.7)$$

For planning year y, the system load shedding loss C_y^{CUT} is represented by the sum of unit load shedding loss $C_n^{Cutloss}$ and the load shedding amount.

$$C_y^{CUT} = \sum_{n\in\Omega_N} \sum_{d\in\Omega_D} \sum_{t\in\Omega_T} N_d^{Day} N_t^{Time} C_n^{Cutloss} D_{CUT,n}^{y,s,d,t} \qquad (15.8)$$

15.3.3 Constraints Conditions

Phase 1: Constraints of the investment decision

(1) Period continuity of unit status variable:

$$I_{y,u} \geq I_{y-1,u} \qquad (15.9)$$

(2) The limit of earliest construction year:

$$I_{y,u} = 0, \forall y < y_{Earliest} \qquad (15.10)$$

(3) The minimum gap year between constructing two generation units:

$$I_{y,u2} = 0, \forall y_{Interval} > y \geq 0$$
$$I_{y,u2} \leq I_{y-y_{Interval},u1}, \forall y > y_{Interval} \qquad (15.11)$$

When the units of a power plant have multiple construction periods, if the construction gap year between two units $u1$ and $u2$ is $y_{Interval}$, the second unit $u2$ cannot be constructed until $y_{Interval}$ is passed after the construction of unit $u1$. If the second unit $u2$ has been put into use in the year y, the first unit $u1$ must be in service in the year $y - y_{Interval}$.

(4) Ensuring adequate power capacity in each year:

$$\sum_{u\in\Omega_{OW}} I_{y,u} Cap_u + \sum_{u\in\Omega_W} r_W I_{y,u} Cap_u \geq (1 + R_y^{Reserve}) D_y^{Max} \qquad (15.12)$$

where the power generation units are divided into non-wind power units (Ω_{OW}) and wind power units (Ω_W). For wind power units, due to the uncertainty of wind power generation, the wind capacity should be multiplied by the capacity credit factor r_W ($0 \leq r_W \leq 1$). $R_y^{Reserve}$ and D_y^{Max} represent the power generation reserve rate and the maximum load of the system for year y.

Phase 2: Operation simulation constraints

(1) Operation capacity constraints. The unit on-line capacity is related to the type of unit. Large thermal power, nuclear power, etc, need to maintain a constant output. These are known as must-operating units. First, for the non-must-operating unit, we have:

$$Cap_{n,k}^{',y,s,d} \geq 0, Cap_{n,k}^{y,s,d} \geq 0$$
$$Cap_{n,k}^{',y,s,d} + Cap_{n,k}^{y,s,d} \leq Cap_{y,s,d,n,k}^{Available} \qquad (15.13)$$
$$Cap_{y,s,d,n,k}^{Available} = \sum_{u \in \Omega_U} K_{u,k} Z_{u,n} I_{y,u} Cap_{U,u}^{y,s,d}$$

where $Cap_{y,s,d,n,k}^{Available}$ denotes the on-line capacity of a k-type unit in scenario s of typical day d in year y, which is related to the investment decision variable.

(2) Reserve constraints. For each determined typical load day, the entire system needs to meet certain reserve constraints so that the peak load and the valley load satisfy the positive and negative reserve requirements, respectively, to ensure definite capacity adequacy and a safe, reliable system.

The conditions that the reserve constraints must satisfy are as follows:

$$\sum_{n \in \Omega_N} \sum_{k \in \Omega_{OW}} (Cap_{n,k}^{',y,s,d} + Cap_{n,k}^{y,s,d}) \geq$$
$$(1 + R_y^{Re\,serve})(D_{y,s,d}^{Max} - \sum_{n \in \Omega_N} D_{CUT,t}^{y,s,d,t})$$
$$\sum_{n \in \Omega_N} \left(\sum_{k \in \Omega_{OW}} \beta_K U_{n,k}^{y,s,d} + \sum_{k \in \Omega_{OW}} \gamma_W Cap_{n,k}^{y,s,d} \right) \leq \qquad (15.14)$$
$$(1 - R_y^{Negative})(D_{y,s,d}^{Min} - \sum_{n \in \Omega_N} D_{CUT,t}^{y,s,d,t})$$

where $R_y^{Positive}$ represents the positive reserve coefficient in planning year y, $R_y^{Negative}$ is the negative reserve coefficient, $D_{y,s,d}^{Max}$ represents the highest load in a typical load day, and $D_{y,s,d}^{Min}$ is the lowest load. In the calculation of negative reserve, the conventional unit can be determined by multiplying the non-start-stop capacity $Cap_{n,k}^{y,s,d}$ by coefficient β_K to indicate its minimum non-start-stop output. For a wind power unit, the non-start-stop capacity $U_{n,k}^{y,s,d}$ is multiplied by the coefficient γ_W to indicate its maximum non-start-stop output.

The above discussion applies to the anti-peak-regulation characteristics of wind power. When the wind power source ratio is too high, extreme wind power scenarios may result in negative reserve constraints that have no solution. This requires the loosening of the constraints to consider the start-stop capacity of a wind power unit in the same way as a conventional unit.

(3) Output constraints. The start-stop output and the non-start-stop output of the conventional unit are each regulated by their corresponding start-stop capacity and non-start-stop capacity. Additionally, the start-stop output

of a conventional unit can take a value of 0 in the minimum-load period. In terms of a wind power unit, its non-start-stop output is 0. The start-stop output does not exceed the value of wind power output in the target scenario at that moment.

$$\beta_k Cap_{n,k}^{y,s,d} \leq P_{n,k}^{y,s,d,t} \leq Cap_{n,k}^{y,s,d}, \forall k \in \Omega_{OW}, t \in \Omega_T$$

$$\beta_k Cap_{n,k}^{',y,s,d} \leq P_{n,k}^{',y,s,d,t} \leq Cap_{n,k}^{',y,s,d}, \forall k \in \Omega_{OW}, t \in \Omega_{PeakT} \qquad (15.15)$$

$$P_{n,k}^{',y,s,d,t} = 0, \forall k \in \Omega_{OW}, t \in \Omega_{ValleyT}$$

where Ω_{PeakT} represents the peak load period, and $\Omega_{ValleyT}$ represents the valley load period.

(4) Line flow constraints. According to the distribution transfer matrix, the flow obtained is as follows:

$$L_l^{y,s,d,t} = \sum_{n\in\Omega_N} g_{l,n} \left[\sum_{k\in\Omega_K} (P_{n,k}^{y,s,d,t} + P_{n,k}^{',y,s,d,t}) - D_n^{y,s,d,t} + D_{CUT,n}^{y,s,d,t} \right]. \qquad (15.16)$$

For line l and area n, $g_{l,n}$ is its corresponding coefficient; when the k-type unit is in area n. Flow can be calculated through the output and load $D_n^{y,s,d,t}$.

Flow is in the following range:

$$-L_{l,y}^{Max} \leq L_l^{y,s,d,t} \leq L_{l,y}^{Max}. \qquad (15.17)$$

(5) Overall power balance constraints. The output and load of the entire system should satisfy the following balance.

$$\sum_{n\in\Omega_N} \sum_{k\in\Omega_K} (P_{n,k}^{y,s,d,t} + P_{n,k}^{',y,s,d,t}) = D_{y,s,d,t}^{System} - \sum_{n\in\Omega_N} D_{CUT,n}^{y,s,d,t}. \qquad (15.18)$$

(6) Ramp-up constraints. The power variation of the conventional unit in the adjacent period has a limit, which is the ramp-up constraint of the unit. Due to the uncontrollability of wind power, conventional units need to make timely power adjustments to complement the wind power. This presents a challenge to the ramp-up ability of conventional units. $R_{n,k}$ is the unit-hour ramp-up limit of a k-type unit, which gives:

$$-R_{n,k} \leq Cap_{n,k}^{',y,s,d,t} - Cap_{n,k}^{',y,s,d,t-1} \leq R_{n,k}. \qquad (15.19)$$

(7) Regional frequency modulation constraints. The uncontrollability of wind power has an influence on regional frequency modulation. The impact is negative, and it is represented by a negative coefficient $(-\alpha_k)$. The conventional-unit frequency modulation coefficient is denoted by a positive value (α_k).

$$\sum_{k\in U_{OW}} \alpha_k Cap_{n,k}^{',y,s,d,t} - \sum_{k\in U_W} \alpha_k Cap_{n,k}^{',y,s,d,t} \geq F_{y,n}^{Min} \qquad (15.20)$$

where $F_{y,n}^{Min}$ represents the minimum frequency modulation capability in area n.

15.3.4   Model Structure

All power generation planning model operating variables have three characteristic dimensions on the time scale: planning year, wind power scenario and typical load day. The specific operating scenario is defined by the given planning year, wind power scenario, and typical load day. The scenario itself contains all the operating constraints, which serve as a computational unit, and different computational units are mutually independent of each other. Therefore, we can denote the operating constraints matrix in the form of a partition matrix, as shown in the following figure. The structure of the partition matrix facilitates storage and computation; the number of computational units for the entire model is $N^{PlanYear} \times N^{Scenario} \times N^{TypicalLoadDay}$. This model is a large-scale composite integer linear planning model.

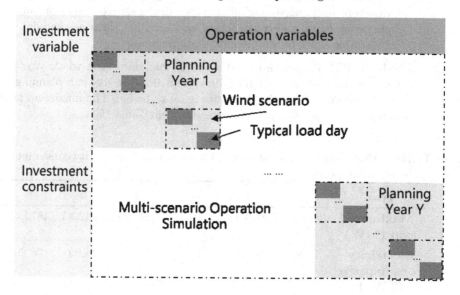

FIGURE 15.1 Overall matrix structure of the model.

15.4   Case Study Analysis

15.4.1   Basic Data

This chapter presents a case study on the IEEE RTS-79 system. The IEEE RTS-79 system contains 24 nodes, 38 lines, and 32 conventional units, with a total installed capacity of 3405 MW and a maximum system load of 2850 MW. Based on the IEEE RTS-79 system data, the following settings were selected:

1. Maintain the original generation and transmission structure and parameters of the IEEE RTS-79 system. All generating units in the original system are assumed to be constructed.

2. Add a certain capacity of wind power to analyze the impact of intermittent renewable sources on the system. Five wind farms are introduced, with a 340-MW capacity for each wind farm. Parameters of these wind farms are identical and mutually independent. All wind farms are already constructed. These five wind farms are connected to node 1 to 5, respectively. The generation output of wind power is simulated using the method in Chapter 6.

3. Except for hydro-power and wind power, replicate the thermal power, nuclear power, and gas turbine units, maintaining all parameters unchanged (including type and scale, construction and operating costs, and construction location). Set them as candidate generation units in the planning optimization.

The IEEE RTS-79 system provides hourly load data for a whole year period. On the basis of original RTS-79 system data, a six-year planning is performed. The annual load increase rate is set to be 5%. The maximum load and electricity demand for each year are shown in Table 15.1.

TABLE 15.1 Expected growth of maximum load and electricity consumption in the planning period

Year	1	2	3	4	5	6
Maximum load (MW)	3819.27	4010.24	4210.75	4421.29	4642.35	4874.47
Electricity consumption (TWh)	19.82	20.81	21.85	22.94	24.09	25.29

15.4.2 Analysis of Basic Planning Results

This section discusses the impact of multiple wind power scenarios in the planning scheme. To highlight the impact of common scenarios on the results, extreme scenarios are not considered in the calculations of this section. For the basic case, only three typical wind power scenarios are selected. The solved planning scheme is shown in Table 15.2.

TABLE 15.2 Analysis of basic planning results

Unit No	Unit Type	Node No	Unit Capacity	Year unit put into operation
1	Single cycle gas turbine	1	20	6
3	Small thermal power	1	76	6
9	Single cycle gas turbine	2	100	6
22	Nuclear power	4	400	1
23	Nuclear power	4	400	1
24	Small thermal power	3	155	1
25	Small thermal power	3	155	2
26	Large thermal power	3	350	1

The planning results indicate the following: (1) In the current system, there is an imbalance between supply and demand in some areas. The power generation deficiency problem is concentrated at nodes 3 and 4. Therefore, in the first two years of planning, large-scale thermal power plants and nuclear power plants are quickly built in nodes 3 and 4, respectively, to meet the demand at a low cost. (2) As load increases, the influence of wind power variability on the system becomes apparent. In the latter planning years, many small thermal power plants and gas turbine units are constructed to address this issue.

According to Table 15.2, a trend chart of the installed capacity structure is drawn; Year 0 on the horizontal axis represents the initial power capacity structure prior to the start of the plan. Figure 15.2 intuitively reflects the rapidly increasing trend of total capacity in early planning years. It also reflects the increasing proportion of flexible power generation including small thermal power plants and gas turbine units.

The analysis of power generation cost can better explain the above phenomenon. Figure 15.3 shows that the start-stop cost proportion gradually increases, occupying a large portion of the annual increasing cost. This indicates an increase in generation share of flexible generation units. As the load continuously increases, wind power output becomes an important form of power generation due to its lower costs, the stochastic generation characteristic of wind power leads to an increase in flexible generation units.

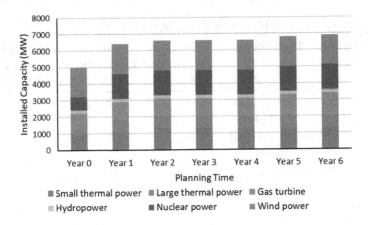

FIGURE 15.2 Installed capacity structure change trend of the basic case study in the planning period.

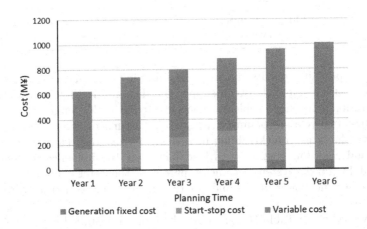

FIGURE 15.3 Power generation cost structure change trend of the basic case study in the planning period.

15.4.3 The Impact of the Number of Wind Power Scenarios

The introduction of multiple wind power scenarios can more accurately reflect the random fluctuation and anti-peak-regulation characteristics of wind power. Compared with the base case, seven typical scenarios are considered. The results are shown in Table 15.3, Figure 15.4 and Figure 15.5. The results show that an increase in the number of scenarios highlights the impact of wind power uncertainty. More flexible units are constructed, and the construction times are earlier. This phenomenon indicates that wind power uncertainty increases the requirement of flexible units.

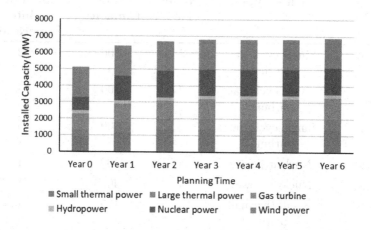

FIGURE 15.4 Installed capacity structure change trend for the multi-scenario wind power case study.

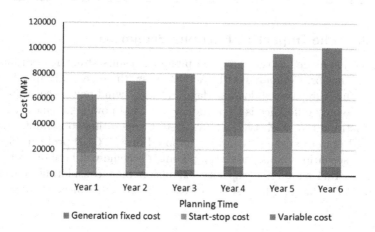

FIGURE 15.5 Power generation cost structure change trend for the multi-scenario wind power case study.

TABLE 15.3 Unit expansion planning in the multi-scenario wind power case study planning period

Unit No	Unit Type	Node No	Unit Capacity	Year unit put into operation
9	Single cycle gas turbine	2	100	2
10	Single cycle gas turbine	2	100	3
11	Single cycle gas turbine	2	100	3
22	Nuclear power	4	400	1
23	Nuclear power	4	400	1
24	Small thermal power	3	155	1
25	Small thermal power	3	155	2
26	Large thermal power	3	350	1

15.4.4   The Impact of Extreme Scenarios

This section examines the change in planning results after the addition of extreme scenarios. Compared with the base case, the case study in this section will still use three typical loading days and three wind power scenarios in each year. The only difference is to introduce the wind power extreme scenarios in the constraints for each typical day. A set of operating constraints considering wind power extreme scenario is imposed on the GEP problem. Since the extreme scenario is considered, the planning scheme will be more robust.

Table 15.4 shows the capacity expansion plan considering the extreme generation scenario. The result indicates the system will need more flexible units earlier.

TABLE 15.4 Unit expansion plan for the extreme scenario case study planning period

Unit No	Unit Type	Node No	Unit Capacity	Year unit put into operation
9	Single cycle gas turbine	2	100	2
10	Single cycle gas turbine	2	100	3
11	Single cycle gas turbine	2	100	3
22	Nuclear power	4	400	1
23	Nuclear power	4	400	1
24	Small thermal power	3	155	1
25	Small thermal power	3	155	2
26	Large thermal power	3	350	1

15.4.5 The Impact of Power Grid Transmission Capacity

In order to analyze the impact of regional transmission connections on the planning results, this section increases and decreases the capacity of each transmission line by 20%, forming a high connection plan and a low connection plan, respectively. The installed power capacity structures in the planning results of the two schemes are compared and shown in Figure 15.6. The results indicate that the total newly installed capacity for the two plans are similar and both satisfy the load increment. However, large differences appear in the incremental installed capacity, leading to power capacity structure difference at the end of the planning period.

For the high connection plan (right figure), due to the large capacity of regional transmission, the overall trend of the system leans toward prioritizing the development of low-cost, concentrated power generation techniques, including large thermal power and nuclear power. Meanwhile, flexible generation units are constructed in some areas and provide support when the generation in other regions fluctuates. In the low connection plan (left figure), since the regional transmission is impeded, each large area must independently accommodate wind power in its region. Therefore, more flexible units should be constructed to increase the regional frequency and peak regulation ability. In the six-year planning period, a large number of small thermal power and gas turbine units are added to the system, distributed at various nodes to meet their respective needs.

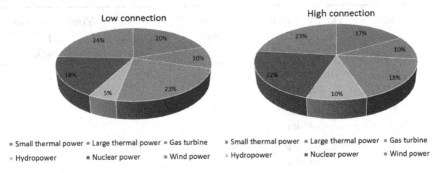

FIGURE 15.6 Installation capacity comparison of regional high and low connection plans.

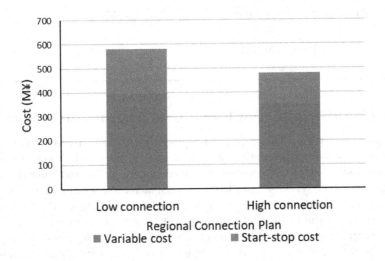

FIGURE 15.7 Comparison of the total power generation cost structure in regional high and low connectivity plans.

The total operating costs of the two plans are compared and shown in Figure 15.7. The results indicate that the larger transmission capacity leads to a lower total cost and a lower ratio of start-stop costs. When the transmission capacity is poor, the impact of uncertainty factors in each region is enhanced, and the flexible unit must frequently change its operating output to compensate for the fluctuations in wind power and load, thereby increasing the total cost and the start-stop cost of flexible units.

15.5 Summary

The uncertainty of renewable energy output requires the future GEP to consider multiple operating scenarios. In order to address this issue, this chapter proposes a multi-region power generation planning model towards high renewable energy penetration. This model has several main features: 1) decoupling investment constraints and operation simulating constraints, 2) group modeling for electricity generators and 3) considering various realistic operation constraints. A case study on the IEEE RTS-79 test system verifies the validation of the proposed method. Flexible generation units are required to handle the uncertainty of renewable energy output. When the volatility of the uncertainty factors in each region increases, the flexible unit will frequently change its operating state to compensate for the fluctuations in wind power and load, thereby increasing the total cost and the start-stop cost of the flexible unit.

Bibliography

[1] K. F. Schenk and S. Chan. Incorporation and impact of a wind energy conversion system in generation expansion planning. *IEEE Transactions on Power Apparatus and Systems*, (12):4710–4718, 1981.

[2] Z. Yu. Planning wind power considering system reliability in peak demand time. *Power and Energy Society General Meeting, 2008 IEEE*, pages 1–6, 2008.

[3] C. Dejonghe, E. Delarue, R. Belmans, and W. D'haeseleer. Determining optimal electricity technology mix with high level of wind power penetration. *Applied Energy*, 88(6):2231–2238, 2011.

[4] J. Nelson, J. Johnston, A. Mileva, M. Fripp, I. Hoffman, A. Petros-Good, C. Blanco, and D. M. Kammen. High-resolution modeling of the western North American power system demonstrates low-cost and low-carbon futures. *Energy Policy*, 43:436–447, 2012.

[5] L. Baringo and A. J. Conejo. Risk-constrained multi-stage wind power investment. *IEEE Transactions on Power Systems*, 28(1):401–411, 2013.

[6] C. Chompoo-inwai, M. Leelajindakrairerk, S. Banjongjit, P. Fuangfoo, and W. Lee. Design optimization of wind power planning for a country with low–medium-wind-speed profile. *IEEE Transactions on Industry Applications*, 44(5):1341–1347, 2008.

[7] A. K. Kazerooni and J. Mutale. Network investment planning for high

penetration of wind energy under demand response program. In *Probabilistic Methods Applied to Power Systems (PMAPS), 2010 IEEE 11th International Conference on*, pages 238–243. IEEE, 2010.

[8] Francois Vallee, François Vallée, Jacques Lobry, and Olivier Deblecker. Impact of wind power integration in the context of transmission systems adequacy studies with economic dispatch: Application to the Belgian case. In *Clean Electrical Power (ICCEP), 2011 International Conference on*, pages 1–8. IEEE, 2011.

[9] L. Baringo and A. J. Conejo. Transmission and wind power investment. *IEEE Transactions on Power Systems*, 27(2):885–893, 2012.

[10] W. Wangdee and R. Billinton. Reliability assessment of bulk electric systems containing large wind farms. *International Journal of electrical power & energy systems*, 29(10):759–766, 2007.

[11] E. Tsioliaridou, G. C. Bakos, and M. Stadler. A new energy planning methodology for the penetration of renewable energy technologies in electricity sector application for the island of Crete. *Energy Policy*, 34(18):3757–3764, 2006.

[12] R. Karki, P. Hu, and R. Billinton. Reliability evaluation considering wind and hydro power coordination. *IEEE Transactions on Power Systems*, 25(2):685–693, 2010.

[13] N. Zhang, C. Kang, D. S. Kirschen, Q. Xia, W. Xi, J. Huang, and Q. Zhang. Planning pumped storage capacity for wind power integration. *IEEE Transactions on Sustainable Energy*, 4(2):393–401, 2013.

[14] T. Senjyu, D. Hayashi, N. Urasaki, and T. Funabashi. Optimum configuration for renewable generating systems in residence using genetic algorithm. *IEEE Transactions on Energy Conversion*, 21(2):459–466, 2006.

[15] H. Bludszuweit and J. Domínguez-Navarro. A probabilistic method for energy storage sizing based on wind power forecast uncertainty. *IEEE Transactions on Power Systems*, 26(3):1651–1658, 2011.

16

Transmission Expansion Planning

A power system with high wind power integration requires extra transmission capacity to accommodate the intermittency that is inherent in wind power production. Storage can smooth out this intermittency and reduce transmission requirements. This chapter proposes a stochastic optimization model to coordinate the long-term planning of both transmission and storage facilities to efficiently integrate wind power. Both long-term and short-term uncertainties are considered in this model. Long-term uncertainty is described via scenarios, while short-term uncertainty is described via operating conditions. Garver's 6-node system and a system representing Northwest China in 2030 are used to illustrate the proposed model. Results indicate that storage reduces the transmission requirement and the overall investment, and allows efficiently integrating wind power.

16.1 Overview

This chapter addresses the long-term planning problem of identifying the best combination of interconnection enhancement and storage installation to achieve the minimum expected cost in the considered power system. For the sake of computational tractability, a static approach is used.

It is important to realize that this long-term planning problem is plagued with both long-term and short-term uncertainty. Long-term uncertainty pertains to the fact that the peak and shape of the demand are unknown throughout the target year and that the available generation facilities in that year are unknown as well. In each day of the target year, short-term uncertainty pertains to the unknown variability of demand and wind power production and to their correlations.

To tackle this long-term planning problem, we use a stochastic programming model that represents long-term uncertainty via scenarios and short-term uncertainty via operating conditions. On the one hand, the considered scenarios represent the uncertain demand level and uncertain installed production capacity in the target year. On the other hand, for each scenario, a number of operating conditions are considered. An operating condition is a representative day including 24-hour profiles of correlated demand and wind production.

A number of works are available in the technical literature regarding the long-term planning of both transmission and storage facilitating for a system with significant wind power integration. Ugranli, *et al.* [1] proposed a transmission expansion planning methodology that minimizes both investment cost and curtailed wind energy under N-1 security. A multi-stage multi-objective transmission expansion planning (TEP) model is proposed in [2] involving investment cost, private investment level, and reliability. In [3], an approach is proposed for determining the optimal location and sizing of an energy storage system (ESS) in a power system. A genetic algorithm is used in [4] to find the location and size of an ESS. Zhang *et al.* [5] proposed a deterministic single-stage TEP model, which considers the deployment of an ESS. A storage deployment scheme to minimize investment and operation costs is analyzed in [6]. This work is based on the assumption that the operation of ESSs follows predefined cycles.

Due to the stochastic nature of wind power, increasing penetration of wind power means more uncertainty in the TEP problem. Several stochastic programming models are available in references [7, 8] to represent the uncertain nature of power systems. Wind power and system reliability are considered as random variables in [7]. A robust TEP model is proposed in [9]. The uncertainties of the load and of renewable resources are analyzed in [10, 11]. Adriaan, *et al.* [12] proposed a two-stage TEP model considering market uncertainties. A two-stage stochastic model is used in [8] to solve the TEP problem considering uncertainties on wind availability and system load. These uncertainties are represented by correlated wind and load scenarios in [13].

Finally, we note that uncertainties of both wind power and load have a critical impact on the TEP problem, and they are not statistically independent. For instance, wind power production is usually high during the night while the demand is low, and vice versa [14]. This trend appears daily and seasonally.

16.2   Problem Description

16.2.1   Assumption

The following assumptions are considered to formulate the proposed model.

1. The model is static as it only considers the investment decisions to supply the demand of a target year. The model can be expanded to a dynamic setting, but at the cost of high computational burden.

2. A DC power flow representation is adopted and line losses are neglected.

3. Disjunctive constraints [15] are used to represent building / not-building decisions for lines.

4. To represent daily storage cycles, and demand and wind production correlations, each operation condition is described by 1 day, i.e., 24 hours. However, the model can be easily extended to represent weekly storage cycles.

5. The storage investment is made by the system operator pursuing social welfare. The cost involved is then socialized among all users.

16.2.2   Model Structure

The planning model presented in this chapter solves a coordinated expansion problem involving transmission and storage. The structure of the proposed model is shown in Figure 16.1. Several scenarios are considered to represent the uncertainty of load growth and available production capacity in the planning year. In each scenario, to consider the correlation between load and wind production, several operating conditions are considered. Each operating condition is constructed by clustering historical data of the daily load and wind power production.

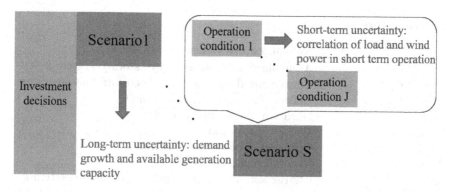

FIGURE 16.1 Structure of the proposed model.

16.2.3   Notations

Indexes

l	Index for transmission lines
w	Index for wind farms
k	Index for storage systems
i	Index for thermal generators
s	Index for scenarios

t	Index for time periods
b	Index for nodes
j	Index for operation conditions

Sets

Ω^I	Sets of thermal units
Ω^W	Sets of wind farms
Ω^K	Sets of storage systems
Ω^L	Sets of all lines
Ω^{LE}	Sets of existing lines
Ω^{LC}	Sets of alternative lines
Ω^B	Sets of nodes
Ω^I_b	Sets of thermal units located at node b
Ω^W_b	Sets of wind farms located at node b
Ω^K_b	Sets of storage systems located at node b
Ω^{L1}_b	Sets of lines whose sending end node is b
Ω^{L2}_b	Sets of lines whose receiving end node is b
Ω^S	Sets of scenarios
Ω^{OP}	Sets of operating conditions in each scenario

Parameters

C^{Line}_l	Annualized investment cost of line l
$C^{Storage}_k$	Annualized investment cost of storage system k
C^{ch}_k, C^{dis}_k	Charge and discharge variable cost of storage system k
C^{Gen}_i	Generation cost of thermal unit i
τ	Duration of one time period
α_s	Weight of scenario s
d_j	Duration of operating condition j
x_l	Reactance of line l
$\eta^{ch}_k, \eta^{dis}_k$	Charge and discharge efficiency of storage system k
$P^{Gen,max}_i, P^{Gen,min}_i$	Capacity and minimum power output of thermal unit i
R^{up}_i, R^{down}_i	Ramp-up and ramp-down limit of thermal unit i
$P^{dis,max}_k, P^{dis,min}_k$	Maximum and minimum discharging power of storage system k
$P^{ch,max}_k, P^{ch,max}_k$	Maximum and minimum charging power of storage system k
S^{max}_k, S^{min}_k	Maximum and minimum energy level of storage system k
L^{max}_l	Capacity of line l
$P^f_{w,s,j,t}$	Forecast power output of wind farm w at time t in operating condition j of scenario s
$D_{b,s,j,t}$	Load demand at node b at time t in operating condition j of scenario s

T	Number of time periods
M	A large enough constant

Variable

$P_{i,s,j,t}^{Gen}$	Power output of thermal unit i at time t in operating condition j of scenario s
$P_{w,s,j,t}^{Wind}$	Power output of wind farm w at time t in operating condition j of scenario s
S_k^0	Initial energy level of storage system k
$P_{k,s,j,t}^{ch}$	Charging power of storage system k at time t in operating condition j of scenario s
$P_{k,s,j,t}^{dis}$	Discharging power of storage system k at time t in operating condition j of scenario s
$L_{l,s,j,t}$	Power flow of line l at time t in operating condition j of scenario s
$\theta_{b,s,j,t}$	Voltage angle of node b at time t in operating condition j of scenario s
X_l	Binary variables that equal 1 if line l is built and 0 otherwise
Y_k	Binary variables that equal 1 if storage system k is built and 0 otherwise

16.2.4   Model Formulation

The objective is to minimize total cost, including the annualized investment cost and the operation cost:

$$\min \; C^{Inv} + C^{Oper} \tag{16.1}$$

where

$$C^{Inv} = \sum_{l \in \Omega^{LC}} C_l^{Line} X_l + \sum_{k \in \Omega^K} C_k^{Storage} Y_k. \tag{16.2}$$

$$
\begin{aligned}
C^{Oper} &= \sum_{s \in \Omega^S} \alpha_s \sum_{j \in \Omega^{OP}} d_j \sum_{t=1}^{T} \tau \sum_{i \in \Omega^I} C_i^{Gen} P_{i,s,j,t}^{Gen} \\
&+ \sum_{s \in \Omega^S} \alpha_s \sum_{j \in \Omega^{OP}} d_j \sum_{t=1}^{T} \tau \sum_{k \in \Omega^K} (C_k^{ch} P_{k,s,j,t}^{ch} + C_k^{dis} P_{k,s,j,t}^{dis}).
\end{aligned} \tag{16.3}
$$

In equation (16.1), C^{Inv} denotes the annualized investment cost in transmission lines and ESSs. C^{Oper} denotes the operation cost of thermal units and ESSs. Operation costs of all thermal units and ESSs are considered linear, as stated in Equation (16.3). The production cost of wind plants is not considered to compute the total operating cost. This is because the marginal cost of wind power production is comparatively very small and generally considered equal to zero. Although the model does not consider a wind power curtailment penalty, the savings of reducing wind power curtailment are reflected as a reduction of the operation cost of the thermal units. Scenarios and operating conditions allow representing the long-term and short-term uncertainties of load and wind production.

The constraints of the proposed model are provided below. Constraint (16.4) represents the power balance at each node in the system.

$$
\begin{aligned}
&\sum_{i\in\Omega_b^I} P_{i,s,j,t}^{Gen} + \sum_{w\in\Omega_b^W} P_{w,s,j,t}^{Wind} - \sum_{l\in\Omega_b^{L1}} L_{l,s,j,t} + \sum_{l\in\Omega_b^{L2}} L_{l,s,j,t} \\
&+ \sum_{k\in\Omega_b^K} (P_{k,s,j,t}^{dis} - P_{k,s,j,t}^{ch}) = D_{b,s,j,t}, \ \forall b, s, j, t = 1, 2, ..., T
\end{aligned}
\tag{16.4}
$$

Constraints (16.5)-(16.7) are DC power flow constraints. DC power flow constraints for existing lines are:

$$
\begin{aligned}
&L_{l,s,j,t} - (\theta_{b1,s,j,t}^l - \theta_{b2,s,j,t}^l)/x_l = 0, \\
&\forall l \in \Omega^{LE}, s, j, t = 1, 2, ..., T.
\end{aligned}
\tag{16.5}
$$

Constraints (16.6) and (16.7) are DC power flow constraints for prospective lines.

$$
\begin{aligned}
&- M(1 - X_l) \le L_{l,s,j,t} - (\theta_{b1,s,j,t}^l - \theta_{b2,s,j,t}^l)/x_l, \\
&\forall l \in \Omega^{LC}, s, j, t = 1, 2, ..., T
\end{aligned}
\tag{16.6}
$$

$$
\begin{aligned}
&L_{l,s,j,t} - (\theta_{b1,s,j,t}^l - \theta_{b2,s,j,t}^l)/x_l \le M(1 - X_l), \\
&\forall l \in \Omega^{LC}, s, j, t = 1, 2, ..., T
\end{aligned}
\tag{16.7}
$$

Constraints (16.8) and (16.9) are the ramp-up and ramp-down constraints of thermal units,

$$
\begin{aligned}
&- R_i^{down}\tau \le P_{i,s,j,t}^{Gen} - P_{i,s,j,t-1}^{Gen}, \\
&\forall i, s, j, t = 2, 3, ..., T
\end{aligned}
\tag{16.8}
$$

$$
\begin{aligned}
&P_{i,s,j,t}^{Gen} - P_{i,s,j,t-1}^{Gen} \le R_i^{up}\tau, \\
&\forall i, s, j, t = 2, 3, ..., T.
\end{aligned}
\tag{16.9}
$$

Constraint (16.10) enforces energy limits for storage systems in each time period, scenario, and operating condition.

$$S_k^{\min} \le S_k^0 + \sum_{t=1}^{t_n} (P_{k,s,j,t}^{ch}\eta_k^{ch} - P_{k,s,j,t}^{dis}/\eta_k^{dis})\tau \le S_k^{\max},$$

$$\forall k, s, j, t_n = 1, 2, .., T \tag{16.10}$$

Note that S_k^0 is the storage level at the beginning of the day, and that $S_k^0 + \sum_{t=1}^{t_n} (P_{k,s,j,t}^{ch}\eta_k^{ch} - P_{k,s,j,t}^{dis}/\eta_k^{dis})\tau$ is the storage level at time t_n.

Constraint (16.11) enforces storage energy balance per day.

$$\sum_{t=1}^{T} (P_{k,s,j,t}^{ch}\eta_k^{ch} - P_{k,s,j,t}^{dis}/\eta_k^{dis})\tau = 0, \forall k, s, j \tag{16.11}$$

Constraints (16.12) and (16.13) enforce transmission capacity limits for existing and prospective lines, respectively.

$$- L_l^{\max} \le L_{l,s,j,t} \le L_l^{\max},$$

$$\forall l \in \Omega^{LE}, s, j, t = 1, 2, ..., T \tag{16.12}$$

$$- X_l L_l^{\max} \le L_{l,s,j,t} \le X_l L_l^{\max},$$

$$\forall l \in \Omega^{LC}, s, j, t = 1, 2, ..., T \tag{16.13}$$

Constraints (16.14) and (16.15) are discharging and charging bounds for ESSs.

$$Y_k P_k^{dis,\min} \le P_{k,s,j,t}^{dis} \le Y_k P_k^{dis,\max}$$

$$\forall k, s, j, t = 1, 2, ..., T \tag{16.14}$$

$$Y_k P_k^{ch,\min} \le P_{k,s,j,t}^{ch} \le Y_k P_k^{ch,\max}$$

$$\forall k, s, j, t = 1, 2, ..., T \tag{16.15}$$

Constraint (16.16) sets the production bound of thermal units.

$$P_i^{Genmin} \le P_{i,s,j,t}^{Gen} \le P_i^{Genmax},$$

$$\forall i, s, j, t = 1, 2, ..., T \tag{16.16}$$

Constraint (16.17) sets the wind production bound.

$$0 \le P_{w,s,j,t}^{Wind} \le P_{w,s,j,t}^{f},$$

$$\forall w, s, j, t = 1, 2, ..., T \tag{16.17}$$

Constraints (16.18) and (16.19) are variable declarations.

$$X_l \in \{0, 1\}, \ l \in \Omega^{LC} \tag{16.18}$$

$$Y_k \in \{0, 1\}, \ k \in \Omega^{K} \tag{16.19}$$

16.3   Illustrate Example

16.3.1   Basic Settings

The proposed model has been tested on the modified Garver's 6-node system
[16] shown in Figure 16.2. This system includes five nodes and six lines. Additionally, a sixth node is considered, which is not connected to the system,
but lines can be built to connect it. Any corridor can accommodate at most
three lines (prospective plus existing). Table 16.1 provides the parameters of
existing and alternative lines. The sixth column of this table provides annualized investment costs. The generating system consists of three generators,
whose parameters are shown in Table 16.2. The generator at node 1 is a wind
farm with a capacity of 400 MW. The wind power output is obtained from
the Wind Integration Datasets of NERL [17]. Loads are located at nodes 1
to 5. Load profiles are based on historical data of the China Southern Grid.
Three different load profiles (scenarios) are considered, high, medium and low,
see Table 16.3. For simplicity, the available generation capacity in the target
year is considered known with certainty. For each load profile, four operating
conditions reflecting different wind production and load in the four seasons
of the year are considered. Representative daily demand profiles (per season)
are shown in Figure 16.3.

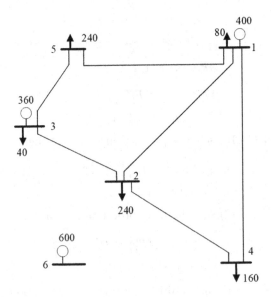

FIGURE 16.2 Garver's 6-node system.

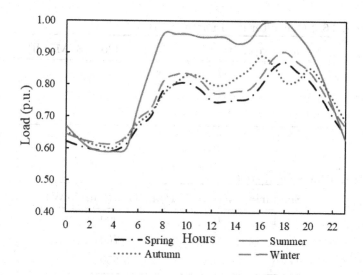

FIGURE 16.3 Representative daily demand profiles.

TABLE 16.1 Line data for Garver's system

From	to	R	X	Limit	Cost	Already built?
1	2	0.10	0.40	100	40	Yes
1	3	0.09	0.38	100	38	No
1	4	0.15	0.60	80	60	Yes
1	5	0.05	0.20	100	20	Yes
1	6	0.17	0.68	70	68	No
2	3	0.05	0.20	100	20	Yes
2	4	0.10	0.40	100	40	Yes
2	5	0.08	0.31	100	31	No
2	6	0.08	0.30	100	30	No
3	4	0.15	0.59	82	59	No
3	5	0.05	0.20	100	20	Yes
3	6	0.12	0.48	100	48	No
4	5	0.16	0.63	75	63	No
4	6	0.08	0.30	100	30	No
5	6	0.15	0.61	78	61	No

TABLE 16.2 Generator data for the Garver's system

Node	Capacity(MW)	Type	Cost($/MWh)
1	400	wind	0
3	360	coal	48
6	600	coal	52

TABLE 16.3 Scenario data

Scenario	Weight	Peak Demand(MW)
1	0.4	900
2	0.3	760
3	0.3	650

Two alternative ESSs are considered at each node. The annualized invest-ment cost of each of them is $60/kWh. Data for the two alternative ESSs are provided in Table 16.4. The model is implemented in GAMS [18] and solved with CPLEX 7.5 [19] on a PC with an Intel Core i5/2.6- GHz-based processor and 8 GB of RAM.

TABLE 16.4 Data of alternative ESSs for Garver's 6-bus system

Type	$P_k^{dis,\max}$ / $P_k^{ch,\max}$ (MW/h)	$P_k^{dis,\min}$ / $P_k^{ch,\min}$ (MW/h)	$S_k^{\min}$ (MW/h)	$S_k^{\max}$ (MW/h)	η_k^{dis} / η_k^{ch}
1	100	0	20	600	0.85
2	50	0	10	300	0.90

16.3.2   Results

Two cases are considered. The first one includes both prospective lines and storage systems. The second one includes no storage. Results are summarized in Table 16.5 and Figure 16.4.

Figure 16.4(a) shows the results of the first case, in which the newly built lines appear in red. The planning outcome includes five new lines with a total annualized investment cost of $140.00 million and two ESSs with a total annualized investment cost of $54.00 million. Among the five new lines, two connect node 6 with node 2, another two connect node 6 with node 4, and the remaining one connects node 3 with node 5. This solution allows the energy produced at node 6 to flow to all loads in the system. An ESS of type 2 is built at node 1 to mitigate the variability of wind power production. And an additional ESS of type 1 is built at node 5, which has the largest demand. In the second case, two additional lines are needed, which are shown in blue in Figure 16.4(b).

TABLE 16.5 Result comparison of two cases with and without an ESS

Scheme	With ESS	Without ESS
Number of new lines built	5	7
Lines to be built	2-6(2),3-5,4-6(2)	2-6(3),3-5(2),4-6(2)
Annualized investment cost in lines (M$)	140.0	190.0
Annualized investment cost in storage (M$)	54.0	—
Operation cost (M$)	207.24	213.52
Total Cost (M$)	401.24	403.52
Wind Curtailment (MWh)	270.37	5307.46
Computing time (s)	40.20	1.39

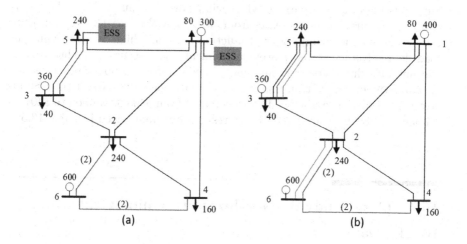

FIGURE 16.4 Result comparison of two cases.

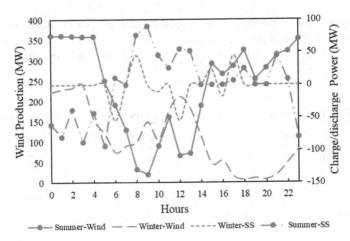

FIGURE 16.5 Hourly output of wind farm and ESS in summer and winter operating conditions.

Figure 16.5 shows the wind power output and the corresponding operation of an ESS in two representative operating conditions. Regarding storage, values above zero represent the discharging power while values below zero represent charging power. The results suggest that the ESS is able to mitigate the variation of wind power on node 1 by charging when wind power output is high and discharging when wind power output is low. The yearly wind power curtailment decreases from 5370.46 MWh to 270.37 MWh if ESSs are available. Since wind power production is usually high at night while the load demand is low, wind power curtailment often occurs at night. Storage can mitigate this waste of clean energy. The decrease in both operating cost and investment cost of transmission lines is large enough to offset the ESS investment cost. Hence, the total cost decreases from \$403.52 million to \$401.24 million if ESSs are considered. These results are summarized in Table 16.5.

16.4   Case Study on Northwestern China Grid

16.4.1   Data

A coordinated planning of transmission and storage in the Northwestern China Grid (NCG) is considered in this section. This system includes 922 GW of generation capacity, with a wind power capacity of 189 GW. The NCG is a synchronized power system that includes five interconnected provincial systems: Shanxi, Qinghai, Gansu, Xinjiang, and Ningxia. These provincial systems are interconnected through 750 kV ultra-high voltage AC lines, shown in Figure 16.6.

The NCG system is modeled as shown in Figure 16.7. Each of the five provincial systems is represented by one node. The current 750 kV lines between these provincial systems are shown in orange. Three external systems (Northern China, Eastern China, and Central China) are represented by three additional nodes. The connections among these external systems are represented in blue. The interconnections between these three external systems and the five regional systems in the NCG are represented in purple.

In this section, we analyze the planning of tie-lines between the five provincial systems and storage facilities in the NCG grid. Interconnections with external systems are considered known.

Three different load scenarios are considered to account for long-term uncertainty, i.e., high, medium and low load scenarios. The weights for these three scenarios are 0.4, 0.3 and 0.3, respectively. The available production capacity is considered known. For each load scenario, 24 typical operating conditions (high and low load operating conditions in each month) are represented. Daily load profiles (including 24 hours) are predicted based on the historical data for each month. Daily wind power production profiles are obtained using the methodology proposed in [20] in which correlations among different wind farms and the load are considered.

Pumped storage power stations are considered as potential storage systems in Northwestern China. Its capital investment cost is estimated to be 600 $/kW. Two identical alternative ESSs are considered at each node with a maximum capacity of 5000 MW. Considering a discount rate of 10% and investment return period of 25 years, the annualized investment cost is 66 $/kW. Data for alternative ESSs are provided in Table 16.6.

FIGURE 16.6 The geographic location of the studied system.

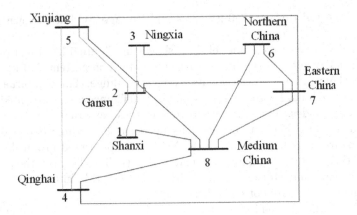

FIGURE 16.7 The configuration of the studied system.

As shown in Table 16.7, five corridors and two lines per corridor are considered. The alternative lines are all 750 kV HVAC with an annualized investment cost of 165 k$/km. Since a large number of thermal units are located in each provincial power system, a clustering method is used to merge units. Units of the same type and capacity located in the same provincial system are merged into an equivalent one.

This case study is also implemented in GAMS and solved with CPLEX 7.5 on a PC with an Intel Core i7/2.6 GHz-based processor and 8 GB of RAM. Two cases are compared below. The first one considers both building storage facilities and transmission expansion, while the second one considers only transmission expansion.

16.4.2 Results

Table 16.8 and Figure 16.8 allow comparing results for the two cases. If storage is available, a saving of $1.13 billion is achieved in transmission investment. In this case, four ESSs are built with an annualized investment cost of $1.32 billion. Two of the newly built ESSs are located in Shanxi and Gansu, while the other two are located in Qinghai. If storage is not available, four additional lines need to be built, two connect Gansu with Xinjiang, and another two connect Qinghai with Xinjiang. The variability of wind power can only be compensated by generation in other regions. However, storage allows effectively mitigating wind power variability and the peak load is supplied with reduced transmission availability.

If an ESS is available, the wind power curtailment decreases by 47%. In addition, the total cost decreases from $207.57 billion to $204.54 billion, rendering $3.03 billion savings. This results from two effects: on the one hand, a lower number of transmission lines is needed and additional wind energy is used. On the other hand, the investment cost of storage increases the total cost. However, the increase of investment cost in the ESS is smaller than the reduction of transmission line investment cost and generation cost. Therefore, the total cost decreases when the ESS is available.

The computational time required to solve the models, which is provided in Table 16.8, is moderate.

TABLE 16.6 Data of alternative ESSs for the NCG

$P_k^{dis,max}$ / $P_k^{ch,max}$ (MW/h)	$P_k^{dis,min}$ / $P_k^{ch,min}$ (MW/h)	S_k^{min} (MW/h)	S_k^{max} (MW/h)	η_k^{dis}/η_k^{ch}
5000	0	1000	30000	0.90

TABLE 16.7 Data of alternative lines for the NCG

No	From	to	Capacity (MW)
1	1	1	7000
2	2	3	7000
3	2	4	4000
4	2	5	4000
5	4	5	4000

TABLE 16.8 Result comparison of two cases with and without an ESS

Scheme	With ESS	Without ESS
Number of new lines built	1	5
Lines to be built	1-2	1-2,2-5(2),4-5(2)
Annualized investment cost in lines (Billion $)	10.13	11.26
Annualized investment cost in storage (Billion $)	1.32	—
Operation cost (Billion $)	203.09	206.31
Total Cost (Billion $)	214.54	217.57
Wind Curtailment (GWh)	5289.05	10044.35
Computing time (s)	1360.78	230.97

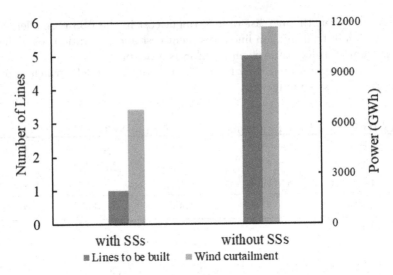

FIGURE 16.8 Results with or without an ESS.

16.5 Summary

This chapter presents a stochastic optimization model for the long-term coordinated expansion planning of transmission lines and storage systems in power systems with a high share of wind power. The proposed planning model characterizes the uncertain nature of load and wind power production. A number of scenarios are used to represent the long-term uncertainty of the demand. Different operating conditions are used to represent daily load, daily wind production, and their correlation.

Simulation results suggest that coordinating the planning of transmission lines and storage systems reduces the total cost of a power system with large wind power integration. This is because storage smooths the variation of the net load (load minus wind power production) and reduces the transmission capacity requirement. A case study based on the Northwestern China Grid system shows that constructing an ESS in Shanxi, Qinghai, and Gansu avoids building four HVAC transmission lines within the Northwestern China Grid. The investment in storage brings about an overall annual savings of $3.03 billion and a wind power curtailment reduction of about 4755 GWh.

Bibliography

[1] F. Ugranli and E. Karatepe. Transmission expansion planning for wind turbine integrated power systems considering contingency. *IEEE Transactions on Power Systems*, 31(2):1476–1485, 2016.

[2] A. Arabali, M. Ghofrani, M. Etezadi-Amoli, M. S. Fadali, and M. Moeini-Aghtaie. A multi-objective transmission expansion planning framework in deregulated power systems with wind generation. *IEEE Transactions on Power Systems*, 29(6):3003–3011, 2014.

[3] P. Xiong and C. Singh. Optimal planning of storage in power systems integrated with wind power generation. *IEEE Transactions on Sustainable Energy*, 7(1):232–240, 2016.

[4] M. Ghofrani, A. Arabali, M. Etezadi-Amoli, and M. S. Fadali. Energy storage application for performance enhancement of wind integration. *IEEE Transactions on Power Systems*, 28(4):4803–4811, 2013.

[5] F. Zhang, Z. Hu, and Y. Song. Mixed-integer linear model for transmission expansion planning with line losses and energy storage systems. *IET Generation, Transmission & Distribution*, 7(8):919–928, 2013.

[6] G. Latorre, R. D. Cruz, J. M. Areiza, and A. Villegas. Classification of publications and models on transmission expansion planning. *IEEE Transactions on Power Systems*, 18(2):938–946, 2003.

[7] M. Banzo and A. Ramos. Stochastic optimization model for electric power system planning of offshore wind farms. *IEEE Transactions on Power Systems*, 26(3):1338–1348, 2011.

[8] H. Park and R. Baldick. Transmission planning under uncertainties of wind and load: Sequential approximation approach. *IEEE Transactions on Power Systems*, 28(3):2395–2402, 2013.

[9] J. Li, L. Ye, Y. Zeng, and H. Wei. A scenario-based robust transmission network expansion planning method for consideration of wind power uncertainties. *CSEE Journal of Power and Energy Systems*, 2(1):11–18, 2016.

[10] A. Marin and J. Salmeron. Electric capacity expansion under uncertain demand: decomposition approaches. *IEEE Transactions on Power Systems*, 13(2):333–339, 1998.

[11] J. Wen, X. Han, J. Li, Y. Chen, H. Yi, and C. Lu. Transmission network expansion planning considering uncertainties in loads and renewable energy resources. *CSEE Journal of Power and Energy Systems*, 1(1):78–85, 2015.

[12] D. W. Van, H. Adriaan, and B. F. Hobbs. The economics of planning electricity transmission to accommodate renewables: Using two-stage optimisation to evaluate flexibility and the cost of disregarding uncertainty. *Energy Economics*, 34(6):2089–2101, 2012.

[13] A. Kalantari, J. F. Restrepo, and F. D. Galiana. Transmission and wind power investment. *IEEE Transactions on Power Systems*, 28(2):1787–1796, 2013.

[14] L. Baringo and A. J. Conejo. Correlated wind-power production and electric load scenarios for investment decisions. *Applied Energy*, 101:475–482, 2013.

[15] R. Romero, A. Monticelli, A. Garcia, and S. Haffner. Test systems and mathematical models for transmission network expansion planning. *IEE Proceedings-Generation, Transmission and Distribution*, 149(1):27–36, 2002.

[16] L. L. Garver. Transmission net estimation using linear programming,. *IEEE Transaction on Power Apparatus and Systems*, Vol. PAS-89:1688–1696.

[17] National Renewable Energy Laboratory. Wind integration datasets. http://www.nrel.gov/wind/integration datasets/.

[18] A. Brooke, D. Kendrick, A. Meeraus, and R. Raman. Gams: A users' guide: Gams develop. 1989.

[19] ILOG. CPLEX 7.5: Users manual. 2001.

[20] N. Zhang, C. Kang, C. Duan, X. Tang, J. Huang, Z. Lu, W. Wang, and J. Qi. Simulation methodology of multiple wind farms operation considering wind speed correlation. *International Journal of Power & Energy Systems*, 30(4):264, 2010.

Index

A

Abnormal Wind Power Forecast (AWPF), 67. See also Numerical Weather Prediction (NWP)

    clustering, 74, 78

    data adjustment engine, 78

    dataset, 70

    feature extraction, 70–73

    identifying, 79

    root mean square errors, 79

AC power flow-based Monte Carlo simulation (AC-MC), 173, 174

Active distribution system (ADS), 161, 163, 167–168

Algorithm HD-DPSO, 20

Algorithm K-means, 74

Algorithm variables-grouping, 22

Artificial Neural Network (ANN), 68–69, 74–75

Autoregressive-moving average (ARMA) models, 36, 141

Available transmission capacity (ATC), 250

Azimuth angle, 120

B

Bayesian network theory, 86

Beta distribution, 4, 35, 39, 85

Binary optimization models, 327–328

Block matrices, 165

C

C-vine copula, 14

Capacity credits, 308–309

    calculating, 278–279, 281–282, 287–289

    defining, 278

    effective load-carrying capability (ELCC), formulating based on, 286

    evaluating, 314

    factors influencing, 290–291

    integrating, 286–287

    overview, 275–278

    wind farm output and spatial correlations, 298–300

Capacity outage probability table (COPT), 17, 18, 287, 289

Central probability interval, 101, 105

Characteristic curve, 38

Cholesky decomposition, 162

Printed in the United States
by Baker & Taylor Publisher Services